Ecology and Ethics

Volume 6

This series is devoted to continuing research at the interfaces of ecology and ethics (embedded in the multiple fields of philosophy and ecology) to broaden our conceptual and practical frameworks in this transdisciplinary field. Confronted with global environmental change, the academic community still labors under a tradition of strong disciplinary dissociation that hinders the integration of ecological understanding and ethical values to comprehensively address the complexities of current socio-ecological problems. During the 1990s and 2000s, a transdisciplinary integration of ecology with social disciplines, especially economics, has been institutionalized via interdisciplinary societies, research programs, and mainstream journals. Work at this interface has produced novel techniques and protocols for assessing monetary values of biodiversity and ecosystem services, as illustrated by the Millennium Ecosystem Assessment. At the beginning of the 2010s, however, an equivalent integration between ecology and philosophy still remains elusive. This series undertakes the task to develop crucial theoretical and practical linkages between ecology and ethics through interdisciplinary, international, collaborative teamwork. It aims to establish a new forum and research platform to work on this vital, but until now insufficiently researched intersection between the descriptive and normative domains. The scope of this series is to facilitate the exploration of sustainable and just ways of co-inhabitation among diverse humans, and among humans and other-than-human co-inhabitants with whom we share our heterogeneous planet. It will address topics integrating the multiple fields of philosophy and ecology such as biocultural homogenization, Planetary or Earth Stewardship.

Luca Valera
Editor

Pantheism and Ecology

Cosmological, Philosophical, and Theological Perspectives

Springer

Editor
Luca Valera
Department of Philosophy
Universidad de Valladolid
Valladolid, Spain

ISSN 2198-9729 ISSN 2198-9737 (electronic)
Ecology and Ethics
ISBN 978-3-031-40039-1 ISBN 978-3-031-40040-7 (eBook)
https://doi.org/10.1007/978-3-031-40040-7

Cover illustration: Cover design by deblik, Berlin.

This Springer imprint is published by the registered company Springer Nature Switzerland AG
The registered company address is: Gewerbestrasse 11, 6330 Cham, Switzerland

Paper in this product is recyclable.

Foreword

Pantheism and Ecology: Cosmological, Philosophical, and Theological Perspectives is the sixth book of the Ecology and Ethics series. It complements the former volumes by introducing a new, and essential, thematic realm: eco-theology. Due to the secularization of global society, religious faith is often left out of public debates in general, and the discussion of environmental issues in particular. Although the theological dimension was not absent in our preceding books, Italian philosopher Luca Valera addresses it in a novel and systematic way in this new volume. Previously, in our series, the first volume integrated ecological sciences and environmental philosophy; the second incorporated the former disciplines into Earth Stewardship; the third presented a conceptual and practical framework to reorient ongoing biocultural homogenization drivers toward biocultural conservation processes; the fourth volume focused on the work of two European women with Mahatma Gandhi and documented their international collaborations to illustrate the value of intercultural dialogue and grassroots actions; and the fifth volume exposed how formal education can often catalyze biocultural homogenization and introduced the field environmental philosophy methodological approach to counteract this trend. Now, in this sixth volume, Valera addresses ontological and ethical dimensions by integrating some key concepts of the previous five volumes into multifaceted theological perspectives with a focus on the relationships between ecology and pantheism and panentheism.

A root interest is how God or the Divine is present in nature. This concern has weighed on the thought of philosopher theologians since deep in Judeo-Christian history and, philosophically, even earlier. They have sought to understand the very structure of nature and, therefore, how humanity is related to it. The introduction of the theological into the ecological discourse opens a new ontological horizon, since religious faith affirms that there is an intentionality in the universe; therefore, also in biophysical entities on our planet. Religious faith – Jewish and Christian and others in their own way – affirms that the cosmos, the planet Earth, and its living beings are the creation of God, who cores them with intentionality, and for pantheists and panentheists, imbues divinity into the very structure of the Earth and its living beings.

This ontological turn has crucial ethical implications. If we accept the divine character of creatures of the biophysical entities with which we cohabit on our planet, then we must formulate a new ethical question: Are we (or are we not) respecting the intention and substance with which the cosmos (including our planet and its biodiversity) was created? The acceptance of God's imminent participation in the cosmos gives a new foundation to affirm the intrinsic value of the natural world and to respect and defend its biocultural diversity. It also (re)dimensions the role of humanity from that of a special creation apart from evolutionary processes to one of kinship with and participation in the rest of creation. As Uruguayan theologian Guillermo Kerber (this volume) states in his chapter:

> An ecotheology of creation, based on the new cosmology, emphasizes that human beings are not above but within creation. The world does not belong to humans, it belongs to God, its creator. But the world is given to humans as a garden to cultivate and tend, and they thus have a relationship of responsibility towards it. In my view, this calls for an ethics of care expanded to the whole creation understood as our common home, as the subtitle of Pope Francis' encyclical evokes. Human beings can only be human and fulfil themselves as they bring the world to fulfilment by becoming involved in it by way of work and by care for it. By contrast with a destructive, dominating involvement in the Anthropocene era, a panentheistic perspective directs to maintaining the balance of creation taking care (i.e., doing justice) to the most vulnerable species (including humans) and environments.

Kerber's eco-theological perspective, undergirded by pantheistic or panentheistic concepts, resonates with a core concept of the biocultural ethic: *co-inhabitants*, which includes diverse human and other-than-human beings (Rozzi 2018). Throughout Valera's volume, the notion of creatures resonates with the concept of co-inhabitants. A particularly relevant historical figure is St. Francis of Assisi for whom the creationist belief implies a view of kinship and universal fraternity among all creatures (including humans and other beings belonging to the animal, vegetable, mineral, or other kingdoms). The idea of a divine immanence leads to ethical implications that are compatible with the biocultural concepts that cross the Ecology and Ethics series.

The concept of creatures as co-inhabitants assumes a continuity and a unity of life that is compatible with both a pantheistic or panentheistic religious worldview and an evolutionary scientific worldview. This convergence leads to a second central concept of the biocultural ethic that applies to both worldviews: *life habits* (Rozzi 2018). We co-inhabit the Earth and the cosmos with a vast diversity of human and other-than-human modes of being, each with its own life habits. From a religious and a scientific viewpoint, we can regard the diversity of forms and their unique life habits as a co-evolving process among creatures as co-inhabitants. German philosopher Konrad Ott (this volume) undertakes in his chapter this dynamic and evolutionary view to affirm that:

> [A] fertile earth brings forth the spheres of the plants which, according to the Hebrew thought, are not living beings yet. Plants cover the dry land and prevent it from becoming a desert. A productive earth and its plants constitute *habitats* for living beings. Three realms (spheres) of life come into being: waters, air, and dry land (emphasis added).

Thus, the understanding of coevolved life habits leads to the notion of cogenerated *habitats*, a third core concept of the biocultural ethic. Indeed, the biocultural ethic is grounded in the conservation of habitats as the condition of possibility for the well-being of the communities of co-inhabitants, and the continuity of complex interrelationships between biological and cultural diversity (Rozzi 2023).

We illustrate the compatibility of the religious and theological concepts with biocultural and scientific ones to stress two complementary values of Valera's volume. First, *Pantheism and Ecology* has essential continuities with concepts examined in other books of the Ecology and Ethics series. Second, *Pantheism and Ecology* challenges and fractures a still prevailing and distorted assessment of the Hebrew land tradition, what Aldo Leopold (1949, p. viii) called the "Abrahamic concept of land" because "we regard it as a commodity belonging to us" and of the Christian religion as "the most anthropocentric religion the world has seen," inherited from Lynn White's (1967) influential work "The Historical Roots of Our Ecologic Crisis." Noteworthily, White (1967, p. 1206) ended his article by alluding to St. Francis who dispossessed humanity "from his monarchy over creation and set up a democracy of all God's creatures."

By taking up pantheism and panentheism, Luca Valera invites the reader of this book to explore and to think again ancient theological and philosophical traditions for their relevance for (re)understanding the rich diversity of values of nature. In his words, "to walk the path that leads from interest or respect for nature (i.e., the ethical issue), to the very idea of nature underlying it (i.e., the cosmological question)" (Valera this volume). Walking this path will help readers to comprehend root ethical causes of our global social-environmental crisis and (re)visualize connections between humans and the biosphere that inspire life habits grounded in social-environmental justice and the appreciation of life in its scintillating diversity.

Cape Horn International Center (CHIC) Ricardo Rozzi
Universidad de Magallanes
Puerto Williams, Chile
Department of Philosophy and Religion
University of North Texas
Denton, TX, USA

Departamento Ecuménico de Investigaciones Roy May
San José, Costa Rica

References

Kerber G (this volume) Panentheism in Christian Ecotheology. In: Valera L (ed) Pantheism and ecology: cosmological, philosophical, and theological perspectives, Ecology and ethics series, vol 6. Springer, Dordrecht

Leopold A (1949) A Sand County Almanac and sketches here and there. Oxford University Press, New York

Ott K (this volume) Genesis 1 as ecosophy. In: Valera L (ed) Pantheism and ecology: cosmological, philosophical, and theological perspectives, Ecology and ethics series, vol 6. Springer, Dordrecht

Rozzi R (2018) Biocultural Homogenization: a wicked problem in the Anthropocene. In: Rozzi R, May RH Jr, Chapin FS III, Massardo F, Gavin M, Klaver I, Pauchard A, Núñez MA, Simberloff D (eds) From biocultural homogenization to biocultural conservation, Ecology and ethics book series, vol 3. Springer, Dordrecht, pp 21–47

Rozzi R (this volume) Biocultural ethics and Leonardo Boff's ecotheology of liberation. In: Valera L (ed) Pantheism and Ecology: Cosmological, Philosophical, and Theological Perspectives, Ecology and ethics series, vol 6. Springer, Dordrecht

Valera L (this volume) Intrinsic values, pantheism, and ecology: where does value come from? In: Valera L (ed) Pantheism and ecology: cosmological, philosophical, and theological perspectives, Ecology and ethics series, vol 6. Springer, Dordrecht

White L Jr (1967) The historic roots of our ecological crisis. Science 1955:1203–1207

Preface

The focal problem of this book is the relationship between pantheism and ecology. As several papers and books have pointed out over the years of research in environmental philosophy and eco-theology, this relationship is fundamental to the perspectives of both ecology and pantheism. Indeed, pantheism offers a consistent cosmological and theological framework for ecology (considered simultaneously as a worldview, philosophy, and political activism). On the other side, ecology can be regarded as one of the most current, vivid, and politically relevant expressions of pantheism. Notwithstanding this latent interrelationship, until now it has not been deeply or seriously examined. For these reasons, it is interesting to propose a book discussing this issue, which is certainly relevant to both environmental philosophy and contemporary theological (and eco-theological) discourse.

The reason for the genesis of this book has been, first and foremost, an interest in Spinoza's cosmology, inherited from the research and insights of the father of deep ecology, Arne Næss. Such cosmology has been overlooked and misinterpreted hastily for too long, and perhaps its importance and relevance have been underestimated. Indeed, I believe many contemporary issues with both practical and metaphysical connotations – such as the ecological approach – can be reconsidered in light of the Spinozian theoretical background.

In this regard, this book simultaneously deals with three lines of research: the line that questioned the cosmology of ecology, overcoming a narrowly defined "applied" environmental ethics, in order to reconsider an environmental philosophy *tout court*; the line that has taken up and reconsidered Spinoza's philosophy, clarifying controversial issues, such as the relationship between the whole and the parts; finally, the line that has explored the issue of pantheism and its differences with panentheism or traditional theism, as well as its permeation into different cultures and religious traditions.

Pantheism is not the only theological – or cosmological – alternative to theoretically ground ecology, evidently. Other schools of thought, philosophies, theologies, or religious traditions have attempted to offer robust ecological and cosmological frameworks and tried to offer convincing answers to the current ecological crisis as well. Part I of this book precisely discusses in-depth and critically analyzes such perspectives, particularly considering some of the main protagonists in the history of philosophy and theology. Accordingly, in Part I, following the brief history of the relationship between the whole and its parts (Chap. 1), the different co-authors of this book analyze thinkers and traditions that have critically considered or challenged pantheistic cosmology, such as Stoicism (Chap. 2); Neoplatonism and Medieval Patristics (Chap. 3); Renaissance Thinkers and Artists (Chap. 4); Spinoza (Chap. 5); Schleiermacher (Chap. 6); Rumi and Tagore (Chap. 7); Jonas (Chaps. 8 and 9); Teilhard de Chardin (Chap. 10); and Næss (Chap. 11). In this sense, Part I offers a historiographical and theoretical substantive contribution to the reflection on pantheism, beginning to introduce some elements of its connection with the ecological question (they will be deepened in Parts II and III). Thus, through this first part of the book, the reader is provided with a comprehensive overview of the historical discussion regarding pantheistic cosmology. An international expert on the subject writes each chapter. It represents, thus, a novelty in the context of the international theological and philosophical debate.

Part II delves more directly into the relationship between pantheism and ecology. In this sense, it is not only to offer a historical perspective to the topic in this part but also a thematic reflection regarding some frontier aspects between ecology and pantheism. Even in this part, several leading authors in the current contemporary debate about ecological thought are considered, such as Panikkar (Chap. 12); Tillich (Chap. 14); and Thoreau (Chap. 15); moreover, it delves into different religious or cultural traditions, such as the Native Americans tradition (Chap. 13); or the Biblical (Chap. 16) and the Christian one (Chap. 17). In addition, Part II addresses a number of metaphysical and theological issues central to the debate regarding the relationship between ecology and pantheism, such as the opposition between pantheism and theism (Chap. 18); the naturalist theology implicit in ecology (Chap. 19); the metaphysical implications of holism and reductionism (Chap. 20); the cosmology implicit in the Anthropocene narratives (Chap. 21); and the possibility of a systemic metaphysics as a consistent framework for a given cosmology (Chap. 22).

Finally, Part III deals more directly with the ethical and political concerns and consequences involved in pantheism in its current intersections with ecology. Some topics addressed in Part II are as follows: the question of boundaries and the possibility of their existence in a pantheistic or ecological cosmology (Chap. 23); the theoretical foundation of intrinsic values – a relevant issue in current environmental ethics (Chap. 24); the possibility of developing a biocultural ethics in dialogue with eco-theology (in particular, with Boff's perspective) (Chap. 25); the eco-feminist question, in relation to pantheism and panentheism (Chap. 26); the need to abandon modern reductionism for an ecological paradigm shift (Chap. 27); and, finally, the discussion of a current utopian project of applied ecology (Chap. 28).

In this sense, this book simultaneously addresses theoretical, historical, and practical issues starting from philosophical and theological viewpoints, with an evident openness to ecology as a field of application. It is also worth noticing that there is a significant presence of scholars from the Ibero-American sphere in the book, mainly due to two reasons. First, my interest in carrying out this text was also motivated by a grant from the John Templeton Foundation ("Supporting Constructive

Research on the Existence of God in Spanish-Speaking Latin America. Project number: 61559-13") as well as my participation in a Chilean research group (Centro Cabo de Hornos – Cape Horn International Center, CHIC – project ANID N° FB210018), which both supported the present research. Second, this book is intended to address an audience that is not exclusively English-speaking, since – as far as I could perceive – there are emerging lines of thought in this area in Ibero-America that need to be emphasized.

Finally, I hope this book may offer insights for further study on the relationship between ecology, environmental ethics, and pantheism. Too often, in my opinion, the environmental philosophical discussion has been reduced to mere applied environmental ethics, or to a simplistic application of principles to compelling ecological dilemmas. In doing so, we have often overlooked the depth of such problems, which can find their meaning only in the light of a broader cosmological and metaphysical vision.

Cape Horn International Center (CHIC)
Puerto Williams, Chile
Bioethics Centre
Pontificia Universidad Católica de Chile
Santiago de Chile, Chile
Department of Philosophy
Universidad de Valladolid
Valladolid, Spain

Luca Valera

Acknowledgments

This volume conveys the work of many researchers, students, and people from different regions of the world. I am grateful to all the authors for their wonderful work and commitment. I am grateful to my institutions in Spain and Chile for their support and engagement, too: the Cape Horn International Center (CHIC) and the Bioethics Centre at the Pontificia Universidad Católica de Chile, in Chile, and the Department of Philosophy of the Universidad de Valladolid, in Spain. Moreover, I am grateful to the John Templeton Foundation (Supporting Constructive Research on the Existence of God in Spanish-Speaking Latin America, Project number: 61559-13) for their support, particularly Professor José Tomás Alvarado. I also would like to acknowledge funding from the National Agency for Research and Development of Chile (ANID/BASAL FB210018) to the Cape Horn International Center (CHIC).

I am especially grateful to Prof. Ricardo Rozzi and Prof. Roy May Jr. for their outstanding work and help with the last version of the papers: they provided priceless support for the production and edition of this volume. I also want to thank Gabriel Vidal for his support. Finally, I would like to thank three anonymous reviewers who generously provided comments on chapter manuscripts.

Part of the inspiration for developing this book came from the online international workshop: "Reflexiones sobre la ecología. Miradas cosmológicas y teológicas," realized with the support of the Department of Philosophy at the Universidad de Valladolid (Spain) and the Bioethics Centre at the Pontificia Universidad Católica de Chile (Chile) on April 26th and 27th.

Contents

Part III From Pantheism to Ethics and Politics

Editor and Contributors

About the Editor

Luca Valera is Associate Professor at the Department of Philosophy, Universidad de Valladolid (Spain). Moreover, he is Adjunct Professor at the Center for Bioethics (School of Medicine), Pontificia Universidad Católica de Chile (Chile), and Associate Researcher at the Cape Horn International Center – CHIC (Chile). He earned his PhD in Bioethics at Università Campus Bio-Medico di Roma, Italy. His areas of research are environmental philosophy, philosophy of technology, and bioethics.

Contributors

Joel Álvarez Department of Philosophy, University of South Florida, Tampa, FL, USA

Rafael Amo-Usanos Department of Theology, Universidad Pontificia Comillas, Madrid, Spain

Emma Lavinia Bon Department of Philosophy and Communication, University of Western Piedmont, Vercelli, Italy

Joseph Bracken Department of Theology, Xavier University, Cincinnati, OH, USA

Alicia Irene Bugallo Department of Posgrade Studies in Philosophy, Universidad Nacional de San Juan, San Juan, Argentina

Fernando Calderón Department of Philosophy, Universidad de Valladolid, Valladolid, Spain

Teresa Calderón Department of English Philology, Universidad de Valladolid, Valladolid, Spain

Sixto Castro Department of Philosophy, Universidad de Valladolid, Valladolid, Spain

Gianluca Cuozzo Department of Philosophy and Educational Sciences, Università degli Studi di Torino, Turin, Italy

Jelson R. de Oliveira Department of Philosophy, Pontificia Universidade Catolica de Paraná, Curitiba, Brazil

Luciano Espinosa Faculty of Philosophy, Universidad de Salamanca, Salamanca, Spain

Lucio Florio Facultad Filosofía y Letras, Pontificia Universidad Católica Argentina, Buenos Aires, Argentina

Juan Jesús Gutierro Carrasco ESCUNI, University Centre for Education, Madrid, Spain
Comillas Pontifical University, Madrid, Spain

Casey Norrington Jackson School of History, Archaeology and Philosophy, University of Winchester, Winchester, UK

Guillermo Kerber Adult Faith Formation, Roman Catholic Church, Geneva, Switzerland

Abubakr Khan Faculty of Humanities and Social Sciences, Information Technology University of the Punjab, Lahore, Pakistan

Graham Lee Candler School of Theology, Emory University, Emory, GA, USA

João Almeida Loureiro Department of Philosophy, Universidade de Lisboa, Lisbon, Portugal

Alfredo Marcos Department of Philosophy, Universidad de Valladolid, Valladolid, Spain

Roy H. May Jr Departamento Ecuménico de Investigaciones, San José, Costa Rica

Konrad Ott Department of Philosophy, Christian-Albrechts Universität zu Kiel, Kiel, Germany

Moisés Pérez Philosophy Department, Universidad Católica de Valencia, Valencia, Spain

Adrián Pradier Department of Philosophy, Universidad de Valladolid, Valladolid, Spain

Zahra Rashid Faculty of Humanities and Social Sciences, Information Technology University of the Punjab, Lahore, Pakistan

Jeyver Rodríguez Department of Applied Ethics, Faculty of Religious Sciences and Philosophy, Catholic University of Temuco, Temuco, Chile

Francesco Vitali Rosati Department of Philosophy and Educational Sciences, Università di Torino, Turin, Italy

Ricardo Rozzi Cape Horn International Center (CHIC), Universidad de Magallanes, Puerto Williams, Chile
Sub-Antarctic Biocultural Conservation Program, Department of Philosophy and Religion and Department of Biological Sciences, University of North Texas, Denton, TX, USA

Jessica Sepulveda Deparment of Applied Ethics, Faculty of Religious Sciences and Philosophy, Catholic University of Temuco, Temuco, Chile
Cape Horn International Center – CHIC, Puerto Williams, Chile

Lisa Sideris Environmental Studies Department, University of California Santa Barbara, Santa Barbara, CA, USA

Luca Valera Department of Philosophy, Universidad de Valladolid, Valladolid, Spain
Center for Bioethics, Pontificia Universidad Católica de Chile, Santiago de Chile, Chile
Cape Horn International Center – CHIC, Puerto Williams, Chile

Gabriel Vidal Philosophy Department, Pontificia Universidad Católica de Chile, Santiago de Chile, Chile

Max Wade Morrisey College of Arts and Science, Boston College, Boston, MA, USA

Part I
Pantheism, Ecology, and Cosmology: Different Perspectives and Traditions

Chapter 1
Brief History of the Organism and the Relationship Between the Whole and Its Parts

Rafael Amo-Usanos

Abstract Organicism is the doctrine that interprets the world and nature by analogy with the organism. This is the starting point of this work and develops the conviction that the various positions of environmental ethics depend on the concept of the organism on which they are based. The studies of the philosophy of biology by Mario Bunge and Rafael Amo support the historiography of the organism. Both detect three stages in this science that coincide with the three great explanatory models of the organism: vitalism, mechanism, and systemism. In this way, the three concepts of organism will be explored, their cosmological repercussions and, above all, their ethical consequences, since the universe and nature, when thought of as an organism, vary and are valued differently. This work shows the importance of the concept of organism in the most widespread models of environmental ethics.

Keywords Organism · Environmental ethics · Mereology · Mechanism · Vitalism

1.1 Why Should We Discuss the Organism When We Are Reflecting on Ecology?

The current work[1] is based on the connection between ecology –in fact, environmental ethics– and cosmology discussed in current literature. It hopes to explain it through the role played by the notion of organism.

> When I first saw Gaia in my mind I felt as an astronaut must have done as he stood on the Moon, gazing back at our home, the Earth. The feeling strengthens as theory and evidence come in to confirm the thought that the Earth may be a living organism. Thinking of the

[1] The following pages are partly indebted to Carlos Giménez Rodríguez who facilitated my access to some works on environmental ethics.

R. Amo-Usanos (✉)
Department of Theology, Universidad Pontificia Comillas, Madrid, Spain
e-mail: rafael.amo@comillas.edu

L. Valera (ed.), *Pantheism and Ecology*, Ecology and Ethics 6,
https://doi.org/10.1007/978-3-031-40040-7_1

> Earth as alive makes it seem, on happy days, in the right places, as if the whole planet were celebrating a sacred ceremony. Being on the Earth brings that same special feeling of comfort that attaches to the celebration of any religion when it is seemly and when one is fit to receive. It need not suspend the critical faculty, nor can it prevent one from singing the wrong hymn or the right one out of tune (Lovelock 1988, p. 205).

We might think that this understanding of the Earth as a living being is a recent idea. However, it is not. Plato had no clearly defined concept of organism but states the same thing: "On this wise, using the language of probability, we may say that the world came into being –a living creature truly endowed with soul and intelligence by the providence of God" (Plato 1961, 30b, 1163).

These two passages witness to the long history of organicism: the doctrine that interprets the world and nature by analogy with the organism (Abbagnano 2007, p. 55). That is the issue of this work. The universe, the world, has always been interpreted as an organism. Understanding the latter means understanding the former. Understanding this link between the universe and the organism will depend on how the link between the parts and the whole in the organism has been understood. I will show that this link between the parts and the whole will account for the various stances on environmental ethics.

This connection between organicism and environmental ethics is double. On the one hand, it is true that the description in Timaeus and some of Lovelock's statements on the Gaia hypothesis have no direct ethical intention.[2] But the connection between cosmology and environmental ethics has already been proved in other fields of ethics (Amo Usanos 2017, pp. 59–65) and requires the inclusion of worldviews, ideas on nature, cosmologies, and research contexts. On the other hand, Lovelock himself as well as other authors, such as Capra, state that their cosmological understanding leads them directly to their ethical commitment.[3]

Briefly, within a consistent system of reflection, ethical stances must coincide with all other aspects of the worldview, particularly the cosmologic elements connected to environmental ethics, so that they influence each other. The following pages, which will go through the history of the organism to gauge its influence on environmental ethics, are backed up by this idea.

---

[2] Lovelock (1987, p. 96): "Gaia is a hypothesis within science and is therefore of itself ethically neutral [...] The environmentalist who likes believe that life is fragile and delicate in danger from brutal mankind does not like what he sees when he looks at the world through Gaia. The damsel in distress he expects to rescue appears as a buxom and robust man-eating mother."

[3] Lovelock (1988, p. xvi): "It would be difficult after spending nearly 20 years developing a theory of the Earth as a living organism –where the evolution of the species and their material environment are tightly coupled but still evolve by natural selection– to avoid capturing views about the problems of pollution and the degradation of the natural environment by humans". Please see: Capra (1996, p. 7) quoted further on.

1.2 The Organism and Its Historiography

The word organism comes from the Greek *organon* which originally referred to an instrument. That is to say, "something composed of well combined unequal parts, made in a way that allowed it to fulfil the function or functions for which it has been designed" (Ferrater Mora 1981, p. 2450). Actually, according to Bertolaso (2017), organism was not used in biology until the end of the seventeenth century. Before that time, the words organisation or organic body were used; that was the criterion to differentiate between the living and the inert.

The features of organisms are "autonomy, freedom, and independence" (Bertolaso 2017). Autonomy has had the greatest impact on organicism and on ecology. Briefly, the issue of "biological subordination: from the parts to the whole, from a whole to itself, from parts to other parts, and the coordination between parts" (Bertolaso 2017), since whichever are the traits of the relationship, it will imply how the connection between the human being –a part of nature– with the whole –the world, nature, should be.

There are several criteria to describe the history of the organism.[4] In this work, I will apply the historiographic criterion that some authors use for the general history of the philosophy of biology. As stated by Mario Bunge (1997, p. 140), there are three models for the definition of life and its causes in philosophy of biology: vitalism, mechanicism, and biosystemic. Rafael Amo (2017, p. 64), who basically agrees with this three-part division of the history of biophilosophy, thinks that these three models are the inheritors of the conjunction between a particular notion of nature, of an image of the universe, and a particular research context.

1.3 History of the Organism and Environmental Ethics

Having established the connection between organism and ecology through the organistic analogy, I will proceed to go through the three stages of the history of the philosophy of biology –vitalism, mechanicism, and systemic thought– with the following items: (a) brief description of the paradigm; (b) definition of organism; (c) cosmologic impact; (d) mereology and ecology.

[4] Marta Bertolaso, for example, suggests as a starting point of the history of the organism the doctoral dissertation of G. E. Stahl in 1684. As from this moment, organism would enter the debate between vitalists and mechanicists which the later will clearly win. The debate then became that of mechanicists and so-called organicists.

1.3.1 Vitalism, Organism, and Environmental Ethics

(a) **Brief description of vitalism**

> In a strictly exact and thorough sense, [...] vitalism describes the doctrine which states that the difference between the living and the non-living is attributed to a particular constitutive and functional principle, the "vital principle," and this is seen as a specific force, the "vital force," ontologically and operationally superior, [...] to the remaining "force" of nature and, consequently, essentially irreducible to them (Albarracín Teulón 1982, p. 243).

(b) **Vitalist definition of organism**

Although vitalism is rooted in the first steps of philosophy, Hans Driesch stands out among its representatives. According to Driesch, "the organism is a specific body, built up by a typical combination of specific and different parts" (Driesch 1908, p. 25). This author, who does not deny the influence of Aristotle, acknowledges that the organism is not a mere roughly organised sum of parts. There is a part that is responsible for organic teleology which, in turn, is responsible for the other characteristics of an organism: entelechy.

Aristotle's thought, though it is not the sole possible vitalism, has left the clearest mark in the history of the philosophy of biology. The entelechy mentioned by Driesch is the soul: the form of the hylomorphic composition that describes organism, responsible for life, teleology, and totality.

(c) **Cosmologic impact of vitalism**

The model of vitalist organicism is embodied in the philosophical tradition of *anima mundi.* Its first systematic philosophical mention is found in Plato's Timaeus and reaches down to Schelling's work, *On the World Soul.* The influence of Timaeus was huge because it became the essential treaty on philosophy of nature. Even Aristotle, who did not share many issues, quotes it on several occasions. Above all, it was spread by the Neo-Platonism of Calcidius and Proclus. The influence of the latter on St. Augustine has been testified and that was how the issue of anima *mundi* entered Christianity.

Schools of thought such as stoicism, which identified *pneuma* with the soul of the world, as well as Paracelsus and alchemy received the idea that the universe, as living beings on earth, had a soul: therefore, it was a living being (Abbagnano 2007, p. 54).

It is well known that the move of *Naturphilosophie* is a reaction to mechanicism. Schelling, who represents that line of thought, recovers the notion of the soul of the world. At bottom, the universe is a great organism that is not made up of the sum of organisms. On the contrary, the latter are the realization of the former.

> In Schelling's view, things do not constitute the beginning of the organism. It is the other way round; the organism is the beginning of things. "The essence of all things [...] is life." For that reason, the cause of life must exist before matter; it must not be searched in living matter but outside of it (Aréchiga 2018, p. 159).

(d) **Vitalist mereology and ecology**

In vitalism the whole is the fruit of the presence of a main part, the soul -or the immaterial element (entelechy in Driesch's words). It is responsible for a body becoming an organism. According to Ramellini's description (2006, p. 89), the vitalist organism is the fruit of the combination of a system of sub-entities vertically interconnected. Thus, the organism is understood as a set of coordinated parts subordinated to the whole that is the fruit of the main part, which works as the vital part.

This element accounts for all the philosophical issues of the organism: autonomy, freedom, independence and, especially, teleology. In the cosmologic level, *anima mundi* is responsible for the universe being an organism. This vitalist model is very close to the way ancient animistic religions understood the world and the universe.

Consequences for environmental ethics may be diverse. If vitalism reaches cosmology, with the aforementioned *anima mundi*, the universe is the greatest living being which must be respected, venerated, and even adored. If vitalism only includes living beings on earth, the most complex living being -the human being endowed with a rational soul- enjoys the highest form of life that must be respected.

1.4 Mechanicism, Organism, and Organicism

(a) **Mechanicism**

In Mario Bunge and Martin Mahner words (1997, p. 140):

> The mechanistic answer comes in two versions, which may be called physico-chemicalism and machinism, respectively. According to the former, organisms are nothing but extremely complex physical or physico-chemical systems: they have no properties or laws of their own. According to machinism, organism are not just extremely complex physical systems, but they are machine-like systems, if just extremely complex physical systems but they are machine-like systems, if not machines proper.

(b) **Mechanistic definition of organism**

It is in the dawn of mechanicism, as has been said, when the biological use of the word organism begins. That is why it has often been taken for granted that organism is a mechanicist term.

One of the first philosophers of mechanicistic biology was Leibniz (1991, p. 25):

> Thus, each organic body of a living being is a kind of divine machine or natural automaton which infinitely surpasses all artificial automata. For a machine made by human artifice is not a machine in each of its parts. For example, the tooth of a brass wheel has parts or pieces which to us are no longer have something recognizably machine-like about them,

> reflecting the use for which the wheel is intended. But the machines of nature, namely living organisms, are still machines even in their smallest parts, ad infinitum.[5]

Loeb represents the physico-chemialist version of mechanicism. In his view,

> scientifically, however, individual life begins (in the case of the sea-urchin and possibly in general) with the acceleration of the rate of oxidation in the egg, and this acceleration begins after the destruction of its cortical layer. Life of warm-blooded animals –man included– ends with the cessation of oxidation in the body (Loeb 1912, p. 14).

Once life has been defined, he states: "We must, however, settle a question which offers itself not only to the layman but also to every biologist, namely, how we shall conceive that wonderful 'adaptation of each part to the whole' by which an organism becomes possible" (Loeb 1912, p. 23). After rejecting the idea that "a part is so constructed that it serves the whole" (Loeb 1912, p. 24), in a clear reference to vitalism, he sets his hope on that when "the structure and the mechanism of the atoms were known to us we should probably also get an insight into a world of wonderful harmonies and apparent adaptations of the parts to the whole" (Loeb 1912, p. 26).

It is physics that can account for the chemical structure of atoms and, by extension, of the organism. The whole is mere appearance, the result of a physicochemical connection between the parts.

(c) **Cosmologic impact of mechanicism**

Mechanistic cosmology reached its summit in the Newtonian universe: a perfect machine, a perfect clock –except the problem of initial movement– ruled by the law of universal gravitation. Descartes had also dealt with the issue. He agreed that physics explained the universe and its movements, but not that it explained everything. For both, force is extrinsic to mass. Objects, even celestial ones, move by a force that is not immanent; that is why the universe cannot appear as living since objects do not have immanent force, self-movement. The universe is a machine which, as all machines, is moved by an external force.

(d) **Mechanistic mereology and ecology**

In mechanistic mereology the whole is more than the sum of its parts,

> in a machine, the parts are causally independent from and temporarily previous than the whole they form. They acquire their function because they are present in the machine as a whole. The whole only reaches existence once all the parts have been appropriately assembled (Bertolaso 2017).

So, the universe is a complex machine –a clock– in which there is no life, and it must be treated in the same way as are machines. If the universe is no longer a

[5] In Capra's opinion (1996, p. 6) this machinist version of mechanicism triumphed: "The paradigm that is now receding has dominated our culture for several 100 years, during which it has shaped our modern Western society and has significantly influenced the rest of the world. This paradigm consists of a number of entrenched ideas and values, among them the view of the universe as a mechanical system composed of elementary building blocks, the view of the human body as a machine."

being with a soul, that is, a living being, then it is an inert being and does not require the respectful treatment due to living beings.

In the face of the argument about the responsibility, or its absence, of Christianity regarding the ecological crisis inspired by Lynn White's article, *The Historical Roots of our Ecological Crisis*, theologian Pannenberg stated:

> The confusion between dominion mentioned in the Bible and despotism does not have its origins in Christianity but in mechanicism and the radical split between *res extensa* and *res cogitans*. The Bible text does not carry the seed of a misdirected anthropocentrism. This has its origins in those who interpret *be fruitful and multiply* based on the notions of materialistic mechanicism post-Descartes that isolates man from God (Amo Usanos 2020, p. 170)

(e) **The crisis of mechanicism**

Mechanicism failed as a system to explain the world for many reasons, but the notion of organism played a special part. At his point and in view of the crucial importance they had, we must quote the words Kant devotes to them in *The Critique of Judgement*. It is Kant who realises that mechanicism is unable to give reasons for the organism. He believes the laws of mechanics are unable to explain the teleology of the organism, in other words, the end that seems to unify the parts in a whole. "For the thing itself is an end, and is thus comprehended under a concept or an idea that must determine a priori everything that is to be contained in it" (Kant 2000, par 65, p. 245). This means that an organism is only seen as such if its end is perceived, something that proves impossible to discover in the laws of mechanics that operate without an end.

Vitalism is equally unable to do so, it appeals to an eternal cause to account for finality. "One says far too little about nature and its capacity in organized products if one calls this an analogue of art: for that case one conceives of the artist (a rational being) outside of it. Rather, it organizes itself" (Kant 2000, p. 246). Kant has only one way out: that organism may be a self-organised being, a being with the capacity to endow itself with its structure: "A thing exists as a natural end if it is cause and effect of itself" (Kant 2000, p. 243).

This being so, in what concerns the relationship between the whole and its parts in understanding the organism, Kant says:

> For a body, therefore, which is to be judged as a natural end in itself and in accordance with its internal possibility, it is a required that its parts reciprocally produce each other, as far as both their form and their combination is concerned, and thus produce a whole out of their own causality (Kant 2000, p. 245).

But at the same time that "the idea of the whole conversely (reciprocally) to determine the form and combination of all the parts: not as a cause –for then it would be a product of art– but as a ground for the cognition of the systematic unity of the form and the combination of all of the manifold that is contained in the given material for someone who judges it" (Kant 2000, p. 245). This is the origin of the notion of self-organisation and, consequently, of system, that will be the clue to the next worldview.

1.5 Systemic Thought, Organism, and Organicism

(a) **Systemic thought**

The basic trait of this paradigm is holistic understanding, as Capra says: "The new paradigm may be called a holistic worldview, seeing the world as an integrated whole rather than a dissociated collection of parts" (1996, p. 6). Now, this holistic understanding goes together with the notion of system applied to the explanation of the whole of reality, but especially of life.

In Bunge and Mahner view, the biosystemic paradigm

> recognizes the bios as an emergent level rooted to the chemical one. More precisely, biosystemism maintains that (a) living systems, though composed of physico-chemical subsystems, have emergent properties, in particular laws, which their components lack, and (b) the units of biological science are the organism-in-this-environment, as well as its various subsystems (molecules, cells, organs) and supersystems (population, community, ecosystem) (Bunge and Mahner 1997, p. 140).

(b) **Systemic definition of organism**

As stated by Bertolaso (2017) "recent literature in biology and philosophy has witnessed a strong growth for the return of the organism." The writings of Kurt Goldstein in 1934 and von Uexküll in 1940 are incipient steps to approach the organism from a holistic perspective which, as Capra said, is the epistemological perspective that characterises this systemic explanatory paradigm.

In this model, the organism recovers a new central role when it relates to the notion of system.[6] Bertalanffy (1960, p. 9), the father of the general systems theory, said that his understanding of reality was organismic. He states that "every organism represents a system, by which term we mean a complex of elements in mutual interaction" (Bertalanffy 1960, p. 11). Briefly, an organism is an open system that shows similar characteristics to those of a system in thermodynamic equilibrium (Bertalanffy 1968, p. 121). Prigogine will enrich this by explaining stability far from equilibrium. On the one hand, in some systems absence of stability does not lead to greater entropy, but to a new reality in a novel form of equilibrium. This is due to the property of self-organisation. We should remember the pages of the *Critique of Judgement.* On the other hand, living beings are not in equilibrium since that would imply thermic death. In view of this, holism and the notion of system have given birth to a new understanding of the organism and the living being. Ramellini (2006, p. 89) discovers in contemporary literature about biology two ways to understand the organism that reply to systemic thought.

The first one understands organism as a system of sub-entities connected in such a way that the organism is the result of some connections between the parts and the whole. Such connections may be vertical or horizontal. When they

[6] Not all the authors of the systemic paradigm can abandon mechanicism that easily. Francisco Varela himself talked about self-poietic machines and he stands midway between mechanicism and systemic thinking (Bertolaso 2017).

are interpreted as horizontal, the organism emerges as such from the relationships between the parts. It could be said that, up to a certain point, the whole is the fruit of the relationships. On the other hand, when connections are interpreted vertically, the organism is understood as a set of coordinated parts that are subordinated to the whole which, at the same time, is the reason for being of its parts. The whole becomes the regulator of life. The philosophy of the process should be understood along this line. For that philosophy, an organism is an organism of hierarchically organised organisms which are in the service of the whole.

The second one is formed by the authors who understand the organism as a system, independently from its sub-entities. It is possible to see the organism as a totality, regardless of its sub-entities. The organism is an isolated entity that is alive due to the activity of the totality, but it is not matter in motion, rather, it is activity that is expressed in individual unities as living organisms. Along this line, we might consider Maturana and Varela's theory of autopoiesis. They state that organism, as a whole, has no parts; these are simply established by the observer. The organism is a self-regulated and homeostatic unity that can remain as such. It is an autopoietic machine whose behaviour results in maintaining its own activity.

(c) **Cosmologic impact of systemic thought**
The different discoveries of astronomy –in particular the contributions of Hubble– proved the expansion of the universe. It was no longer understood as static and began to be seen as dynamic (Doblado 2019). It spelt the death of the clock-metaphor of the universe as a huge machine composed of small machines. With the discovery of a universe in evolution, dynamism returns to it and to nature. And this dynamism also comes hand in hand with the notion of system. The universe, and all its elements, are systems. Some of them are open systems, living organisms: vegetables, animals, ecosystems and, why not, the Earth. One of the clearest examples of organicism in the systemic paradigm is Lovelock's work, The Gaia Hypothesis. He states: "To me it was obvious that the Earth was alive in the sense that it was a self-organizing and self-regulating system" (Lovelock 1988, p. 31).[7]

(d) **Mereology and ecology**
Holistic epistemology –typical of systemic thought– leaves its trace in mereology by proposing the organism from the whole and not from the parts. The consequence is that "this monumental paradigm shift is bringing with it a new perspective on human beings and their role in the great drama of nature" (Davies and Gribbin 1991, p. 4). This leads Lovelock (1988, p. 236) to say: "It all depends on you and me. If we see the world as a living organism of which we are a part –not the owner, nor the tenant; not even a passenger– we could have

[7] Næss's (2008, p. 92) statement is quite similar to Lovelock's one: "Now is the time to share with all life on our maltreated earth through the deepening identification with life-forms and the greater units, the ecosystems, and Gaia, the fabulous, old planet of ours."

a long time ahead of us and our species might survive for its 'allotted span.'" Capra (1996, p. 7) states in similar terms:

> Deep ecology does not separate humans –or anything else– from the natural environment. It sees the world not as a collection of isolated objects, but as a network of phenomena that are fundamentally interconnected and interdependent. Deep ecology recognizes the intrinsic value of all living beings and views humans as just one particular strand in the web of life.

1.6 Conclusion

In the light of the road travelled I have proven the heuristic capacity of the notion of life and living beings (organism). It is able to account for the ultimate reasons of many stances in an array of sciences. As well as for ecology and environmental ethics.

This work pretended to show that the environmental sensitivity of our times is not only the result of the environmental crisis or of a mutation in the values system. Rather, it is the opposite. The traits that make up modern ecology and the concern for it are a consequence of a paradigm change –marked by holism and the systems concept applied to living beings– that impacts on the notion of nature, life, and organism.

Much time elapsed from *Timaeus* to *Gaia* hypothesis. And, besides their similarity when considering that the universe or the Earth are alive, they are similar concerning the role of organicism in understanding reality. The Earth and the universe have always been considered an organism. What has changed is the explanatory model of the organism. With the notion of systemic paradigm, ecosystems or planet Earth or the universe itself are considered to be alive. And life is a value that must always be cared for and respected. That may be one of the ultimate reasons, or the ultimate reason, of the current ecological sensitivity.

Nevertheless, research should go on studying something that is beyond the purpose of this chapter: the link between the whole and the divine, both in pantheism and in panentheism. The organism is always a whole, be it in vitalism, in mechanicism, or in the systemic paradigm. But in the latter, the possible excess of holistic epistemology turns it into something that either undervalues the parts or annuls them. This opens the way to a crushing version of the whole towards the minor units of life.

References

Abbagnano N (2007) Diccionario de filosofía. Fondo de Cultura Económica, Mexico DF

Albarracín Teulón A (1982) La teoría celular, paradigma de la biología del siglo XIX. Dynamis 2:243–262

Amo Usanos R (2017) Vida y ética. Síntesis, Madrid

Amo Usanos R (2020) Una nueva síntesis humanista para un desarrollo humano integral. In: Larrú JM (ed) Desarrollo humano integral y Agenda 2030. Aportaciones del pensamiento social cristiano a los Objetivos de Desarrollo Sostenible. BAC, Madrid, pp 143–175
Aréchiga V (2018) El concepto de vida en el alma del mundo de Schelling. Meta 8:159–168
Bertolaso M (2017) Organismo. In: Diccionario Interdisciplinar Austral. Available at: http://dia.austral.edu.ar/Tiempo
Bunge M, Mahner M (1997) Foundations of biophilosophy. Springer, Heidelberg
Capra F (1996) The web of life. Anchor Books, New York
Davies P, Gribbin J (1991) The matter myth. Beyond chaos and complexity. Penguin Books, Harmondsworth
Doblado A (2019) Física de partículas y cosmología. In: Garrido M, Valdés LM, Arenas L (eds) El legado filosófico y científico del siglo XX. Cátedra, Madrid, pp 753–762
Driesch H (1908) The science and philosophy of the organism. Adam and Charles Black, London
Ferrater Mora J (1981) Diccionario de filosofía. Alianza, Madrid
Kant I (2000) Critique of the power of judgment. Cambridge University Press, Cambridge
Leibniz G (1991) Monadology. Routledge, London
Loeb J (1912) The mechanistic conception of life. Biological essays. The University Chicago Press, Chicago
Lovelock J (1987) Gaia. A model for planetary and cellular dynamics. In: Thompson WI (ed) Gaia. A way of knowing. Political implications of the new biology. Lindisfarne Press, Hudson, pp 83–97
Lovelock J (1988) The ages of Gaia. A biography of our living earth. Oxford University Press, Oxford
Næss A (2008) Self-realization: an ecological approach to being in the world. In: Drengson A, Devall B (eds) Ecology of wisdom. Counterpoint, Berkeley, pp 81–96
Plato (1961) The collected dialogues of Plato, including the letters. Princeton University Press, Princeton
Ramellini P (2006) Il corpo vivo. La vita tra biologia e filosofia. Edizioni Cantagalli, Siena
von Bertalanffy L (1960) Problems of life. Harper and Brothers, New York
von Bertalanffy L (1968) General systems theory. George Braziller, New York

Chapter 2
Stoic Pantheism and Environmental Ethics in Pliny the Elder

Max Wade

Abstract This chapter explores the relationship between two themes of this volume, pantheism and ecology, as it is present in Pliny the Elder's *Natural History*. Specifically, it examines the influence of Stoicism on Pliny's cosmology, as he adopts and challenges philosophical positions associated with the Stoa, particularly in his engagement with the view that the entire cosmos is a single living being which governs the variety of natural phenomena in both the heavens and the sublunar world. Additional themes, such as the role of natural teleology and providence are explored, as these provide the intellectual foundation for what I consider to be Pliny's "environmental ethics." I argue that Pliny, despite often expressing an anthropocentric outlook toward nature, holds that, in virtue of the divinity of the Earth, there are certain human duties toward maintaining natural order. These are justified both in terms of the benefits they provide for humans as a species, given that the Earth is a source of food and other material goods, but also in terms of harms and benefits for the Earth itself. Recognizing the difference between goods that exist for our sake and that which exists for the benefit of the non-human world is a crucial distinction according to Pliny and one that underlies many of his comments about the duties humanity has toward Nature.

Keywords Pliny the Elder · Natural history · World soul · Pantheism · Ecology

---

M. Wade (✉)
Morrisey College of Arts and Science, Boston College, Boston, MA, USA
e-mail: wademf@bc.edu

L. Valera (ed.), *Pantheism and Ecology*, Ecology and Ethics 6,
https://doi.org/10.1007/978-3-031-40040-7_2

2.1 Introduction

Pliny the Elder remarks that one of humanity's great signs of ingratitude toward the life-giving earth is our ignorance of her nature.[1] This ignorance is made manifest in many ways, but most clearly of all in our reckless plunder of the riches of the earth which we use not for our betterment, but instead to wage war and kill our fellow living creatures (Gibson and Morello 2011, p. 14). This human attitude, Pliny says, is a kind of "madness" [*insania*] and "crime" [*scelus*] (Gibson and Morello 2011, p. 102ff) which his *Natural History* serves to remedy by drawing his reader's attention to the beautiful and miraculous aspects of the natural world to instill both respect and admiration for the order underlying it.

Pliny, although not a philosopher himself,[2] draws heavily from philosophical resources as he attempts to capture the entirety of the world in all its parts, including everything from the distant motions of the heavens to the natures of gems buried deep within the earth. His project, therefore, should be seen as eminently philosophical in nature, as it is not merely a catalog or index of various natural phenomena, but is an attempt to explain the connections between them and the purposes they serve (Gibson and Morello 2011, p. 57). This connection, ultimately, is founded upon the assumption that the world is a single, divine being that has purposively designed the order of Nature; all things are connected as parts within a whole and are expressions of this cosmic design. In doing so, Pliny presents a distinct worldview which, although not expressed in a typically philosophical manner, nonetheless contains intellectually refined views on physics, ethics, and the place of humanity within the whole of Nature.[3]

This chapter explores the interrelation between Pliny's theological commitment to the divinity of Nature and the consequences this has for the relationship between humanity and the non-human world. I begin by exploring Pliny's particular form of pantheism, first in terms of his cosmology and secondly as regards the divinity of the Earth itself. I will then examine the ways in which Pliny describes humanity's place in the world in both anthropocentric and non-anthropocentric terms. Finally, I conclude by examining what can be called Pliny's "ecological ethics," using the case study of his prohibitions on mining as an example of how he believes that

[1] *Inter crimina ingrati animi et hoc duxerim quod naturam eius ignoramus* (*HN* 2.63); translations of the *Natural History* (abbreviated *HN*) are based on the Loeb edition (Pliny 1938–1962) unless otherwise noted, often with some modification. Passages are cited based on the book number, followed by the chapter.

[2] Beagon (1992) views Pliny as operating with a degree of skepticism toward philosophy, as he considers it foreign wisdom that may be useful to Roman thought, but not entirely compatible with it. For more on the relationship between Pliny and philosophy, see (Beagon 2005, p. 15ff), (Gibson and Morello 2011, ch. 6), (Gábli 2011, pp. 49–50), and (Griffin 2007, pp. 88–91). See also (Paparazzo 2005) and (Paparazzo 2008).

[3] Beagon (1992, pp. 13-15) emphasizes the philosophical structure of the *HN*, saying that in it, "philosophy is united with practical philanthropy" and that "his acquaintance with philosophical ideas molds his thought in a general way and colours his attitude to everyday life."

humans possess ethical duties to preserve the life-giving flows and processes of the non-human world.

2.2 Pliny's Pantheism

Pliny's description of the cosmos opens in the second book of the *Natural History* with an openly pantheistic declaration: "the world [*mundus*] and this—whatever other name men have chosen to designate the sky whose vaulted roof covers everything [*cuncta*]—is rightly believed to be divine [*numen*], eternal, immeasurable, ungenerated and imperishable" (*HN* 2.1). This *mundus* encompasses all things; there is nothing outside of it since it is "all within all" [*totus in toto*].[4] Therefore, the world, understood as a single, spherical body (*HN* 2.2), is identified as a god and, as such, all things within the world can be viewed as parts or manifestations of this cosmic divinity. Nature then, "is, in Stoic fashion, both rational and divine… Nature's supreme power is proven through her supreme variety" (Beagon 2005, pp. 24–25). The divine cosmos, as the totality of all things,

> is sacred, eternal, immeasurable, wholly within the whole [*totus in toto*], nay rather itself the whole, finite and resembling the infinite, certain of all things and resembling the uncertain, holding in its embrace all things that are without and within, at once the work [*opus*] of Nature and Nature herself (*HN* 2.1).

For Pliny, therefore, "*Natura* is the world, both as a whole and as its separate components; she is both the creator and the creation… *Natura* is everything" (Beagon 1992, p. 26).

Viewed from the perspective of the world taken as a whole, all of reality is a single living being engaged in constant, dynamic activity. On a celestial level, this activity consists in the rotations of the heavens, which, in conjunction with a variety of terrestrial processes, work together to facilitate the diversity of life on earth.[5] The world in its entirety is the product of this divine Nature, as all things come about as a result of the work of Nature; but, simultaneously, there is never anything outside of Nature, since Nature encompasses all things as parts within a unified whole. All parts work together in light of the design of the whole, much in the same way that

[4] In *HN* 2.5, Pliny writes that "for whatever God be, if there be any other God, and wherever he exists, he is all sense, all sight, all hearing, all life, all mind, and all within himself." This leads Pliny to reject conventional Greco-Roman polytheism, as the various gods fail to rise to the level of supreme divinity that Pliny attributes to Nature. This final condition, the "all within himself," appears to be to the crucial condition which both eliminates the possibility of genuine polytheism and grounds the other attributes of omniscience and omnipresence. For more on Pliny's rejection of polytheism for pantheistic Nature, see (Beagon 1992, p. 92).

[5] The Stoics held the view that the cosmos, as a living being, was united by *sympathy* between its parts, much in the same way that terrestrial organisms constitute an organic unity that organizes its constituent parts toward a certain end. For the influence of this view on Pliny's cosmological considerations, see Gábli (2011, p. 46).

an animal's body possesses a variety of organs that in conjunction facilitate the health and longevity of the entire organism.

This view of the universe as a living being, articulated by Plato in the *Timaeus*,[6] is a core tenet of Stoic philosophy that Pliny is clearly drawing upon (Beagon 1992, p. 30) in his picture of the cosmos:

> Both Pliny and Seneca espoused an essentially Stoic view of the universe, which emphasized a pantheistic notion of divine power dispersed through all parts of nature. Theirs was not the dichotomy between an incorruptible, eternal heaven and a corruptible and transient earth which Christianity, inspired in part by Platonism, promoted (Gibson and Morello 2011, p. 74).

God, as an immanent principle, is what animates the whole cosmos according to the Stoics; as Sedley (2002, p. 41) succinctly notes, "the Stoic god is the single cause of everything, himself a body and immanent throughout all the world's matter," and it is this same divine power which directs all things toward their natural ends (Baltzly 2003, p. 13ff). Additionally, Pliny speaks of Nature as having a pervasive vital spirit [*spiritus vitalis*] that is present in all things. This divine *spiritus* (understood not as something immaterial or ghostly, but as the corporeal element of air), which "penetrates all the universe" (*HN* 2.4), is likewise drawn from the Stoic *pneuma* and serves a similar function (Gábli 2011, p. 56; Beagon 2007, p. 25). Without this, life would be impossible and the order of the cosmos would be impossible to maintain. Nature, then, by means of this pervasive vital spirit, is spoken of by Pliny as the "parent of all things" (*HN* 37.78).[7] It is this vital spirit that is present in animal and plant life (*HN* 12.1), accounting for its growth as well as the various other life processes that allow them to function. It is by means of this vital spirit that the cosmos is able to be a single living being, as something akin to a soul permeates throughout all the parts of its "body" and serves as the active principle that accounts for the life and activity of all things.

At the center of this cosmic body lies the earth (*HN* 2.69). While the stars and other heavenly bodies operate somewhat at a distance from ordinary human life, the

---

[6] Specifically, *Tim.* 30c: "divine providence brought our world into being as a truly living thing, endowed with soul and intelligence" (trans. Donald Zeyl). Sedley (2002) sees Plato as the ultimate origin of Stoic pantheism, but acknowledges the ways in which the idea developed in ways distinct from that of Plato and other prior philosophers. Similarly, Beagon (1992, p. 27) notes that "his divine, eternal, immeasurable, but finite *mundus* is directly derived from the cosmological theories of Plato, Aristotle and the Stoics." Pliny, however, is also willing to draw on other sources –such as the pre-Socratics and Epicureans– indicating a lack of allegiance to any particular cosmological theory, preferring instead to develop his own views independently of the positions of any particular philosophical school. For more on how Pliny diverges from his philosophical predecessors, see Robertson and Pollaro (2022).

[7] *Parens rerum omnium.* Pliny's invocation of this phrase comes at the end of the *Natural History*, given as a parting remark in praise of Nature, that he alone among all men had "praised thee in all thy manifestations." The *Natural History* itself, therefore, can be seen as a devotional text; by examining Nature in every manifestation and understanding each as praiseworthy, one is able to adopt a properly reverential attitude toward the divine source of all things.

earth is in many ways the most immediate and imminent source of the divine manifestations of Nature. After all,

> it is the earth which literally bubbles with the divine power of nature (*uis naturae*), portrayed in 2.207-208 as situated deep within her, rather than concentrated in the heavens, and continuously bursting forth unexpectedly and eternally as precious minerals, medicinal springs, volcanic fires and mephitic gasses (Gibson and Morello 2011, p. 75).[8]

This constant "bursting forth" of the divine power of Nature contained within the earth (*HN* 2.95) leads Pliny to speak about the earth as a divinity in itself –not numerically distinct from the divine cosmos, but a divine body infused with the many powers of Nature worthy of a "maternal veneration" [*maternae venerationis*] (Merchant 1989, pp. 29–41) by the human beings that live and die on its surface. The earth "receives us at birth" and "nurtures" us, giving us food, shelter, and all sorts of medicinal herbs for our benefit; even poisonous plants serve the benefit of allowing us to commit suicide by means of a quick and (comparatively) painless death (Beagon 1992, pp. 238–239).

The earth, therefore, represents a particular place in the universe –one which is characterized by a cornucopia of different sorts of living beings and natural phenomena, all of which are the product of the living, divine earth. As Manorlaraki (2018, p. 220) notes, this is tied up with a strongly sensorial experience: "Both cultivated and grown in the wild, the bounty of the earth offers its human consumers a sensual overload." Many of these things are produced directly from the body of the earth itself: gems, precious stones, and metals, but also many species of plants (*HN* 21.50) and animals (*HN* 9.74)[9] are spontaneously generated out of the body of the earth. In speaking about the generation of fungi, for example, Pliny describes the formation of the organism within the "womb" [*vulva*] of the earth itself (*HN* 22.46), which nourishes the mushroom like the yolk [*luteum*] of an egg, and "spring[ing] up spontaneously and [not] grown from seed[s]" (*HN* 19.11). The earth, far from a mere conglomerate of rocky and dense matter, is a living being constantly producing life, both directly and indirectly.

These products of nature all speak to the power of their creator, who is itself manifested in the diversity of its various creations, and they all serve a purpose within the broader whole. Everything, from miniscule insects and plants to great sea monsters and bizarre human sub-species, is a sort of microcosm of Nature, provided

---

[8] Beagon (2007, p. 20) makes a similar observation in relation to the earth's connection to *pneuma*: "It is from the depths of the earth that nature's creative force, *pneuma*, literally bursts forth, creating miraculous springs and mysterious exhalations, and causing nature's basic elements to exhibit unexpected properties or to coalesce and react together in unusual and bizarre ways."

[9] The generation of animals out of muddy and slimy water was a commonly held view among most ancient philosophers (see Aristotle, *Generation of Animals* 3.11). According to Manning and Wilberding (2021, p. 179), for many Platonists "it is universal Nature, i.e., the world soul, that is responsible for providing both the souls (principles of life) and the forms of [spontaneously generated] creatures. What this means is that what came to be known as "spontaneous" generation turns out on this world soul-based theory not to be spontaneous at all." Pliny's account of spontaneous generation has significant overlap with this Platonist view, particularly insofar as the spontaneous generation of life out of "non-living" material is ultimately due to the animating power of Nature.

one is able to view them in light of the whole. It is this central importance Pliny places on the earth due to its role in the generation of life –both human and nonhuman– that will serve as the basis for the various normative claims he makes about human duties toward the natural world, particularly the earth itself. However, before we can speak about these duties and ethical views, we must first examine Pliny's understanding of the place of humanity within the natural order.

2.3 Humanity and Nature

As we have seen in the previous section, Nature's power is made manifest in the variety of its products; its presence is felt in everything from the stars in the heavens to the most mundane insects and vegetation. Yet, as Pliny often reiterates to his reader, these things are seemingly provisioned for the sake of mankind, as the "produce she fosters [*alit*] for our benefit" (*HN* 2.63). Humanity is given the "first rank" in the natural order [*principium iure tribuetur homini]* (*HN* 7.1) and, as such, receives the bounties of nature as "gifts of the gods" (*HN* 27.1–2). Nature, the parent of all things, seems to have a favorite child.

As many have recognized (Millar 2021, p. 256; Parejko 2003), Pliny's worldview is extremely anthropocentric in its orientation. Given humanity's place at the top of the terrestrial hierarchy, Nature's products appear to exist for our sake and our sake alone: "the centrality of the human race in the Plinian universe and the teleological emphasis imply that, without an understanding of man in nature, we cannot hope to understand nature itself" (Beagon 2005, p. vii). In point of fact, one simply can look at the structure and organization of the *Natural History* itself, as the vast majority of the text speaks about the *mirabilia* of Nature not in its own terms, but as things best understood in terms of their use for humans and the history of humanity's interaction with them.

> Scholars point out that Pliny's presentation of nature in inventory form reveals the perspective of a conqueror, which sees the material world as filled with items to be catalogued and administered. The main significance of the *Natural History's* form, it has been thought, lies in the way it alludes to an instrument of government, and therefore to Roman military power (Gibson and Morello 2011, p. 35).

The animal, plant, and mineral worlds are all fundamentally useful for humans, typically in the form of medicine and remedies, but are also obtained for reasons that are far more ornamental.[10] It is fair to say, therefore, that "the history of Nature is thus simultaneously a history of Culture. The *Natural History* of the earth is by

[10] Pliny expresses frustration at the tendency for humans to search for luxury goods while simultaneously ignoring the plethora of things actually produced for our benefit. The earth has not distributed these goods carelessly; far from it, as the things that grow on the land for us to immediately make use of are some of the most immediately beneficial things for us. As he notes in *HN* 33.1: "Although medicines also earth bestows upon us on her surface, as she bestows corn, bountiful and generous as she is in all things for our benefit! The things that she has concealed and hidden under-

inversion the unnatural History of Man" (Wallace-Hadrill 1990, p. 81). One cannot make sense of Nature without doing so through the eyes of humanity, as without us, Nature would have no need of many of its various flora and fauna.

This anthropocentrism is a strong mark of the Stoic influence on Pliny. For the Stoics, all entities in the cosmos are directed toward the same end – namely, the benefit of rational beings: "God exercises providence in relation to rational creatures: everything in the world is for them…far from being congenial to a deep ecology ethic that locates divinity and thus value in nature, Stoic pantheism is breath-takingly anthropocentric" (Baltzly 2003, p. 15). The existence of various species of plants and animals are for the sake of humans, serving as food or medicine. This leads Chrysippus to speculate that the souls of animals are analogous to salt, as both are used by Nature to preserve animal meat until it is ready for human consumption (Baltzly 2003, p. 18). Pliny's view of non-human life echoes this same sentiment, as he exhaustively lists the various healing properties of other creatures and frequently speaks about their usefulness in relation to particular human endeavors.

However, to paint Pliny in broad strokes as someone who is entirely anthropocentric in his outlook would be a disservice to the nuances of the *Natural History*.[11] Crucially, there are many points in which this anthropocentrism breaks down and I wish to highlight three particular ways in which Pliny emphasizes aspects of Nature existing for her own sake, rather than for humanity.

First, Pliny singles out certain aspects of the terrestrial world as being produced by Nature for her own use and needs, explicitly indicating that these are *not* meant for human use. This is most clearly seen when he speaks about things closely tied to the earth, such as mountains:

> For everything that we have investigated up to the present volume may be deemed to have been created for the benefit of mankind. Mountains, however, were made by Nature for herself to serve as a kind of framework for holding firmly together the inner parts of the earth, and at the same time to enable her to subdue the violence of rivers, to break the force of heavy seas and so to curb her most restless elements with the hardest material of which she is made (*HN* 36.1).

This passage is particularly interesting because of the way in which the effects of mountains –mitigating and repelling the destructive force of watery bodies and containing the earth's "inner parts"– allow the earth to achieve a sort of harmonious balance (or, at the very least, a tenuous stalemate) between its various elements and ecosystems. The ratio of sea and land on the earth, along with the particular arrangement of the various zones and biomes, seems to be valued by Nature because it

ground, those that do not quickly come to birth, *are the things that destroy us* and drive us to the depths below." Emphasis added.

[11] Wallace-Hadrill (1990, p. 85), for example, speaks about Pliny as a "proto-environmentalist." Beagon (2007, p. 19) views many things produced by Nature as being "deviations" from the otherwise human-centric universe and considers Pliny to challenge an anthropocentric outlook to a certain degree (see Beagon 1992, p. 38ff). Denson (2021, p. 147), on the other hand, sees Pliny as fully "post-anthropocentric," at least when speaking about certain parts of Nature.

results in much more diversity, allowing Nature to make its power manifest in a variety of creatures and phenomena that are strange and unfamiliar compared to the things within the more human-friendly regions. This balance may benefit humans, as it is a necessary condition for our existence as a species, but Nature itself seems to desire this geographical arrangement first and foremost for its own sake.

When Nature's order is disrupted in instances of mining, in which humans strip away mountains or threaten to destabilize the integrity of the earth by removing its mineral veins, the conditions that give rise to life are threatened:

> We trace out all the fibres of the earth, and live above the hollows we have made in her, marvelling that occasionally she gapes open or begins to tremble—as if forsooth it were not possible that this may be an expression of the indignation of our holy parent! (*HN* 33.1).

The sea, without the confinement imposed by its mountainous barriers, will forcefully overpower the flat and less rocky terrain on land; likewise, earthquakes and fissures in the earth may very well reflect the earth's "indignation" at our unscrupulous treatment of its mineral deposits. Nature, however, seems to have built-in defense mechanisms to prevent this from occurring, as mines become infested with poisonous snakes and other dangerous beasts to "protect her and keep off our sacrilegious hands" (*HN* 2.63). The anthropocentric teleology of the Stoics is rejected here, as these poisonous plants and animals are speculated to serve as a sort of natural immune system that exists to protect Nature from the violence inflicted upon it by human beings.

Second, certain domains are inhospitable to humanity and are presented as territory which we should be wary of intruding upon. The extreme regions of the earth –the poles and the equatorial middle band of the earth– are "taken from us" by the heavens, as the climates of these places are too extreme to allow human life to flourish. The poles are described as freezing cold and cast in darkness, while the equator is "parched and burnt" by the heat of the sun (*HN* 2.68). Likewise, Pliny expresses a similar degree of trepidation when speaking about the ocean (Beagon 1992, pp. 159–161), which occupies much of the earth, even within the temperate bands within which human habitation is possible. Even though the sea, by virtue of its watery nature, is associated with the generation of a plenitude of life and is the source of a variety of meteorological phenomena necessary for climatic regulation (*HN* 2.66), its tendency to produce monstrous creatures and violent storms makes it a realm upon which humans are only temporary visitors who must tread with caution.

Even within the terrestrial areas that we do inhabit, plenty of land is crisscrossed with rivers and mountains, forests and deserts, all of which are not conducive to human habitation.[12] So, while Pliny originally sets off the earth as the part of Nature "belonging to man" (in contrast to the heavens, which "belong to God" [*sic*

[12] Nature encourages humans to farm in certain places, but not others. Beagon (1992) encourages us to see these areas as ones "set aside" for us by the Providence of Nature, speaking to the way in which Pliny's admiration for the grandeur of Nature mitigates the otherwise anthropocentric tendencies of his philosophical sources.

hominum illa ut caelum dei]), he goes to great lengths to emphasize the way in which the scope of the earth pales in comparison with that of the heavens, just as the scope of land inhabitable to humans pales in comparison with the entirety of the earth. This all serves to emphasize the way in which our place within Nature is precarious and small in comparison with the entirety of the cosmos. In our hubris we take our tiny sliver of the world to be the most important when, in reality, we are but a small part of a much larger whole.

The third and final aspect that must be noted is the way in which all parts of Nature, when viewed properly, are microcosmic representations of the whole and, as such, occupy a particular place and are manifestations of Nature's divinity. Beagon (1992, p. 38) notes that "other animals, besides man, can to a certain extent be seen as microcosms. It is this, together with the interest in the range of species in the overall totality of Nature, which helps balance the stress on man's supremacy." After all, Pliny intends to emphasize in his cataloging of Nature's various manifestations that "in the contemplation of Nature nothing can be deemed superfluous" [*in contemplatione naturae nihil possit videri supervacuum*] (*HN* 11.1). As such, all things need to be viewed within the context of being parts of a broader natural order, as "the power and might of Nature lacks credibility at every point unless we comprehend her as a whole rather than piecemeal" (*HN* 7.7).

Small things are just as important as things that are more easily appreciable by humans as serving some obvious purpose in the natural world. The mineral world, for example, rather than being the lowliest and least significant domain of things, is instead "the supreme and absolute perfection of Nature's work" (*HN* 37.1).[13] Even tiny insects, which through a limited human perspective are a mere nuisance, are a testament to the amazing power of Nature: "The enlightened response to the mosquito's touch, suggests Pliny, should not be humancentric irritation, but nature-centric wonder at its gustatory endowment" (Manolaraki 2018, p. 211).

Therefore, we ought to see Pliny as having an ambiguous relationship with the variety of anthropocentrism characteristic of Stoic philosophy. On one hand, he clearly appreciates the purposiveness that it grants to all things –nothing in Nature is produced in vain– but nonetheless it seems to lead to the unfortunate consequence of making Nature exist for humanity's sake, diminishing the autonomy and sovereignty of the divine. Clearly, for Pliny, there are aspects and domains of Nature which are produced for its own sake and its sake alone; these aspects may inadvertently benefit humans, but their ends are nonetheless for the sake of Nature, not humanity. In fact, Pliny notes Nature's ambivalence in this regard, stating that it "asks a cruel price for all her generous gifts, making it hardly possible to judge

[13] In *HN* 36.23 and 36.25, Pliny speaks of Nature directly giving powers to stones as another form of *mirabilia* that speaks to the variety and power of Nature. Nature has "given to rocks a voice" [*dederat vocem saxis*], as well as imbuing magnets with "senses and hands" [*sensus manusque*] and iron "feet and instincts" such that it moves on its own accord to the magnetic stone. Pliny is more than willing to attribute a baseline degree of agency and awareness to all things in the cosmos, even rocks and stones, given the way in which Nature's power animates them and gives them particular powers.

whether she has been more a kind parent to man or more a harsh stepmother" (*HN* 7.1).

It is these domains, therefore, that steer Pliny toward the anti-anthropocentric aspects of what can be called his "ecological ethics"; a picture of duties towards Nature that are in place for the sake of Nature, not humanity (even if humanity benefits from following these same duties). In turning now to these particular duties, it seems, following our previous discussion, that they must all be understood as being invoked for the sake of encouraging a particular ethical attitude. Namely, the idea that human activity must take place within certain limits designated by Nature. It is in the crossing of these boundaries that harm is done – first to Nature and the earth, and then, secondly, to ourselves.

2.4 Our Duty to the Earth

To begin with a particularly evocative example, Pliny bemoans the lack of laws prohibiting the mining and importation of marble. Humans quarry mountains and "haul them away on a mere whim" (*HN* 36.1), which should be met with a response of "blushing prodigiously with shame" (*HN* 36.2). The arbitrariness of this is particularly emphasized for Pliny. Mountains, which, as we have previously seen, exist to serve a particular purpose within Nature's order, are flattened out and the natural geographical boundaries dissolved. The harvesting of these mountains comes at great cost, both environmental and economical, and only produces the reward of a kind of superficial luxury (Carey 2000). The natural purpose of mountains is being disregarded solely because humans take marble (and other stones) to be beautiful, which is only possible if it is considered apart from its role within the whole of Nature.

Not only is this unnecessary –after all, these ornamentations made from marble are alleged to serve no utilitarian function apart from any other stone– but produces a net-negative result. Pliny encourages his reader to "think how much more happily many people live without [precious stones]" (*HN* 36.1) and the way in which this sets a "bad moral precedent" [*malo exemplo moribus*] (*HN* 36.2); accordingly, there is a consistent disappointment on Pliny's behalf that there have not been laws put in place to prohibit this practice (Merchant 1989, p. 30ff). If anything, it reflects a deep lack of gratitude toward Nature, as we pursue these goods blindly, ignoring things immediately at our disposal in search of rare and exotic luxuries. Pliny distinguishes between these utilitarian goods provided by Nature and those pursued solely for the sake of luxury, noting that "we penetrate her inner parts and seek for riches in the abode of the spirits of the departed, as though the part where we tread upon her were not sufficiently bounteous and fertile" (*HN* 33.1).

But why such strong moral indignation? Viewed abstractly, we can recognize that this is a transgression of the boundaries Nature has laid down for humanity: "Man's vice is a result of his use of *ratio* [reason] to pervert *Natura*" (Beagon 1992, p. 154). But Pliny wants to ensure that there is a visceral emotional reaction against mining the earth and to this end he employs some of his strongest language to evoke an explicitly bodily sense of horror. In exploiting the earth, we "torture" [*cruciatur*] her body and, in digging deep into the earth in search of luxuries, we "drag out her entrails [*viscera*] to seek a jewel merely to be worn upon a finger" (*HN* 2.63). The kind of person who delights in precious gems and other luxuries must be some sort of monster, gaining pleasure from the pain they inflict not just on other people, but on their very own mother (Pepper et al. 2003, p. 340).

Without the particular pantheistic formulation provided by Pliny, this argument would not hold water. But since divinity does not have a transcendent component[14] and is instead most associated with the life-giving, generative aspect of Nature, an extractive, exploitative attitude toward the earth and its products is a direct affront to the rational moral order laid down for us to follow. As Wallace-Hardrill (1990, p. 83) notes, gratitude toward the earth [*terra*] is necessary because everything it provides for humanity is good. And, in violating this natural order, we commit grave harms that result in pain and suffering inflicted upon Nature itself.

Proper human activity, therefore, operates in line with the ebbs and flows of Nature; working to enhance and promote these flows, resulting in the proliferation of life and human happiness.[15] Actions contrary to this –particularly mining and the wanton pilfering of exotic flora– disrupt these Natural flows, causing pain and suffering for nearly everyone and everything involved. Pliny's ecological outlook has rightly been characterized as deeply agrarian in orientation: the proper human agent is like a farmer who, in tilling the fields and creating the proper conditions for Nature to flourish,[16] is able to work with Nature to turn uncultivated land into fertile soil lush with plant and animal life. Similarly, Pliny's disdain for magic can be seen in the way in which it seeks to circumvent the natural order, disrupting the flow of life in order to yoke Nature to human needs, rather than having humanity and Nature operate in tandem (Beagon 1992, p. 100; Gibson and Morello 2011, p. 174).

Pliny's worldview, therefore, shares with the Stoics a close intertwining between physics and ethics (Gibson and Morello 2011, p. 103ff). By understanding Nature in all its parts and as a unified whole, we are able to understand our place within this broader unity and thus act accordingly. Viewed from a purely human, anthropocentric perspective, the violence we commit against the earth comes back to harm us as

---

[14] Pliny, additionally, rejects the existence of an afterlife (*HN* 7.55), considering the belief an indication of human vanity (Beagon 1992, p. 106). Beagon (2005, p. 114) notes that "Pliny was unusually sceptical by Roman standards, pouring scorn not only on all forms of popular religion, magic, and astrology but even on the traditional Olympian religion," all of which stems directly from his pantheism.

[15] Nature is most properly viewed as *artifex*: the creative, productive craftsman of all things in the world. Nature's wild products are more abundant and plentiful than cultivars, as the variety of all life, both plant and animal, can be traced to the glory of Nature as *artifex*. On this point, see Beagon (1992, pp. 131–133) and Manorlaraki (2018, p. 218). Humans, as a microcosm of Nature itself, reproduce Nature by the agency of art (*HN* 36.68).

[16] On Pliny's agricultural outlook, see Beagon (1992, p. 162ff). Likewise, "Pliny's criticism of medicine is, in fact, criticism of the way man's activity can misuse Nature's most basic gifts" (Beagon 1992, p. 203).

well –Nature always gets the last laugh– as we not only exploit our fellow humans in our pursuit of Nature's luxuries, but also are forced to contend with the resulting disruptions of the environment that make our own lands less habitable as a result.

But Pliny encourages his reader to go beyond this and see things from Nature's perspective to imagine the pain and suffering we inflict upon the earth in our shallow pursuits. As Beagon (2005, p. 46) notes, "to the variety and universality are added the versatility of mind, albeit with a twist; human powers of rational and moral deliberation allow for the possibility of perverting the natural law", but it is this same power of the mind that lets us appreciate Nature in all its glory as well. Humans are unique in our potential for understanding how all of reality is a unified whole and it is in this grasping of the totality of Nature that we are able to break out of our anthropocentric outlook, treating Nature with kindness and respect for its own sake.

So, while Pliny is still nonetheless deeply anthropocentric compared to contemporary perspectives that challenge the assumption that humanity is the "pinnacle" of the natural world, he nonetheless is able to reconcile this with a form of ecological ethics that insists upon respect and careful consideration given toward Nature. As Beagon (1992, p. 47) describes, "rather than seeking to improve on Nature, man is invited to stand back and admire the truly perfect." Our place within the world is unique, as we are especially well positioned to engage in activities that enhance and multiply the innately generative power of Nature, but this comes with a great burden of responsibility, as this same human power is what allows us to inflict deep cuts within the earth itself, ripping open mountains and letting its entrails spill out so we can decorate our houses and bodies with its riches.

Ultimately, it is this pantheistic religiosity that is supposed to direct our attitude toward the whole of Nature. From the smallest of insects to the tallest of mountains, each thing is both a product of Nature and its immanent embodiment; as such, its particular place within Nature must be respected and honored, rather than ignored or dismissed for the sake of a particular human pursuit. By working with Nature, not contrary to it, not only are we able to flourish, but Nature as well.

Acknowledgments I would like to sincerely thank Jack Bagby, James Taylor, and Cameron Pattison for their contributions to my research and critical suggestions on my initial formulations of this chapter, particularly in regard to their suggestions for various eco-critical readings of the *Natural History*.

References

Baltzly D (2003) Stoic pantheism. Sophia 42:3–33

Beagon M (1992) Roman nature: the thought of Pliny the Elder. Oxford University Press, Oxford

Beagon M (2005) The Elder Pliny on the human animal: natural history book 7. Oxford University Press, Oxford

Beagon M (2007) Situating Nature's wonders in Pliny's natural history. Bull Inst Classical Stud 50:19–40

Carey S (2000) The problem of totality: collecting Greek art, wonders and luxury in Pliny the Elder's natural history. J Hist Collect 12:1–13
Denson R (2021) Divine nature and the natural divine: the marine folklore of Pliny the Elder. Green Lett 25:143–154
Gábli C (2011) The terminology of Pliny the Elder's cosmology. Acta Antiqua 51:45–58
Gibson R, Morello R (2011) Pliny the Elder: themes and contexts. Mnemosyne supplements: monographs on Greek and Latin Language and Literature. Brill, Leiden
Griffin M (2007) The elder Pliny on philosophers. Bull Inst Classical Stud 100(50):85–101
Manning G, Wilberding J (2021) Reflection II: the world soul and spontaneous generation. In: Wilberding J (ed) World soul: a history. Oxford University Press, Oxford, pp 177–185
Manolaraki EH (2018) Senses and the sacred in Pliny's natural history. Ill Classical Stud 43:207–233
Merchant C (1989) The death of nature: women, ecology and the scientific revolution. HarperCollins Publishers, San Francisco
Millar JE (2021) Roman climate awareness in Pliny the Elder's natural history. Class Antiq 40:249–282
Paparazzo E (2005) The Elder Pliny, Posidonius and surfaces. Br J Philos Sci 56:363–376
Paparazzo E (2008) Pliny the Elder on metals: philosophical and scientific issues. Class Philol 103:40–54
Parejko K (2003) Pliny the Elder's Silphium: first recorded species extinction. Conserv Biol 17:925–927
Pepper D, Webster F, Revill G (2003) Environmentalism: critical concepts, vol 2. Taylor & Francis, New York
Pliny (1938) Natural history. Harvard University Press, Cambridge
Robertson P, Pollaro P (2022) Ancient Greco-Roman views of ecology, sustainability, and extinction: Aristotle, stoicism, Pliny the Elder on Silphium, and the modern legacy in Cuvier, Humboldt, Darwin and beyond. In: Hufnagel L (ed) Ecotheology – sustainability and religions of the world. IntechOpen, London, ch. 7
Sedley D (2002) The origins of stoic god. Trad Theol 41:41–83
Wallace-Hadrill A (1990) Pliny the Elder and Man's unnatural history. Greece Rome 37(1):80–96

Chapter 3
The Presence of God in Creation: Medieval Motifs of Ontological Continuity, Light and Sympathy for Creatures

Adrián Pradier

Abstract The difference between Creator and Creation is strict in medieval thought. We can confirm that pantheistic doctrines were not only considered philosophically erroneous, but theologically inadmissible. However, the distinction between both dimensions was not always clear. The influential doctrine of an ontological continuity between God and created things was a source of controversy, if not of pantheistic or panentheistic risk. Such is the case of some thinkers who, although located in different periods, were influenced by the same Neoplatonic sensibility. In this chapter I present a balanced review of this structure of thought inherited by these authors. This led them, on critical occasions, to bypass or even to overcome the problem of pantheism, but not without difficulties. My aim is not to take stock of avowedly pantheistic medieval positions, but to analyze the risks faced by one of the most powerful and influential currents of thought of the period through one of its most celebrated doctrines. This contribution is divided into three parts. In the first one, I present a brief approach to the "ontological continuity" Neoplatonic doctrine, and its close link with the theology of light, which was especially successful for a long time in the framework of Christian thought, both Greek and Latin. Then, in the second, I review the foundational position of Dionysius Areopagite in order to overcome the problems derived from this use of light allegory and also the threats of its pantheistic drifts. Finally, in the third, I present two medieval authors, John Scotus Erigena and St. Francis of Assisi, which have been labelled pantheistic in the first case, pan-animistic in the second case, because of certain propositions which, no doubt, can generate controversial positions on the matter.

Keywords Ontological continuity · Medieval Neoplatonism · Light · Presence · Sympathy

A. Pradier (✉)
Department of Philosophy, Universidad de Valladolid, Valladolid, Spain
e-mail: adrian.pradier@uva.es

L. Valera (ed.), *Pantheism and Ecology*, Ecology and Ethics 6,
https://doi.org/10.1007/978-3-031-40040-7_3

3.1 The Neoplatonic Doctrine of Ontological "Continuity" (συνέχεια)

The Neoplatonic approach to the shaping of reality is built on the basis of an "unfolding" or "procession" (πρόοδος) of the One. This unfolding is limited by two elementary conditions. On the one hand, the absolute transcendence of the One, that is, of its source. On the other hand, the persistence of its unity beyond its hypostases and all ontic diversity (Piñero 2022, pp. 61–84; Bales 1982, pp. 40–50), which are completely traversed by the unity nascent in the One (O'Daly 1974, pp. 159–169). This consideration, difficult to conceive, was thus explained in different allegorical forms. First, as a river without origin, which supplies water to all the rivers, without exhausting itself, so that "the rivers flowing from it confuse their waters at first, before each one takes its particular course, but each one already knows where its flow will take it." Plotinus also identified it with an immense tree whose vital principle remains immobile, inasmuch as "it is not dispersed throughout the tree but has its seat in the root," which allows it to provide the plant with "life in its multiple manifestations." For Plotinus it was clear: the principle of all things "is not manifold," even if it is "the principle of this multiplicity" (Plot. *Enn.* VI, 9, 9).

But undoubtedly the most influential rhetorical device appropriate to the nature of the One was the use of the metaphor of light. First, from a comparison between the One and the Sun, insofar as the latter is "he whom the Sun imitates" (Plot. Enn. V, 5, 8, vv. 7–8) in its activity. What in the intelligible realm resembles the Sun is not the One, but the Intelligence, the first of the hypostases in which the procession of being is concretized; and this One illuminates the Soul -second hypostasis-, who resembles "the lunar star, which takes its light from the Sun" (Plot. *Enn.* V, 6, 4, vv. 15–16; I, 6, 6, v. 27). However, ultimately, the one who provides (παρέχον) Intelligence with the capacity (τὴν δύναμιν) to be what it is, is the One, "pure light" (Plot. *Enn.* VI, 4, 4, 18–20). Such is the case that, if the reality of light "is like an activity" (ὡς ἐνέργεια) (Plot. *Enn.* IV, 5, 7, v. 43), the activity emanating from the One is according to Plotinus "like the light of the Sun" (ὡς ἀπὸ ἡλίου φῶς) (Plot. *Enn.* V, 3, 12, v. 40).

The scheme of reality proposed by Plotinus required the Parmenidean notion of ontological "continuity" (συνέχεια) as an elementary principle of his metaphysical approach. This idea was especially emphasized by Proclus, for whom this continuity, in the ontological unfolding of the One to the territories of the manifold, traversed different hierarchical levels of reality, which facilitated, in turn, a criterion of dignification and ordering in the progressive multiplication of intermediating entities. Such entities gained or were lost in invisibility and perfection depending on whether they were closer or farther away from the source of being itself, that is, higher or lower on the scale of transcendence. This commitment implied, among other things, the non-existence of emptiness in the physical world; the persistence of that same continuity not only within the sensible world, but between the sensible and the suprasensible dimensions of intervals of qualitative nature (Proclus 1963, p. 216). This also meant the establishment of a tendency towards similarity, in accordance with the consequent ontological diminution, from the posterior to the anterior order. This latter process is, moreover, carried out according to the two laws of continuity: "Every producing cause brings into existence things similar to itself before different things," and "every procession is achieved through a resemblance of the secondary to the primary" (Proc. *Inst. Theol.* §28 [32, 10–11] & §29 [30, 12–13]).

The doctrine of continuity thus permeates all mature Neoplatonic philosophy and is, at the same time, one of the keys to understanding the success of the theology of light in later patristic thought. The principle is evident: the best allegorical strategy for expressing God's activity in relation to the created world is to think of Him in the form of an unattainable source of light, which radiates throughout his creative self-display. However, the difficulties that sensible light itself poses in establishing the boundaries between its activity and its source lead, almost irremediably, to a consideration which, without being exactly pantheistic —inasmuch as there is no full identification between an immanent God and created nature— is nevertheless close to positions that could pass for panentheistic —inasmuch as all that is, is *literally* in God. The theology of light, therefore, carried in its bosom an evident danger of pantheism or, at best, panentheism.

3.2 The Ontological "Continuity" in the Patristic Framework: Dionysius the Areopagite

The most representative case of this panentheistic drift is that of Ps. Dionysius Areopagite, whose works would become, with the passing of time, an authentic "manual of the Middle Ages of the symbolism of light" (Álvarez Gómez 2004, p. 533). According to his theological principles, the Divine Scriptures deploy an extensive repertoire of images, figures, and elliptical resources that, in his own words, are "figures and appearances of that which has neither form nor appearance" (*De Coel. Hierar.* II [PG 3, 140a]). In this sense, the most important work of his oeuvre is the treatise on the names of God, in which he goes through each of the appellatives collected in the biblical texts as if they were units of transcendent meaning, whose intelligibility collaborates in the "return" (ἐπιστροφή) from the multiple and varied towards the indistinct Unity. One such name is Light, inasmuch as it expresses exactly the mechanics of the "unfolding" or "procession" (πρoόδος) of God as an excellent form of extroversion or "giving-away Himself" (Schäfer 2006, 179).

This circumstance fits perfectly with his own consideration of the ontology of the created as possessing two elementary notes, on which their respective beauty rests. On the one hand, the criterion of the "good fit" or "fitness" (εὐαρμοστία) of its parts. On the other hand, the presence of a certain "brightness" or "radiance" (ἀγλαΐη) (*De Divin. Nom.* IV, 7 [*PG* 3, 701c]) that renders them beautiful and charming. Therefore, if things are also, under this ontological mode, "lights" or "depositaries of light," which receive their ontic condition from a superior Light, this not only legitimizes theologically the name of Light for God as Light-giver, but also its symbolic use both in the scriptural, as well as in the specifically material, realm.

A paradigmatic example of the tropism linked to the use of "sensible light" to shed "intelligible light" on the background of a theological mystery can be seen in his formula to explain the indissoluble character of the Trinitarian Persons. The particle that articulates the whole is καθάπερ, "as if," which enables a translational exercise where the vehicle is the sensible light and the tenor is the Trinity. The allegory is certainly illustrative: like the lights of the lamps in a house, each enjoys distinction in unity and unity in distinction, "they are unified in their difference and differenced in their unity." However, when there are many lamps it ends up resulting in "a light unifying all the lights into one light which is one, clear, undifferentiated illumination" (*De Divin. Nom.* II, 4 [*PG* 3, 641a-b]), in such a way that no one would be able to determine exactly the limits of each one, even if one persisted in the idea that they are, indeed, distinct lights.

However, the recourse to light as the most appropriate symbol to reflect the activity of God generates a problem that is repeated in other figures of similar structure has illustrate its tropism. Dionysius writes, using the metaphor of the circle, that all things lose their individuality in the original source of being, as happens with the radii of a circle and its center: "All the lines of a circle subsist together in the center of a circle in one unity: the point uniformly has all the straight lines in itself both among themselves and with respect to the one source from which they proceed." Thus, "if they stand away from it a small distance, they are slightly differentiated; if they stand away a great distance, they are more greatly differentiated" (*De Divin. Nom.* V, 6 [PG 3, 821a]). It is true that apparent undifferentiation does not imply its ontological equivalent, and, in this sense, the analogical structure itself could operate here in terms of a check against pantheistic risk. As far as the allegories of light are concerned, the matter will not be resolved until the appearance of Arabic philosophy, and in particular the proposals of Avicenna and Alhazen, who introduce a triple distinction that makes it possible to better articulate the phenomenological mechanics of light and, consequently, to better refine theological considerations. Thus a distinction will be made between (1) *lux*, i.e., illumination as observed in incandescent objects, such as fire, stars or the Sun; (2) *lumen*, equivalent to the immediate brightness that arises around any point of light, in the manner of a "sphere of light" (*sphaera lucis*) that formally allows the visibility of objects; and (3) *radius* or *radiositas*, i.e., the irradiation of light that allows the brightness of objects (Lindberg 1978, p. 356) at a distance from the light. Thanks to these distinctions the metaphor of light came into common use, especially from the thirteenth century onwards, in thinkers such as Grosseteste, Roger Bacon, or Bartholomew of Bologna, without thereby incurring the original problems of pantheism.

Both the metaphor of light and those with which it shares the problem of the lack of distinction between the source of the unfolding and what is properly unfolded share the same panentheistic risk, whereby things participate to such an extent in the divinity that His presence in them makes them, in effect, also divine. Some authors have pointed out that the key to overcoming this problem lies in one's understanding of the concept of "emanation." First, according to Christian Schäfer, the fact that the unfolding of being takes place in the form of an emanation does not imply that the activity of God, on the one hand, and His accomplished or realized acts, on the other, are not distinguishable, at least *in intellectu*. Second, a key element of emanation and that distinguishes the One from the rest of the hierarchical levels consists, precisely, in His exclusive capacity to create, which allows Him to make Himself present in the different levels of being, but as "a decreasing and multiplying mediated echo in Neoplatonic mediation" (Schäfer 2006, p. 57). If this approach is correct, any pantheistic risk is overcome when this principle is assumed as the basis for a more than evident discontinuity between God and the rest of radical creatures (Pseudo-Dionysius 1987, p. 142). At the same time, the problem of panentheism is also overcome, although in a more subtle and controversial way, since the ontological identity is resolved on the basis of the notion of divine presence: God is present in everything created, without being confused with the created, which, nevertheless, dwells in Him. In other words, about unity and plurality, "as all things *without distinction*, God is neither any one thing nor all things in their plurality" (Perl 2007, p. 33).

A problem still remains, nevertheless, in Dionysius' characterization of the Creator, in that it involved the appellation, admittedly difficult to translate, of "overflowing" (ὑπερπλήρες). In his own words, "the Cause of all is over full of all in one excess beyond having all, the names of Holy of holies and the rest must be celebrated of the overflowing cause and separate preeminence, as one might say" (*De Divin. Nom.* XII, 4 [*PG* 3, 972a]). This and other similar expressions pose an unavoidable difficulty in drawing exact boundaries between the transcendent principle of the divine and that which springs, flows, or emanates from it, which necessarily implies a veiled acceptance of a panentheistic commitment, since, in effect, there is an ontological diminution as one descends from the source of being to the diversity of things that are. Precisely because of this deficit of perfection in being, one cannot speak of pantheism: creatures participate in the Light as far as they are permitted by their own nature, but it is evident that this is diluted and softened at the furthest limits of ontological unfolding (*De Divin. Nom.* XI, 2 [*PG* 3, 952a]). Again, this circumstance implies a degradation of the emanative power and, consequently, a foreseeable ontological discontinuity. Pantheism, inasmuch as it involves precisely an ontological continuity from the original One, is necessarily alien to this approach.

3.3 The Case of John Scotus Erigena: Pantheist, Panentheist... or neither One nor the Other?

Although the risks of pantheism and the specifically panentheistic issues were overcome, not without controversy, by Ps. Dionysius, this was not the case with some of his most eminent commentators. The Irishman John Scotus Erigena, a determined promoter of Latin Neoplatonism in the High Middle Ages, seems to have surpassed such limits, especially if we take into account the understanding of his work by its main interpreters during the 19th and 20th centuries. His most controversial doctrine is based on a conception of nature as divided into four basic instances:

1. as "nature that creates and is not created," i.e., God the creator, as the cause of the universe;
2. as "nature created and that creates," equivalent to the archetypal ideas of God, created in the mind of the Word and serving as the model of all things;
3. as "nature created and that does not create," corresponding to created things, which receive their ontological notes from (2);
4. as "uncreated and uncreating nature," where God is thought as final cause of the whole process.

This last idea imbues all reality with a clearly Neoplatonic dimension by an ontological procession from Unity to things and, also, teleological, when this procession is from things to God. In synthesis, and in his own words, "it follows that we ought not to understand God and the creature as two things distinct from one another, but as one and the same" (*De Div. Nat.* III [*PL* 122, 678c]). Erigena's approach has thus tended to be understood from a monistic and, at the same time, pantheistic perspective, which earned him the definitive condemnation of his works in 1225 by Pope Honorius III.

Dermot Moran suggested in the late 1980s that there were at least three keys to approaching this issue. First, Erigena balanced his more clearly pantheistic propositions with "assertions of the absolute difference between God and creation" (Moran 1989, 89). Second, some of his more polemical considerations concerning the identity of God and nature were made in the light of an evangelical doctrine according to which God will be all in all (I Cor. 15: 28). This versicle can be interpreted under the idea that God is present and that all things depend totally on God's being, independently of other more radical readings of pantheistic orientation —incompatible, on the other hand, with the negative theology inherited from St. Dionysius and which, according to Moran, Erigena develops into a "negative dialectic" (Moran 1989, p. 89) that crosses his vision of the whole system of reality. Third, and finally, Erigena himself distinguished God as genus and his creatures as species (*De Div. Nat.* III [*PL* 122, 677c]), in such a way that surreptitiously the Platonic notion of "participation" seems to play an important role in all his thought (Moran 1989, pp. 88–89). This would move the approach away from pantheistic positions and bring it closer to moderate panentheism, hard to bypass, even under Moran's complex argumentation. In any case, it seems advisable to review more carefully the subtleties of Erigena's thought, especially if the author of Periphyseon himself expressly rejected the accusations of pantheism.

The idea that John Scotus Erigena's pantheistic influence lasted until his final condemnation in 1225 can be traced to the Synod of Paris in 1210. The problem lay in certain assertions concerning the idea that (1) God is the essence of all things (*essentia omnium*) and, on the other hand, (2) the form of all things (*forma omnium*), an idea that we can trace in other thirteenth-century authors who are not suspected of any pantheism, but are deeply neoplatonic, such as Robert Grosseteste. The meeting of Paris dealt with three issues: the problem of the Amalrician heresy (Capelle 1931), the management of the Aristotelian texts translated by David of Dinant, and, finally, the express order that his theological texts, written in the vernacular, should be sent to the bishops of the individual dioceses for their assessment. What we know of David's doctrine comes from the commentaries of his contemporaries, especially St Albert the Great and St Thomas of Aquinas. Following the latter, David identified universal matter with God (*Sum. Theol.* I, q. III, a. 8). This proposition was articulated on the basis of a reduction of reality to three elementary categories: separate eternal substances —God— minds or spirits —angelic realities and human souls— and bodies —matter—. However, these categories were reduced to a single one, namely, the divine essence, which was defined as the formal principle of all things, without our knowing exactly what dimension this expression pointed to.

In any case, if this identification of matter with God was true, it implied at least two lines of problems. First, the difficulties of overcoming the realm of material pantheism, once the God-matter identity has been solved. Second, the closure of any transcendence and, consequently, the need to assume the ontological continuity and coincidence of the divine with the material, which places us in a position where it is not possible to explain the existence of separate entities. However, it is worth noting that there is insufficient evidence to support either Erigena's influence on David of Dinant or Dinant's own pantheism, to the extent that we cannot absolutely confirm this pantheistic drift —regardless of specific studies such as Thery's classic one (Théry 1925), the question about the deep meaning of that *forma omnium* expression remains unanswered.

3.4 The Case of St. Francis of Assisi: The Alleged "Pananimism" in Franciscan Theology

The theological perspective of St. Francis of Assisi is certainly a unique case, due, above all, to a double biographical and bibliographical circumstance. The first refers to the testimonies of some of his most celebrated biographers, who frequently delve into his extraordinary way of relating not only with the living forms of his environment -animals and plants-, but also with inert forms, such as water, fire, the sun, or the stones themselves (Rivera de Ventosa 1955, pp. 210–215). In line with this approach, around 1225 St. Francis of Assisi composed the *Canticle of the Sun*, one of the two texts of his production that he wrote in Umbrian, his native language and dialect that was spoken in the central region of the Italian peninsula. The text echoes his more than evident conviction of the value that creatures play in his basic theological approach, lacking in subtlety, but of great theological depth. In this respect, there are two considerations: the first defends the existence in the thought of St. Francis of, if not a pantheism, then of a pan-animist perspective, by which all creatures —even those that are properly inert, such as water, fire, or stones— enjoy a soul and, therefore, participate in some way in the immortality of the divine. The second position, on the contrary, defends the revelatory value of the loving and disinterested contemplation of creatures, in line with the more orthodox approaches of the previous century.

The first position finds its most canonical statement in environmentalist Lynn White, Jr.'s classic and influential work, *The Historical Roots of Our Ecologic Crisis*. White opposes two ways of understanding humanity's relationship with nature from the Christian, medieval point of view in particular. On the one hand, the dominant perspective, according to which Christianity "is the most anthropocentric religion the world has seen," to the point of affirming that the dismantling of "pagan animism" was the condition of possibility to abusively exploit nature "in a mood of indifference to the feelings of natural objects" (White Jr. 1967, p. 1205). On the other hand, the vision of Francis of Assisi is presented as an alternative in the treatment of humanity's relationship with nature, where the human being is, by way of humility as a species, dispossessed "from his monarchy over creation and set up a democracy of all God's creatures" (White Jr. 1967, p. 1206). This consideration is fundamentally supported by St. Francis' famous Canticle of the Sun, where the beauty of the creatures, as well as their mere presence, constitute the minimum necessary conditions for an express fraternal recognition of them. This could lead to a conclusion of "pan-psychism of all things animate and inanimate, designed for the glorification of their transcendent Creator" (White Jr. 1967, p. 1207).

In this sense, White did not contemplate a pantheistic commitment in St. Francis' natural theology, for, after all, the treatment of each thing as "a brother or a sister, since each had been created, together with himself, by God" does not reach the ontological depths of pantheism, which implies an identification or, at least, the denial of divine transcendence in favor of his total immanence. In this sense, "the case of St. Francis may be rather exceptional in Christianity" (Nishitani 1983, p. 281), but only insofar as his love for creatures was capable of leading him to go beyond strictly human limits to reach out to all things and trace —or, if one prefers, re-signify— our relationship with them.

It is worth noting that, although the hypothesis of Franciscan pantheism is not supported, neither by the texts nor by attitudes of Francis himself, White's idea of a pan-animist perspective in the theological thought of St. Francis could be related to probable panentheism, "meaning that he believes that God includes the world as a part of his being" (Moloney 2013, p. 83). After all, the extolling not only of living creatures —that is, effectively endowed with souls under the canons of thirteenth-century psychology— but also stones, water, or the Sun, as recorded in the Canticle, might give rise to such a thought.

However, White's position is doubly problematic: firstly, because this "is curiously devoid of evidence" (Hoffmann 2014, p. 89); hence, secondly, it is situated in a perspective that is more intuitive and prone to clichés than properly scientific. In reality, the main reason why numerous environmentalists turn to the conduct of St. Francis of Assisi as exemplary does not rest on an exaggerated pananimist conception, but because of his incombustible love for the created and for nature as the basis for a new relationship of human beings with their natural environment. To this circumstance responds the Encyclical Letter *Laudato Si'* of the Holy Father Francis on care for our common home (2015): "Saint Francis of Assisi reminds us that our common home is like a sister with whom we share our life and a beautiful mother who opens her arms to embrace us."

A conception of the order of being in which every creature, just because it comes from the hand of God, necessarily has a bond of familiarity with the rest of creatures. But from the fact that there is a certain ontological familiarity between all things -based on a certain monistic conception of reality- one cannot infer an ontic equality between all of them. Thus, the position of Max Scheler is interesting, for whom, far from being a pantheist or a panentheist, what St. Francis simply does is to extend the love of God through uncontainable love for his creatures. It is an "expansion of the specifically Christian emotion of love for God the Father, and for our neighbour and brother in God, to include all the lower orders of Nature" (Scheler 2017, p. 87). In short, this broadening of the human capacity to extend love toward all other creatures, living and inert, increases the scope of love for God *in* his creatures and not only *from* his creatures. This allows for the evolution of Christian theological thought toward a more dignified consideration of the created environment. This in no way implies either a pananimistic consideration of the created or an immanent consideration of the Creator.

3.5 Conclusions

In general terms, pantheistic and panentheistic positions were hardly echoed in medieval thought, except in very specific cases such as those represented by John Scotus Erigena, under due caveats, or Meister Eckhart, whose more extreme positions can be explained, to a large extent, by his deliberately ambiguous use of mystical language. There are specific accusations against some authors, such as David of Dinant or Amalric of Bena, but in both cases, in the light of the existing bibliography, it is not possible to determine exactly whether they maintained pantheistic positions, panentheistic ideas, or rather interpretations excessively close to the limits of theological decorum but inserted within orthodoxy.

In any case, it is obvious that the Neoplatonic matrix, starting from the thesis of "ontological continuity," facilitated the explanation of the intimacy of creatures with God, while at the same time problematizing the status of this relationship, forcing the exegesis of the biblical texts and the patience of the commentators of the Fathers of the Church, particularly those who remained in the orbit of the original Neoplatonism, such as Dionysius himself, but also St. Gregory of Nyssa or Maximus the Confessor. In this sense, it is difficult to avoid the tensions that animate their respective positions. Medieval Neoplatonic thought presents a very complex and subtle stratigraphy, which, to this day, continues to generate an abundant and profound philosophical and theological literature.

As for other relevant figures, such as St. Francis of Assisi, it is evident that a feeling of universal fraternity, under the creationist paradigm, of the human being towards the rest of creatures, regardless of their belonging to the animal, vegetable, or mineral kingdom, is perfectly compatible with the idea of a divine transcendence and does not imply, in any way, the assumption of pantheistic or panentheistic doctrines, or pananimistic ones. The case of St. Francis is, moreover, particularly critical because of the role he is given in the history of environmentalist thought. Therefore, it is very important to establish his position in order to avoid misunderstandings both from an ethical or practical point of view, and especially from a theological perspective.

References

Álvarez Gómez M (2004) El poder de una metáfora. In: Álvarez Gómez M (ed) Pensamiento del ser y espera de Dios. Sígueme, Salamanca, pp 527–546
Bales EF (1982) Plotinus' theory of the one. In: Baines Harris R (ed) The structure of being: a Neoplatonic approach. SUNY Press, Albany, pp 40–50
Capelle C (1931) Autour du décret de 1210: III. Amaury de Bène. Étude sur son pantheism formel. Vrin, Paris
Hoffmann R (2014) An environmental history of medieval Europe. Cambridge University Press, Cambridge
Lindberg DC (1978) Science in the middle ages. The University of Chicago Press, Chicago
Moloney B (2013) Francis of Assisi and his "canticle of brother sun" reassessed. Palgrave Macmillan, New York
Moran D (1989) The philosophy of John Scotus Erigena: a study of idealism in the middle ages. Cambridge University Press, Cambridge
Nishitani K (1983) Religion and nothingness. University of California Press, Berkeley
O'Daly GJP (1974) The presence of the one in Plotinus. In: Atti del Covegno internazionale sul tema Plotino e il neoplatonismo in Oriente e in Occidente (Roma, 5–9 ottobre 1970). Academia Nazionale dei Lincei, Rome, pp 159–169
Perl ED (2007) Theophany. The neoplatonic philosophy of Dionysius the Areopagite. SUNY Press, New York
Piñero R (2022) No hay palabra alguna… una estética de la contemplación. Sindéresis, Madrid
Proclus (1963) Proclus. The elements of theology. Clarendon Press, Oxford
Pseudo-Dionysius (1987) Pseudo-Dionysius. The complete works. Paulist Press, Mahwah
Rivera de Ventosa E (1955) San Francisco y el panteísmo naturalista. Naturaleza y gracia 2:209–227
Schäfer C (2006) Philosophy of Dionysius the Areopagite. An introduction to the structure and the content of the treatise on the divine names. Brill, Leiden
Scheler M (2017) The nature of sympathy. Routledge, New York
Théry G (1925) Autour du décret de 1210: I. David de Dinant. Étude sur son pantheism matérialist. Vrin, Paris
White L Jr (1967) The historical roots of our ecological crisis. Science 155:1203–1207

Chapter 4
Nature, *Venustas*, and Harmony

Gianluca Cuozzo

Abstract The present essay investigates aspects of Renaissance philosophy, taking as its starting point a selection of pictorial images that can be understood –because of their rich theoretical references– as true treatises on the philosophy of nature. These traits contribute to a present-day proposal of a plausible ecological paradigm, which builds on what Botticelli, Giorgione, and Leonardo were able to communicate through their pictorial works. Concepts such as harmony (as an ontological balance among existing things); limit (which can control a human planning aimed at the transformation of the datum of nature); unemendable human ignorance (regarding the long-term effects of our actions); and the relationship between the preservation of the beauty of creation and the survival of our species are constitutive elements of what can be called the paradigm of a "new humanism." This paradigm can be inferred both from the studies by Arne Naess and from those by Aurelio Peccei: it is a new philosophical *koinè*, aimed at a respectful knowledge of the world, modeled on the one proposed by the multifaceted genius of the philosopher-painter-architect-engineer-literate Leonardo da Vinci.

Keywords Nature · New humanism · Environment · Beauty · Ecology

4.1 Images of Nature

This chapter primarily highlights features of Renaissance philosophy based on a selection of paintings that can be understood, because of their rich theoretical references, as authentic treatises on the philosophy of nature. Three Renaissance artists offer examples that might be useful for proposing a possible ecological paradigm.

G. Cuozzo (✉)
Department of Philosophy and Educational Sciences, Università degli Studi di Torino, Turin, Italy
e-mail: gianluca.cuozzo@unito.it

L. Valera (ed.), *Pantheism and Ecology*, Ecology and Ethics 6,
https://doi.org/10.1007/978-3-031-40040-7_4

First, Sandro Botticelli's *La Primavera* [*Spring*] (1477–1478). Figure 4.1 in this painting, we may point out the concept of *venustas* (beauty, grace, gracefulness): this should be assimilated both to the representation of nature by the cornucopia of the nymph Amalthea (Plutarch) and "to those places most fertile with every kind of *specieria* [spices], and gems and precious stones" (Chiarelli 1992, p. 15). These fertile places were explored by navigators who landed in the New World, particularly Columbus and Vespucci (Cuozzo 2021, pp. 51–64). The Botticellian ideal of beauty, however, is primarily to be understood as a principle of moral exhortation to the contemplation of the harmony and proportionality of creation. This principle is evidenced by the conception of the didactic painting as a true *icon pulchritudinis* [icon of beauty] prepared for the education of the young Lorenzo of Pierfrancesco de' Medici, the second cousin of Lorenzo the Magnificent, whose instruction was entrusted to Ficino, Poliziano, and the poet Naldo Naldi: "*Agite igitur, ô amici, divinam ipsam virtutis ideam formamque semper ante oculos habeamus*" [Let's act therefore, my friends, that we may always have the divine idea and form of virtue before our eyes] (Ficino 1561, p. 807).

Second, Giorgione da Castelfranco's *I tre filosofi* [*The three philosophers*] (1506–1508) (Fig. 4.2). This work should be understood in connection with the Neoplatonic conceptions of the time (particularly Ficino's) and in consonance with some of Leonardo da Vinci's reflections about the science-natural world

Fig. 4.1 Sandro Botticelli's *La Primavera* –Uffizi Gallery (Florence, Italy). This is a faithful photographic reproduction of a two-dimensional, public domain work of art. (Source: Wikimedia Commons). The work of art itself is in the public domain for the following reason: this work is in the public domain in its country of origin and other countries and areas where the copyright term is the author's life plus 100 years or fewer

Fig. 4.2 Giorgione da Castelfranco's *I tre filosofi* –Gemäldegalerie, Cabinet (Berlin, Germany). This is a faithful photographic reproduction of a two-dimensional, public domain work of art [Source: Wikimedia Commons]. The work of art itself is in the public domain for the following reason: this work is in the public domain in its country of origin and other countries and areas where the copyright term is the author's life plus 100 years or fewer

relationship. In this regard, it is worth noticing the three characters, from right to left: from the Ptolemaic savant to the young Copernicus (or one of his disciples), and in the middle, the oriental wiseman (Hermes Trismegistus?). It is also worth noticing the representation of the cave in semi-darkness, which attracts the absorbed gaze of the young researcher, as well as the realization of the painting according to the *duo lumina* technique: one natural (with incidence from left to right), the other otherworldly (affecting the rocks in the upper left part of the painting and giving them that warm gilding that attracts the gaze of the young scientist equipped with a square and compass, "symbolic attributes of the science of measuring the Earth" – Hirdt 2004, p. 78). The resulting concept of nature is the layered one of a physical world to be measured by both scientific procedures based on "mathematical demonstrations" (Leonardo da Vinci 2002, p. 20), and by the appreciation of a divine force, "spiritual, incorporeal, invisible" (Leonardo da Vinci, Cod. Ar. II, 151r). Regarding the latter, Leonardo writes that the *impetus* can be seen "as a spiritual being in every way similar to a soul" (Duhem 1906–13, p. 223). In the case of Giorgione and

Leonardo, however, it is a "weak" pantheism that rests on a mathematical axiom. In Leonardo's case, this axiom is deduced from the studies of Luca Pacioli of the proportionality and harmony of the structure of existence. The latter consists of the relationship between the individual members "together composed" in relation to the whole and of each part with each other. This mutual relationship is such that the various elements "simultaneously compose the divine harmony of this set of members" (Leonardo da Vinci 2002, p. 40).

Third, Leonardo's *Monna Lisa* [*Mona Lisa*] (1503–1516) (Fig. 4.3), *La dama con l'ermellino* [*Lady with an ermine*] (1488–1490) (Fig. 4.4) and *Ginevra de' Benci* (1474) (Fig. 4.5). Taken together, these three paintings represent a real revolution in the conception of nature, far from the ancient concept of *physis*, insofar as it is impregnated with temporality, entropic flux that consumes everything with the "hard teeth of old age" (Leonardo da Vinci, Cod. Atl., fol. 71 r.a –new 195r). In this regard, a concept that would have substantial success in the following centuries' *Naturphilosophie* should be highlighted: the relationship of dependence of human artifacts on natural processes of transformation. This is the central theme of the *vanitas* of architectural constructs, or *artificiata*, predisposed by human ingenuity, as stressed by the slender bridge on our right, close to being swept away by the violent motion of the waters collected in the two asymmetrical basins placed in the background of the mountain landscape. The isomorphism between human virtue and animal grace, to which the portrait of Cecilia Gallerani bears vivid witness, with reference to Leonardo's poetic bestiary, according to which the candid animal, the ermine, would rather "die than be soiled" (Leonardo da Vinci, Ms. H, fol. 48v). In this way, it acts as a physiognomic mirror of the virtue of the female character portrayed, full of kindness of soul and capable, by reason of this natural grace, of remaining in harmony with the whole. The Leonardian axiom "*virtutem forma decorat*" [Beauty is the ornament of virtue] is portrayed in the verse on the panel held by Ginevra de' Benci. This motto seems to be inspired by Ficino's *Convito sopra lo divino amore* [Convite over divine love], according to which morality, truth, and beauty are fused together in an organic vision of nature. It might be further inspired by the vision of the three Graces offered by Botticelli's Primavera embody youth and beauty, mirth, and elegance: "Orpheus calls *Splendor* the grace and beauty of the soul, shining forth the clarity of science and morals; *Viridity* is greenness, the suavity of figure, and color: for these maxims in green signal the flourishing of youth; and he calls Joy, that sincere, useful, and continual delight, which Music brings us" (Ficino 1544, p. 91; emphasis added).

Some corollaries of Leonardo's thought, such as the assumption of the prodigies of technique in an eminently imitative paradigm, according to which scientific applications, rather than altering "*artifiziosa natura*" [artifactual nature] assume its features and prolong its formative virtues at the level of invention (or second nature). should be added to those concepts. These are precise limits to the technical abilities of human beings in their admirable operations on the "*universa macchina di questo mondo*" [whole machine of this world] (Ficino 1544, p. 128). According to this statement it is legitimate to ask how far one can go altering the existing harmony in an attempt to realize the "paradise of science" (Leonardo da Vinci, Ms. E, fol. 8).

Fig. 4.3 Leonardo's *Monna Lisa* –Musée du Louvre (Paris, France). This is a faithful photographic reproduction of a two-dimensional, public domain work of art [Source: Wikimedia Commons]. The work of art itself is in the public domain for the following reason: this work is in the public domain in its country of origin and other countries and areas where the copyright term is the author's life plus 100 years or fewer

Fig. 4.4 Leonardo's *La dama con l'ermellino* –Czartoryski Museum, Kraków (Poland). This is a faithful photographic reproduction of a two-dimensional, public domain work of art [Source: Wikimedia Commons]. The work of art itself is in the public domain for the following reason: this work is in the public domain in its country of origin and other countries and areas where the copyright term is the author's life plus 100 years or fewer

Fig. 4.5 Leonardo's *Ginevra de' Benci* –National Gallery of Art, Washington, DC (US). This is a faithful photographic reproduction of a two-dimensional, public domain work of art [Source: Wikimedia Commons]. The work of art itself is in the public domain for the following reason: this work is in the public domain in its country of origin and other countries and areas where the copyright term is the author's life plus 100 years or fewer

Regarding this last point, it is worth noticing that Leonardo defines the human being, even with the use of technological aids, as the most "cruel and despised monstrosity": in fact, there was no thing "above the earth, or below the earth and water, that has not been persecuted, removed or broken [by humans]" (Leonardo da Vinci 1974, p. 129). Some of Leonardo's inventions, not surprisingly foreshadowed in futuristic sketches, would have had to remain secret in order to prevent human beings from giving vent to their destructive energies amplified by these technological prostheses. For example, the design of a diver's diving suit, once realized, would have prompted humans to lower themselves into the seas in order to rip open, completely unnoticed, the hulls of ships, thereby increasing their potential for annihilation in the event of a conflict and causing the already significant sacrifice of human lives in the bloody wars with which human history is studded to rise exponentially. Better, then, not to divulge such inventions "for the evil natures of human beings, who would use assassinations in the bottom of the seas by breaking the ships at the bottom and submerging them together with the human beings who are in them" (Leonardo da Vinci, *Cod. Atl.*, fol. 237 v.b. –new 647v).

4.2 Towards a New Humanism

At the end of this brief iconological review, I would like to justify some statements by two of the founders of modern ecological thought, Arne Næss and Aurelio Peccei. I maintain that these statements are helpful to address two tasks. First, to cope with current global challenges, such as pollution, depletion of nonrenewable energy sources, global food problem, shortage of water resources, and so on. Second, to re-propose a "new Renaissance" (Næss) or a "new Humanism" (Peccei) in current times. The center of this new paradigm of thought, capable of effectively reacting to the twentieth century alarmism, interweaves heterogeneous knowledge: *Naturwissenschaften* [natural sciences] and *Geisteswissenschaften* [humanities]. These disciplines should work hand in hand in order to stem those systemic imbalances introduced by human productive activities, systemic alterations that cannot be addressed unilaterally by the natural sciences alone. Indeed, such solutions require a new philosophical *koinè*, starting from the multifaceted genius of the philosopher-painter-architect-engineer-literate Leonardo da Vinci. Leonardo's definition of *homo artifex* achieved a happy balance between "the rehabilitation of scientific *curiositas* [curiosity]" (Blumenberg 1992, pp. 393–394) and a religious sense of wonder at the immense cosmic spectacle revelated –"in its height, breadth, and depth" (Cusanus, *De vis.* 8: h VI, n. 30, lin. 17–18)– to the modern natural scientist (Blumenberg 1992, p. 488).

A biographical fragment of Leonardo (who is represented plastically in the process of scouring a cavern never before explored) is particularly relevant. At an existential level, we can evoke both the image of Giorgione and those of Leonardo's own words:

> Drawn by my eagerness, eager to see the great copy of the various and strange forms made by nature's artifice, I turned a little way among the cloudy cliffs, and came to the entrance of a great cavern, before which I remained somewhat stupefied. Unaware of this, I bent my kidneys in an arc. I stopped my weary hand above my knee, and with my right hand I made darkness to my lowered and closed eyelashes, and often bending this way and that to see inside I discerned something, and this I was forbidden to do because of the great darkness that was there. And being a while, immediately two things rose in me: fear and desire, fear for the threatening and dark cavern, desire to see if there within was any miraculous thing (Leonardo da Vinci, *Cod. Arundel* II, fol. 155r).

Regarding Peccei's thought, this would be a rebirth of knowledge "on a human scale," capable of prompting a human transformation that would elevate human capacities and qualities to match new responsibilities in the global context. This philosophy of human life, at the basis of an adequate consciousness of the species (based on eminently biocentric theses), should have been groundbreaking, overthrowing "untouchable principles and norms, and fostering the rise of new motivations and spiritual, ethical, social, aesthetic, and artistic values, responsive to the current imperatives" (Peccei 1976, p. 155). In the face of the danger of the disappearance of humankind, such a revolution should induce the entire political, social, and economic system to prevent its structural collapse by bringing collective institutions as a whole "to a higher level of understanding and organization, based on a stable internal balance and fruitful communion with Nature" (Peccei 1976, p. 155).

Since the time of Aurelio Peccei, overcoming of the *predicament of humankind* that is both erosive and forgetful of the world, continues to be the main challenge for the future of civilization. With acumen and foresight, Peccei hoped for the rebirth of new humanistic knowledge based on quite different assumptions than the current approaches promoting the simple grafting of the *humanities* on digital technologies (the so-called ICTs), accepted uncritically. The one most widely recognized to have forcefully re-proposed the renewal of such a humanistic perspective (in an original synthesis based on biblical revelation, enriched by the legacy of the classical tradition as well as by the reflections on the person and the various contemporary sensitivities present in the most diverse cultures aimed at "a search for harmony within and with creation") has been Pope Francis.

Pope Francis states that today, in the face of a catastrophe, "there can be no renewal of our relationship with nature without a renewal of humanity itself. There can be no ecology without an adequate anthropology" (Bergoglio 2015, n. 118). Therefore, for an integral ecology (as an interaction of scientific, humanistic, sociological, economic, political, and juridical approaches), "we are asked to rethink our presence in the world in the light of the humanist tradition: as a servant of life and not its master, as a builder of the common good with the values of solidarity and compassion" (Bergoglio 2021). We need to rethink this in the face of what is called the "sacralized mechanisms of the prevailing economic system" (Bergoglio 2013, n. 117), due to which "the earth, our home, is beginning to look more and more like an immense pile of filth" (Bergoglio 2015, n. 21).

Peccei (1986b, p. 71) has proposed "a philosophy of human life." This philosophy stems from the combination of ethics, politics, technology, science, and ecology, and seeks to confront the hopeless illusion that techno-scientific developments are the only "bearer of human progress, despite their anarchic character and the bad use made of them" (Peccei 1984, p. 8). Nevertheless, in the face of a proposal based on a sound ethical view of reality, more needed to be done in the past 50 years, while environmental problems have become enormous. So, it is worth remembering that "unless we elevate and develop our quality of life, in harmony with the changes we cause in the outside world, ours will be a race toward catastrophe" (Peccei 1986b, p. 72). In his words:

> The task incumbent upon us is to orchestrate, with a keen sense of reality, different ideals spreading from these various sources [spiritual, philosophical, ethical, social, and aesthetic] in order to generate in ourselves both those moral tensions and that will that are necessary for creative activities while setting for ourselves a set of goals consonant with our times (Peccei 1976, p. 151).

4.3 Ecological Intersections

Besides the need to combine very different methods of inquiry –e.g., economics, technical knowledge, physics, chemistry, ethics, philosophy, and cybernetics– some guiding theoretical principles emerge from the standpoints of Arne Næss and Aurelio Peccei. They take up the considerations highlighted in the commentary on the images of Botticelli, Giorgione, and Leonardo, whose artistic works represent an imaginal condensation of the doctrines of the philosophy of nature of that time. I identify five core guiding principles that emerge from the work of Næss and Peccei.

Principle 1. *The critical claim of an insuperable limit to be assigned to the purposes of human praxis, which tends to exponential growth and imperious transformation of the natural world.*

According to Næss, the fragility of the human situation, constantly exposed to the risk of negative feedback that might overcome the tipping point (upsetting the application of the "principle of revisability" –Næss 1989, p. 69), is such that there is a need for a prior assumption of responsibility, even at the expense of our cognitive uncertainties –concerning any possible intervention that might irreversibly alter natural balances. Attempting reliable conjectures about the spillover effects of our actions in the medium and long term, Næss (1989, p. 90) states that "whenever an ecological policy decision is made, it is necessary to make a synthesis: we take responsibility for all aspects of the problem." Indeed, Næss recalls the Pascalian bet about the existence or non-existence of the Absolute, transformed into a principle of environmental ethics: again, if I were to be wrong about the health of my planet, saying (out of over-optimism) that catastrophe is unreal, I would have everything to lose. Better to do something here and now: in the face of life in general, there is no point in "the recommendation and instigation of bold, radical conservation steps justified by the statements of our lack of knowledge." (Næss 1989, p. 27). As Hans Jonas (2009, p. 46) announced: "Never make the existence or essence of man globally understood a stake in the wagers of action." The loss, in this case, is too significant to rely on chance.

Principle 2. *The mutual interdependence among action, the given world as a whole, and individual natural phenomena (according to the Anaxagorean principle, "omnia in omnibus et quodlibet in quolibet").*[1]

[1] "*Omnia in omnibus et quodlibet in quolibet*" can be translated as "everything in everything else and whatever in whatever else."

As Næss (1989, p. 83) writes, the motto "everything depends on everything expresses the need to develop global worldviews, where in principle everything is relevant to any choice." To this end, to better understand the ecological relevance of this interdependence (avoiding "culpable negligence" [Næss 1989]), it is worth emphasizing the unity between *oiko*-logy and *eco*-logy, such that we should overlap between *οἶκος* (i.e., the home or environment constituted by organisms in a given external context, organic and inorganic (according to Ernst Haeckel's 1868 – classic definition), and *'Ηχώ* (i.e., the Greek nymph Echo, daughter of Air and Earth, the personification of the physical phenomenon of echo). According to this *krasis* of meanings, ecology would be concerned not only with the community of living things in their possible mutual relationships within their common living context, but also with their ability to react/respond (individually and collectively) to the systemic consequences produced there by their transformative activities.

A responsible community, in this sense, is that living consortium of living organisms (called by Næss 1989 as "complex interaction") that starts a virtuous, two-way correspondence with the variations produced by practice in the surrounding world. Acting responsibly, therefore, means virtuously interacting with the context and with the consequences of one's actions (feedback theory), knowing how to make appropriate and innovative responses from time to time to the phenomena of perturbation determined therein, even questioning the intentions of one's actions.

Today, the response (*responsum*) par excellence should internalize the principle "act in such a way that the consequences of your action are compatible with the permanence of authentic human life on earth" (Jonas 2009, p. 16). Otherwise, as in Ovid's fable, the nymph Echo, instead of fostering the bio-social community of the living, will resonate as the mournful wail of our belated repentance, hauntingly echoing from the depths of that erosive "hole-world" from which nothing can return (Dick 2007, p. 139): our defaced Earth, transformed into the tomb of our intentions for growth to the bitter end. The symbol of this new cybernetic knowledge, based on a theory of feedback (Meadows et al. 1973, p. 38), is not so much the Hegelian owl of Minerva, which takes flight when the time is ripe, but that funny and ingenious nocturnal mammal that is the bat, capable of perceiving the world in the form of reflected acoustic iteration. I believe that reconstructing the surrounding reality on the basis of ultrasonic impulses returned from the environment in the form of echoes (a perceptual prodigy referred to as *echolocation*, the basis of the invention of sonar) can take us beyond the erosive solipsism of our times: I am referring to the acosmism of a subject who no longer takes pleasure in earthly things (Plotinus 1992, 1363), locked among the referentless images of his evanescent pseudo-identity, whose fate is exemplified by the unwary Narcissus, the victim of his own empty and seductive *imago*. To borrow from Hegel: *the bat takes flight when postmodernism –a veritable kaleidoscope of seductive, de-emphasizing, referentless images– is accomplished, reaching its twilight*. Its circumspect hovering, free from the pleonastic seductions of an ever-available world, points out the recapture of the real world.

Principle 3. *A profession of ignorance about the long- and medium-term consequences of altering natural balances (which recalls the need for consequential ethics* (Næss 1989, 4).

Here it is appropriate to recall Nicholas of Cusa. First, his *docta ignorantia* [learned ignorance], according to which the possibility of mistake is never entirely averted in our scientific projects. In this sense, Ernst Cassirer argues that Cusanus "has grasped the whole of the fundamental problems of the age from a single methodical principle" (Cassirer 2019, p. 9). Second, his *scientia ignorationis* [knowledge of ignorance] (forerunner of falsificationism), for which the conjecturality of human knowledge is the inevitable corollary (*ars coniecturalis* [conjectural art] which is disclosed, from the politico-religious point of view, in the ideal of universal *concordantia* [concordance], tolerance, and peace among peoples [Mancini 2015, pp. 18–19]). It is, after all, the awareness of human fallibility, captured by Cusanus in the paradoxical formula "*nihil scire, nisi quod ignoraret*"[2] (Cusanus, *De docta ign.* I, 1: h I, p. 6, lin. 11), which is at the heart of every scientific theory, indicating humility and uncertainty as the constitutive basis of all rational knowledge. This criterion, Karl Popper would say, is thus given by the falsifiability and revisability of every cognitive hypothesis. This principle –still present in Leonardo– becomes a true corollary of Næss's (1989, p. 27) thought:

> Should we proceed with the project or not? The burden of proof rests with those who are encroaching upon the environment. Why does the burden of proof rest with the encroachers? The ecosystems in which we intervene are generally in a particular state of balance which there are grounds to assume to be of more service to mankind than states of disturbance and their resultant unpredictable and far-reaching changes. In general, it is not possible to regain the original state after an intervention has wrought serious, undesired consequences. And intervention, ordinarily with a short-sighted gain for some minor part of mankind in view, has a tendency to be detrimental for most or all forms of life.

This dubitative method, as noted, in Cusanus takes the form of a critical principle that subordinates knowledge "to the method of learned ignorance. Without this, the attention paid to the worldly reality of sensible char acter would expire into idolatry" (Thurner 2001, p. 86). The celebration of fact, which has as its inevitable corollary those hyper-specialized experts who are "intoxicated' by facts" (Næss 1989), and who contemplate no strategy of repentance. Popper (2012, p. 57) writes:

> The more we learn about the world, the more conscious, specific, and articulate will be the knowledge of what we do not know, the knowledge of our ignorance. For this is the main source of ignorance: our knowledge can only be finite, while our ignorance can only be, of necessity, infinite.

Principle 4. T*he deep intertwining of harmony (i.e., equilibrium, homeostasis, resilience principle of natural systems), on the one hand, and survival of the human species (including its health or "mental ecology"), on the other.*

[2] *"Nihil scire, nisi quod ignoraret" can be translated as* "to know nothing more than what he did not know."

A prerequisite of this awareness, according to Peccei, would be the titanic endeavor concerning the readjustment of historical memory (by which humans construct their progressive narratives and measure their brief successes) with geological memory (the *tempus naturae*), consequently reconfiguring goals and values of their strategies of production/transformation of the material data; this is what Næss (1989, 45) calls "a *substantial reorientation of our whole civilization.*" If we were to measure the history of *homo sapiens* over the entire twenty-four hours by which we measure the history of the world, it would occupy barely a few seconds. As Benjamin points out, drawing on the studies of J. Rostand (1939, p. 104):

> The miserable fifty thousand years of homo sapiens represent, concerning the history of organic life on earth, something like two seconds at the end of a twenty-four-hour day. The history of civilized humankind, reported on this scale, would also occupy one-fifth of the last second of the last hour (Benjamin 1997, 55).

To illustrate the problem of the exacerbation of this hiatus between the two temporalities (natural and historical), Peccei juxtaposes the times of exploitation of recourses with the cyclical times by which nature regenerates itself, highlighting the enormous gap existing between the respective scales of measurement (for which the assertion *nulla proportio* [no proportion] seems to hold true). In his own words:

> To manufacture an inch of *humus*, nature needs an average of 1 to 4 centuries. Having said that, everything we destroy now, all the species we wipe out, will not return to life at the time of our children, grandchildren, and great-grandchildren. Deserts are spreading, and tropical forests –the habitat of the largest number of animal and plant species– are being felled, burned, and cut down at a dizzying rate. All this will have uncountable repercussions on human ecology [...]. I believe that this phenomenon of destruction of nature is the most dangerous factor because it is advancing quietly. We do not notice what we are losing, what we are far poorer of than our fathers and what will be lacking to our children even more than us: the natural resources essential to life (Peccei 1986a, p. 201).

The problem has been further exacerbated; think of the unbridgeable hiatus that has resulted from the fact that we have very rapidly consumed enormous amounts of resources (coal, gas, and hydrocarbons) that were formed in geological processes of hundreds of thousands of years: "When I start my car I am setting liquefied dinosaur bones on fire" (Pellegrino 2021, p. 19). The consequences of this acceleration, in addition to a devastating input of CO_2 into the atmosphere (the cause of *global warming*), are not yet fully known to us; thus, our existence seems to float over an abysmal void, which is endangering human ecology itself –i.e., our psychic balance and mental health.

Principle 5. *The relationship between natural beauty (to be understood in an ontological sense, as harmony, proportionality, and intrinsic consonance of the parts into which reality is articulated) and the sustainability of complex political-economic systems.*

The Planet's health, both from the natural point of view and from that of the so-called *mundus hominum* (by which human being shapes both the technical-productive structures and those of the socio-political organization of his/her existence), depends on the balance achieved between the first and second nature;

when this latter is highly entropic, exceeding the immunization threshold provided by the natural cyclical processes, a dangerous systemic instability is created. Indeed, Patrik Artus and Marie-Paul Virard (2008, p. 8) argue that global capitalism is configured more and more as "a formidable unequalitarian machine, stoking the fires of all manner of disorder: financial, economic, social and environmental." This makes the socio-historical world resemble a "pot that burns up scarce resources," encouraging the politics of hoarding and accelerating the greenhouse effect, as well as the insane production of waste. Additionally, it reduces the world to "a casino where all the excesses of financial capitalism are expressed" (Artus and Virard 2008, pp. 8–14). This process involves the sharpening of the discrepancy between capital incomes (constantly increasing) and the standard of living of the middle class (drastically impoverished). All of this, between excesses and bets, has been called the systemic trap of our production model, "which adds misfortune to misfortune": it keeps the poor poorer and continues to grow the income and capital of the few, "damaging soils, waters, forests and ecosystems" (Meadows et al. 2006, pp. 71–72) – anything but the dream of a *mondialisation heureuse* [happy globalization] (Artus and Virard 2008, p. 59).

4.4 Magic Constraints and Ecological Knowledge

The features analyzed through principles one to five fall within the framework of a philosophy of nature inspired by a neo-Platonic tradition. This tradition is central to much of Renaissance thought, such as emphasizing the holistic and integral character of our relationship with the world. It is a philosophy of nature in an eminently critical sense due to the intertwining of emerging knowledge. While assigning a normative value to nature, this philosophy recalls the immense human responsibility towards the aforementioned challenges, that is, a responsibility dealing with his/her sciences (and techno-practical applications) and his/her ultimate values. Indeed, humankind, by virtue of its peculiar position in the world, results at once inside and outside the natural world, immanent to the whole of the world precisely in its irreducible eccentricity. In this regard, the Renaissance's figure of the wizard-philosopher, from Ficino to Bruno (recalling the Hermetic tradition), embodied a twofold character. First, the dimension of goodness deals both with the world, in its axiological fullness (whereby ethics must always start from the axiom *primum vivere* [first live], as Anders recalled since it must guarantee the biotic basis of the subsistence of the human species[3]). Second, our distancing ourselves from it (in the sense of our *ek-sistere* [ex-sist] and our constituted *Welthoffenheit* [world hope]).

[3] We may find this normative fullness of nature, Valera points out, in Næss: Following Spinoza, whose immanent and integral view of natural processes had so much influence on the Norwegian's thought, nature not only has moral, utilitarian, or aesthetic perfection, but it is perfect "in itself" (Valera 2015, p. 16).

Timothy Morton, taking up Bruno's *De vinculis in genere* (1591), reminds us that ecology (as a dark science translatable into an eco-gnostic), is that "weird knowledge" that magically connects –in the sense of a winding and entangling (loop)– people, things, machines, and natural elements; all beings, ultimately, "are suffused and enveloped in mysterious hermeneutic clouds of the unknown" (Morton 2021, p. 54). These range from the tiniest detail to the biosphere (and the solar system itself), where every gesture and step forward in the control of the natural process corresponds to "a recoil, a contrary and symmetrical movement" involving ever larger spheres of reality (Morton 2021, p. 54). In this regard, what could be more magical (and supernatural) than the seemingly inconsequential fact that as I turn the key to my car, I am helping to "cause the sixth mass extinction of a four-and-a-half billion-year-old planet?" (Morton 2021, pp. 55–56).

Indeed, as Næss already noted, linking Gestalt ontology and the theory of the "total relational field" as the objective structure of reality (the environment as *Lebenswelt* [Lifeworld]), "the individual acquires consistency only within a relational context, which continually nurtures and generates him. Thus, to know for human beings does not so much mean to abstract from their life context as to interact with their surroundings" (Valera 2015, p. 24), to be aware of this fundamental interconnectedness.

The ecological topic of limitations, then, could be retranslated into the question posed by Karl Löwith in the 1950s. He denounced in the purely historical paradigm of technical manipulation, the modern "*Denaturierung des menschlichen Lebens*" [Denaturation of human life] (Löwith 1983, p. 288), whose nihilistic effects threaten our very subsistence on the planet. How far are we allowed to distance ourselves from the world, as an integrated system of interdependencies, in order to still be human without compromising our own cultural and biological existence?

Rethinking our tradition of the 15th and 16th centuries, as I have tried to point out, may help us to reflect both on the subject of nature (as *venustas* et *harmonia*, the ontological presupposition of any ethical-philosophical problematization) and the *humanitas* of a full responsibility towards our actions. We design, impact, and transform our environment within a framework of references, both human and extra-human ones, characterized by a network of dense mutual connections. In the encounter between these two vectors –passivity and activity, material basis and power of conjectural abstraction, life, and existence– the meaning of a renewed philosophy of nature and philosophically grounded ecological knowledge remains.

References

Artus P, Virard M-P (2008) Globalisation: le pire est à venir. La Découverte, Paris

Benjamin W (1997) Sul concetto di storia. Einaudi, Turin

Bergoglio JM (2013) Esortazione apostolica Evangelii Gaudium. Tipografia Vaticana, Vatican

Bergoglio JM (2015) Lettera enciclica Laudato si'. Sulla cura della casa comune. Tipografia Vaticana, Vatican

Bergoglio JM (2021) Videomessaggio del Santo Padre Francesco in occasione dell'Assemblea plenaria del Pontificio Consiglio della Cultura. https://www.vatican.va/content/francesco/it/messages/pontmessages/2021/documents/20211123-videomessagggio-plenaria-pcc.html
Blumenberg H (1992) La legittimità dell'età moderna. Marietti, Bologna
Cassirer E (2019) Individuo e cosmo nella filosofia del Rinascimento. Bollati Boringhieri, Turin
Chiarelli B (1992) Paolo da Pozzo Toscanelli. In: VVAA (ed) La carta perduta. Paolo dal Pozzo Toscanelli e la cartografia delle grandi scoperte. Alinari, Florence, pp 13–24
Cuozzo G (2021) Cusano. L'arte e il Moderno. Morcelliana, Brescia
da Vinci L (1974) Scritti letterari. Garzanti, Milan
da Vinci L (2002) Scritti. Tutte le opere: Trattato della pittura, Scritti letterari, Scritti scientifici. Rusconi, Milan
Dick PK (2007) Ubik. Fanucci, Rome
Duhem P (1906–13) Études sur Léonard de Vinci (II). F. De Nobele, Paris
Ficino M (1544) Sopra lo amore o' ver Convito di Platone. Néri Dorteláta, Florence
Ficino M (1561) Epistolarium. Opera omnia, vol I. Henricus Petri, Baseline
Haeckel E (1868) Natürliche Schöpfungsgeschichte. Gemeinverständliche wissenschaftliche Vorträge über die Entwicklungslehre im Allgemeinen und diejenige von Darwin, Goethe und Lamarck im Besonderen, über die Anwendung derselben auf den Ursprung des Menschen und andere damit zusammenhängende Grundfragen der Naturwissenschaft. G. Reimer, Berlin
Hirdt W (2004) I tre filosofi di Giorgione. Società Tipografica Fiorentina, Florence
Jonas H (2009) Il principio responsabilità. Einaudi, Turin
Löwith K (1983) Sämtliche Schriften. Metzler, Duisburg
Mancini S (2015) Congetture su Dio. Singolarità, finalismo, potenza nella teologia razionale di Nicola Cusano. Mimesis, Milan
Meadows D, Meadows D, Randers J, Behrens WW III (1973) I limiti dello sviluppo. Rapporto del System Dynamics Group MIT per il progetto del Club di Roma sui dilemmi dell'umanità. Mondadori, Milan
Meadows D, Meadows D, Randers J (2006) I nuovi limiti dello sviluppo. La salute del pianeta nel terzo millennio. Mondadori, Milan
Morton T (2021) Ecologia oscura. Logica della coesistenza futura. Luiss University Press, Rome
Næss A (1989) Ecology, community and lifestyles. Outline of an Ecosophy. Cambridge University Press, Cambridge
Peccei A (1976) La qualità umana. Mondadori, Milan
Peccei A (1984) Il mondo di domani. In: VVAA (ed) Verso il Duemila. Mondadori, Milan
Peccei A (1986a). Introspezione nel problema dell'apprendimento e dell'educazione. In: Peccei A, Lezioni per il ventunesimo secolo. Scritti di Aurelio Peccei. Presidenza del Consiglio dei Ministri, Rome
Peccei A (1986b) L'imperativo di un nuovo umanesimo. In: Peccei A (ed) Lezioni per il ventunesimo secolo. Scritti di Aurelio Peccei. Presidenza del Consiglio dei Ministri, Rome
Pellegrino G (2021) Dopo la natura. La New-Wave post-ecologista di Timothy Morton. In: Morton T (ed) Ecologia oscura. Logica della coesistenza futura. LUISS, Rome
Plotino (1992) Enneadi. Rusconi, Milan
Popper K (2012) Congetture e confutazioni. Il Mulino, Bologna
Rostand J (1939) Héredité et racism. Gallimard, Paris
Thurner M (2001) Theologische Unendlichkeitsspekulation als endlicher Weltentwurf. Der menschliche Selbstvollzug in Aenigma des Globusspiels bei Nikolaus von Kues. Mitteilungen und Forschungsbeiträge der Cusanus-Gesellschaft 27:81–128
Valera L (2015) Arne Næss. Introduzione all'ecologia. ETS, Pisa

Chapter 5
Spinoza: Ecosystemic Consequences of the Intersections Between Pantheism, Panentheism, and Acosmism

Luciano Espinosa

Abstract This chapter shows how Spinoza's metaphysical approach has a variety of applied consequences since it always operates within an integrated whole in a formal and material, intensional and extensional sense. Indeed, certain paradoxes must be assumed, such as obviating the impossible dialectic between the infinite and the finite, while moving from the indeterminacy of the divine substance to the determination of the modal realm. This requires a speculative loop and indirect approach, rejecting both the isomorphism being/thinking and anthropomorphism. This approach emphasizes the univocity of cause and potency when understanding the relationship between substance/finite modes without failing to distinguish between the natural and natural realities. This needs to: (a) consider the ontological binomial *implicatio/explicatio*; (b) understand the infinite modes as keys of organization (i.e., articulation, composition) of the real; and (c) understand intuition as transversal knowledge that goes beyond the categories of reason. Such presuppositions give rise to two consequences: in the absolute, different facets respond to the enunciated concepts so that a certain perspectivism can be used when facing the richness of the ontological expression of the divine/natural; both the insufficiency and the intersection of such approximations allow considering "cosmotheism," insofar as it affirms the physical and intelligible world where human life takes place, which is here interpreted according to a general key called ecosystemic, and here the practical consequences make sense. Indeed, we can speak of a metaphysics that is neither materialistic nor idealistic, crossed by an awareness of limits (ontological heteronomy), from which the autonomy of the human being emerges.

Keywords Spinoza · Cosmotheism · God · Immanentism · Metaphysics

---

L. Espinosa (✉)
Faculty of Philosophy, Universidad de Salamanca, Salamanca, Spain
e-mail: espinosa@usal.es

L. Valera (ed.), *Pantheism and Ecology*, Ecology and Ethics 6,
https://doi.org/10.1007/978-3-031-40040-7_5

5.1 Meaning and Purpose

The comparison between the various ontological expressions listed in the title of this paper witnesses to the complex character of Spinoza's tenet and avoids all reductionism, beyond the obvious criticism of theological monotheism. Insofar as they limit and complement each other, the enriching aspects of these perspectives come to light, both from the point of view of their contribution and of what remains undealt with. Only at the end of that analysis we will see why it is convenient to end up in a different framework, here called *ecosystemic*, where the lives of individuals –human or other-than-human– are carried out with all their practical implications.

The point is to go from the *macro-logical* level to a *micro-physical* one. This poses some difficulties which must be examined –such as the relationship between the whole and the parts, eternity and time, or infinitude and finitude. We face the problem of how to connect the need of the real with the duration of what happens or, if preferred, what tradition calls *nunc stans* (the now that remains), and *nunc fluens* (the now that passes). It is known that everything is in God and that without God nothing can either be or be conceived of (E1P15).[1] At the same time, there are two reciprocal levels –with no dualism involved: the difference between singular things (and their ideas, including those of non-existent things) which are always "included" among God's attributes (and in God's infinite Idea). Yet, they also "imply" existence insofar as they are said to last (E2P8C). Therefore, the implicit and the explicit –as different, yet not dissociated– must be combined.

The focus is on the consideration of the real alien to time and space; in other words, understood in terms of need and what is common, which is truly rational. Though there is another vision immersed in them insofar as it is diverse and contingent, typical of imagination (E2P44C1 and C2). A substantial part of the contents of pantheism, panentheism, and acosmism consists in making them comprehensible and somehow compatible. We will not deal with the matter in its full scope, but two facts are undeniable: that reason is not limited to quantity, number, and size (Ep 12), which only apply as regards duration and are alien to the eternal. While, at the same time, humans must cope with plurality, variation, and fluctuations typical of finitude and time. In other words, there are two basic types of concepts –abstract and sensorial– linked to movement (we will see to intuition further down). However, this does not curtail discussing a peculiar *experience* of the eternal, starting by ourselves (E5P23S), which seems to ground and cut across everything.

There are several loose ends still, but we must note that ambivalences (if not ambiguities) included in Spinoza's system prove fertile. That is where crucial

[1] I am using the internationally accepted abbreviation (E) for Ethics, followed by the n° of the part and the proposition (P), completed in its case with the demonstration (D), the scholium (S), corollary (C), lemma (L), definition (def), axiom (Ax), appendix (A); TTP is used for Theological-Political Treatise, followed by the n° of the chapter and the page in Gebhardt's (G) edition (Spinoza 1972), vol. 3; and for Correspondence or group of epistles Ep is used and the corresponding n° plus the page in Gebhardt's edition, vol. 4.

issues -theoretical as well as practical- are discussed, both as regards the global notion of reality and its consequences on human life, individual and collective. It should be kept in mind that Spinoza stressed the close connections among metaphysics, ethics, and politics (Ep 27 and TTP 4, Spinoza 1972, G. 47). That is why it is convenient to emphasize that the Dutch author's philosophical revolution begins in the foundations of the system.

5.2 On Pantheism

It must be kept in mind that *pantheism* is a notion that was coined after Spinoza's time. Consequently, it must be used cautiously when interpreting his work. In this work, I select some lines that coincide with the pantheist approach, insofar as they help to reply to the issues posited. Firstly, God, as *causa sui* (cause of itself)[2] (E1def1) and absolute nature (constituted by infinite attributes or truly different essences), all-inclusive, is, at the same time, efficient cause, first cause, and by itself (E1P16) of the essence of all things (E1P 24). This means that singular things follow from divine need, they depend on God for all effects and are inherent to God by definition. It is necessary to define the kind of causality and note that it has a univocal character, because God is cause of Godself "in the same sense" as of them (E1P25S). This excludes the equivocal or pluri-vocal character that would separate God from his productions/affections, be it in terms of eminence and/or analogy. That is why it is also stated that he is not a "remote cause." There is a formal community between cause and effect, in other words, between God, attributes, and modes (E1P28S and Ep 4). Moreover, it is the "immanent, and not the transitive cause" of things (E1P18), which eliminates transcendence and emanation.

Ultimately, there is a direct causal link between the divine and the singular, within a *horizontal* ontological framework. Consequently, although things are heteronomous because they depend essentially and existentially on God, it is also true that God is unthinkable without them. It happens altogether and at once if I may say so. On the other hand, substance is absolute (E1P11) insofar as it includes the infinite attributes (dimensions of the real), of which only extension and thinking are known. All these express truly different essences and have nothing in common (E1P10S): because of that, they are independent and equal, that is why no hasty materialistic or idealistic readings should be deduced from them. No doubt, it is not the personal God of monotheisms, guided by moral will: "It cannot be called a free cause, but can only be called necessary," since in Spinoza will and understanding are equal. Will and understanding are related to the nature of God as motion and rest (E1P32 and C2) in order to produce things in the only possible manner (necessary means complete, perfect or finished) and not in view of achieving the good (E1P33

[2]The property possessed only by God, of being his own cause, i.e., independent of any other ground, yet containing within himself a sufficient explanation of his own being.

and S) or, in general, of the best, as was an issue since Plato's time (Timaeus 28c). This kind of *amoral neutrality* –things are as they are, without extrinsic qualifiers or global ends– assumes that God is not a subject (as is a king or a judge) and eludes all anthropomorphism, which, at the same time, permits human autonomy.

Jointly with that *static* view of the system, there is another one that might rightly be termed *dynamic*, linked to the equally absolute potency of God or Nature (E1P32–34); that is, *essentia actuosa* (E2P3S) of which all things participate in some degree and thus become causes that generate effects. The principle of causality orders and avoids chaos by connecting things and events of that vast activity, but there is no way to grasp that unlimited production: it is intelligible, but the infinite causal sequences that move reality are the same that turn it into something indeterminate and indeterminable from the human point of view. Reality cannot be apprehended, let alone unravel its details. That may be the reason why Spinoza searches once more for a direct link between God and finite beings through potency when he refers to "the nature itself of the existence" (not duration in a quantitative and abstract sense) of individual things:

> [I]n so far as they are in God. For although each individual thing is determined by another individual thing to existence in a certain way, the force nevertheless by which each thing perseveres in its existence follows from the eternal necessity of the nature of God (E2P45S).

It is as if there was an intimate transfusion of strength (*vis*) or potency, away from the multiple causal connections in between (the so-called proximate causes) that define the essence or *conatus* of the thing which thus perseveres in existence. It is the immediate, *intensional*, link besides the *extensional* way due to the many interconnections of singular things.

The Dutch author refers everything "actually existing" to the "eternal and infinite essence of God" as its source. This allows him to posit the relationship inversely and in most revealing terms in the following proposition: the idea of an object "involves" the divine essence, as is obvious from everything stated, and "that which gives a knowledge of the eternal and infinite essence of God is common to everything and are equally in the part and in the whole" (according to E2P38) "this knowledge therefore will be adequate" (E2P46D). Now it is the thing that remits to God, and it does it in an epistemic key: the divine essence is the common ground of reality, where rational knowledge is grounded (the *common notions* that connect singulars), according to the conceptual game of whole/part and implication/explanation, since knowing something *implies* appealing to the divine essence (everything), since it *explains* its existence (part), leaning on something common to all things and which, consequently, is true.

Immediately after, another element is added to this that leads to the essence of God, by introducing intuition or the third type of knowledge: as the soul (*mens*) knows itself and its body, as well as when knowing other bodies "actually existing," it sends to or places everything in God. Because of that, it possesses an adequate knowledge of the essence of God (E2P47D). In addition to that, the Scholium says that "the infinite essence and the eternity of God are known to all; and since all things are in God and are conceived through Him, it follows that we can deduce from this knowledge many things which we can know adequately, and we can thus form that third sort of knowledge" (E2P47S). Therefore, access to God is relatively simple, it is enough to see God as the background of permanent reality to which things belong (connection whole-parts), although in order to know the singular essence of the large plurality of things that are inherent to it (not only what they have in common, which is a partial aspect) intuition must be applied. This allows to understand the particular fully and immediately (E2P40S2).

Briefly, pantheism provides three keys that *ecologically* place the finite and temporal in the heart of eternal infinitude, fully and directly: connection through the cause (immanent), participation of potency (internal) and a double gnoseological way, first relational (common notions according to the whole-parts scheme) and after that, the one that singularises (intuition): all of it in terms of implication and explanation. Things are in God, and God is present in them with no mediation (besides causal interconnections that cannot be grasped and are indiscernible). So, understanding them leads to a one-way journey to the essence of divine nature (or reality), which is totality that includes them as parts, and from there it is possible to understand them in a return journey in their particularity. This means that they can be considered inherent, be it as affections, forces or potencies, and singular essences.

5.3 On Acosmism

With these tools it is possible to heed Hegel's criticism when he states that there is no pantheism in Spinoza, but acosmism, insofar as singulars are not properly acknowledged and stated, in view of the fact that immanent and participated potency finally turns into "empty potency" (Hegel 1987, p. 150; p. 155). In other words, the finite reaches the edge of "annihilation" because it lacks its own identity, since it does not establish a dialectic connection with the infinite. The deep reason is that the Dutch thinker posits an "abstract substance," inert, lacking "subjectivity" and "spirituality." Consequently, there is not the necessary inner movement toward the gradual reconciliation of levels along history (Hegel 1987, pp. 413ff.). That absence of the self-realizing impulse of the Spirit, with the entailing mediations would prevent the true grounding of multiplicity, compounded by the absence of the historical dimension as constitutive of reality.

Oddly, Hegel is right in one sense, but he is wrong in another. Substance, God or Nature, is not a subject nor does it become one along time, that is correct; as is also that there is no dialectic connection with the singular. However, that does not mean that this level lacks protection, as we have seen. The point is that Spinoza's conception is absolutely affirmative in the macro and microscopical levels. And that requires emphasizing eternity, which does not include any sort of ontological denial nor change over time. The Dutch thinker does not deem it necessary to overcome antagonisms in the ontological level (maybe it proves necessary in the political level, which corresponds to human autonomy). And there is no awareness of historical progress or cognitive accumulation either, let alone through the famous *cunning of reason* mentioned by the German thinker. Nor does he interpret history as theodicy since there is no evil or subject to be justified. It is obvious that he lacks Hegel's historical sensitivity, but the point is that his perspective is different. There is no Spirit that is unfolded and known in history. The absolute is already complete and does not need to have a *sense*: be it scatologic, moral, or epistemic. Everything is as it should be from the start.

This impersonal neutrality, however, and criticism of all finalisms does not cancel the possibility of grounding the singular: establishing univocity and immanence of the cause, implies understanding the fact that modes "come from eternity" (Ep 12, G.58); in other words, they are grounded in substance. Just as potency-essence that constitutes them as a participated expression of God's implies affirming then, without the fact that finitude is "partly negation" (of what is not that thing itself) leading to any mistake in the face of the "absolute affirmation" of the infinite substance that includes everything (E1P8S1). In the opposite sense, it must be kept in mind that "the universal power of all nature is nothing but the power of all individuals together" (TTP 16, G. 189), which leaves out any kind of submittal to an alien and independent power. As we can see, each level has its statute, and both are reciprocal.

Once again, it is a two-way journey, acknowledged ontologically and epistemically by the fact that existing things "imply and express the concept of God in proportion with his essence" (TTP 4, G. 60). In other words, they acknowledge and involve God under all aspects, according to the corresponding degree in view of their particular nature. And this happens with no need to connect metaphysical essence with moral "fatality," which allows Spinoza to defend ethical behavior be it as regards freedom or need (Ep 43, G.222s). It is not possible to go into details here about these issues, complex and delicate. However, we must complete the reply to presumed acosmism saying that finite beings have a certain amount of autonomy, and that panentheism provides new arguments to guarantee the consistency of finitude.

5.4 On Panentheism

The stance is enriched when it states explicitly that things are in God. This completes what was stated about the cause: "I state [...] that God is immanent cause [...] of all things and non-transitive; furthermore, that all things are in God and they move in God," as many ancient philosophers as well some Hebrews and Paul of Tarsus have said. This has nothing to do with identifying God with a "certain mass or corporeal matter" (Ep 73, G. 307). Spinoza insists on dependency as regards God, but now he emphasizes *insertion* of all existent beings in absolute reality. He highlights belonging and vicinity, including a metaphorical (spatial) and symbolic touch which might be interpreted in Gestalt style: it contrasts background and figure. Individual things find full backing and location in God, which does not imply a confusion of levels. That also accounts for distinguishing between *natura naturans* (nature acting), and *natura naturata* (nature acted upon) (E1P29S): the *thickness* of the real includes those two aspects, clearly distinguished, but not dissociated (e.g., in terms of activity and passiveness).

They are two poles of the same thing, that might point to a mystic background, maybe similar to that of authors such as Meister Eckhart, just to mention an example. It is not easy to establish a frontier in that subtle combination of union and difference, but what is clear in the whole setting is the face-to-face struggle against anthropomorphism and that God cannot be *demarcated* in any way. Indeed, it is possible to posit the hypothesis that the absolute is as amoral as it is, in the long run, *a-logic*: the fact that reason can understand it under the causal perspective that connects substance-attributes-modes and that definitions may be set does not mean it may be fully explained. Not only on account of ontological non-determination (infinite unknown attributes and inapprehensible causal sequences), but also because reason is only useful for and corresponds to human nature, and it cannot be stated whether there is rationality in God or not.

There is true knowledge for human beings, but they are still a particle in the universe. The valuations he attributes to it depend solely on human nature and usefulness (TTP, 16, G. 190s). Consequently, it is intensionally (according to its nature) and extensionally situated, as if an imaginary worm were in the blood, with a certain view of the world and ignoring how the parts and the whole coincide because even the "philosophical mind" errs when it views things as vain, absurd, or disorderly (Ep 32, G. 170–173). In a few words, the human perspective is partial and inevitably biased, in view of the very limits of its nature. It is not possible to develop this issue here either but suffice it to say that for humans the whole "common order of nature" is equivalent to chance, and carrying out a full criticism of anthropocentrism and anthropomorphism actually leads to some sort of *perspectivism of reason*. Not even the language of mathematics, the most trustworthy, fully grasps and expresses the real, as opposed to Galileo's words (*Il Saggiatore, 6 and 48 [The Assayer]*), something Spinoza knows and never repeats. This does not imply an anti-rational stance. Rather, it means being aware of limitations and opening to the meta-rational, insofar as it avoids reification of the divine.

When Spinoza insists that there is no plan, end, global order, or harmony (E1A and 4Pref), he seems to state very clearly that human rationality is one thing and *rationalisation* of the real is another, very different thing. Modifying Hegel, here everything rational is real but not the other way round. Because of that and because God is not a subject, as has been seen, there is no room either for *imitatio Dei,* in view of that unsurmountable gap between the human and the divine, since there is no model to follow nor personal relationship of image and likeness (TTP 13, G. 171). Nor is there community in the will that would supposedly draw them near (according to Descartes' words in *Letters to Chanut 1-2-1647 and 6-6-1647*). In Spinoza's view, it is possible to know – applying a famous analogy with the aspects of language – the structure or basic *syntaxis* of reality, which accounts for philosophical discourse, but there is no *semantics* in common between God and humanity. And that is also why free *pragmatics* in social and political affairs is opened. In a few words, the real does not end once it has been understood from a human perspective, no matter how valuable it may be.

Lastly, panentheism makes it possible to frame the situation within bounds, the status of singulars in general and humans in particular –among other reasons, because they are incardinated in the absolute, not the opposite. Nature is not manufacture nor is it *mindfacture*. That also accounts for the absence of will of power and exploitation in Spinoza. Such will inspires a good part of Western modernity. Spinoza's metaphysical premises are different; they can combine praxis and contemplation with some balance. The thinker trusts reality, while not turning it into a projection of his desires, and establishes a relationship of belonging, not of property. It is summed up in a formula that subjects the principle of sufficient reason to the *sufficient reality principle* (Rosset 1994, pp. 13–37), insofar as the real eludes control and is expressed free of conditionings, lacks or additions, far from demanding any justification.

5.5 Ecosystemic Consequences

Of course, discussion of *Deus sive Natura*[3] provides the discursive groundings and when analysed from the intersection between pantheism, acosmism, and panentheism there is a closer assessment of the dialogic content –if not dialectic– between infinitude and finitude, eternity and time, whole and parts. We have seen that the absolute is comprehensible while incommensurable, it is closely connected to the finite while it cannot be managed in any sense, it is the living source of being and knowing all existence but does not impose any model. Therefore, the point lies in showing where those bases lead or what they permit, what is their framework for life, and what practical applications derive from it, along the line of neutrality mentioned above to open the way to properly human autonomy. Seen from a different perspective: according to what has been said, there is no natural moral or right along conventional lines, nor is there a globalizing *Sense*. On the contrary, the door to the creation of *senses* is opened.

A good example of this distinction of levels is the differences among universal laws of nature, the laws corresponding to human nature, and the positive laws of right (TTP 4, G. 57). About the former, he refers to the example of physical communication of movement among bodies (in TTP 6, he adds that laws admit no exceptions or miracles). As to the second ones, which are a subset of the previous ones insofar as God expresses Godself in human nature, he mentions the association of memory with two concomitant things. As to the third ones, they depend on intelligence and human agreement when it comes to settle life in community. Well, the challenge lies in generating a way of life that will connect rationally the three levels

[3] For Spinoza, God and Nature are interchangeable –i.e., there is no distinction between the creator and the creation.

as much as possible. To do that, it is necessary to achieve the relevant knowledge, both about sets of connected things (through common notions) as well as singular things in themselves (through intuition).

The first step is to stand in the appropriate ontological site and say that the infinite immediate modes of extension and thought are not mediators in any sense. Rather, they are more concrete keys of organization and reference frameworks: motion and rest (M/R) that permit composing bodies, and Infinite Understanding/Idea of God that perceives the intelligible dimension of all existence (ideas of bodies, etc.). At the same time, the only mediate infinite mode is *Facies totius universi*, understood as the unifying expression of the previous ones (Ep 64, G. 278). This seems to become the closest and most effective natural environment for the human individual. The point is to place it within the framework of the immutable total or maximally composed individual (E2P13L7 and S), which includes all bodies (and their corresponding ideas). In other words, particular individuals (couple idea-body) do vary and articulate by means of M/R proportions in multiple scales of composition. That is why there are individuals of first, second, third… way, ending in the notion of nature as a great global individual that includes uncountable assembled parts; humans among them.

This new physical and logic presentation of the whole and the parts, defined by relationships fixed by means of M/R proportions and levels of composition, is what we term *ecosystem* or neutral playing field for human life. This matrix of physical origin that articulates everything is also reflected in other instances as possible gatherings of individuals when acting in different fields, including politics, because the State is an individual composed by many minor individuals. The eco-systemic reading chosen allows to relate and articulate open systems (individuals) through an interchange of all kinds (matter-bodies, energy-potency, and information-knowledge), always at the service of a flexible perspective, rich in contents and transversal, which places human beings in nature better (Espinosa Rubio 1995, ch. 2). That is where the outcomes of previous perspectives meet, though now they are completed with the modifications and interactions that spell reality: to causal and potency connection, plus the insertion of each thing in God (something similar to *latitude*), ties and contacts between things are added (*longitudes*), and it all definitely inserts the parts in the whole. There are causes and effects, forces, and perspectives… that are eventually situated and translated into ideas and affections.

Spinoza starts with the affective and thinking experience *in media res*, not with the *cogito* (E2Ax2, E2P13Postul, E2P23). Additionally, thanks to some process of observation and induction (TTP 7, G. 99) he formulates the great concepts from which he later deduces the rest. It is a fruitful circle between both ways: we might say that the eco-systemic reading responds to the empirical vector, while pantheism and panentheism express the other way, more speculative. In this sense, joining both analytic stances, we may talk about *cosmotheism* (Assmann 2006, pp. 50 s; pp. 87 s), understood as highlighting the mundane which has an affirmative character in and by immanence, though never *taming* the divine. Disentangling it is in charge, as we said above, of rational knowledge of common notions and the particular one of intuition. As it is impossible to know the causal particular sequences (or *proximate causes)* in each case, it is necessary to appeal to the common (which may gather more or fewer individuals) and to the singular intuited essence insofar as it is immediately linked to God. The human being can thus understand not only the metaphysical framework, but the physical and social contexts where they live, to guide themselves pragmatically and cognitively. Also, the fact that they participate in *Amor Dei intellectualis*[4] leads them to understand with maximum qualification and transparency (E5P36S) their intimate and *ecological* connection with God or Nature.

References

Assmann (2006) La distinción mosaica o el precio del monoteísmo. Akal, Madrid
Espinosa Rubio L (1995) Spinoza: naturaleza y ecosistema. UPSA, Salamanca
Hegel GWF (1987) Lecciones sobre filosofía de la religión. (trans. Ferrara R). Alianza, Madrid
Melamed Y (2020) The enigma of Spinoza's amor Dei intellectualis. In: Naaman-Zauderer N (ed) Freedom, action, and motivation in Spinoza's ethics. Routledge, New York, pp 222–238
Rosset C (1994) El principio de crueldad. Pre-textos, Valencia
Spinoza B (1972) Opera. C. Gebhardt (ed.), Carl Winters Universitäts Buchhandlung, Heidelberg

[4]According to Yitzhak Y. Melamed (2020, p. 222) "the notion of divine love was essential to medieval Christian conceptions of God. Jewish thinkers, though, had a much more ambivalent attitude about this issue. While Maimonides was reluctant to ascribe love, or any other affect, to God, Gersonides and Crescas celebrated God's love. Though Spinoza is clearly sympathetic to Maimonides's rejection of divine love as anthropomorphism, he attributes love to God nevertheless, unfolding his notion of *amor Dei intellectualis* at the conclusion of his Ethics."

Chapter 6
Schleiermacherean Panentheism and Ecology

Graham Lee

Abstract Although Friedrich Schleiermacher is relatively well known as "the father of modern liberal theology," not widely known are the ecological implications of his thought despite its considerable influence, that of contemporary ecological concerns, and the evident contiguity of the two. Nor is it well known how exactly his thought is panentheistic. In the chapter I address this question: what Schleiermacher's overall metaphysical-cosmological (MC) account has to offer to ecology particularly inasmuch as the account is panentheistic and it is supposed that the ecological ontology is monistic. This question is best answered once Schleiermacher's MC account is understood to be a certain form of panentheistic post-Kantian neo-Spinozism. The answers to the question and two questions from which it is derived are drawn out in three steps taken in this chapter. The first step involves surveying some important observations made in recent sources addressing the relevance of Schleiermacher's MC account to ecology. The second step is a delineation of in which ways Schleiermacher's MC account, those pertaining to ecology in particular, is panentheistically post-Kantian neo-Spinozistic. There I rely on helpful points concerning the topic drawn from two groundbreaking studies. The chapter concludes with the third step, where takeaways from the preceding two parts are summarily connected, and the authors' insights thereby synthesized in addressing the initial question. Through all this I attempt to show how Schleiermacher's MC account's being panentheistic in the way it uniquely is has significant import for contemporary ecology.

Keywords Ecology · Neo-Spinozism · Panentheism · Post-Kantian · Schleiermacher

G. Lee (✉)
Candler School of Theology, Emory University, Emory, GA, USA
e-mail: glee27@CougarNet.UH.EDU

L. Valera (ed.), *Pantheism and Ecology*, Ecology and Ethics 6,
https://doi.org/10.1007/978-3-031-40040-7_6

Friedrich Schleiermacher's title as "the father of modern liberal theology" (or alternately, "the father of modern Protestant theology") is relatively well known based on his profound influence on modern hermeneutics, the development of biblical higher criticism, and thereby subsequent Christian thought by and large. Not widely known are the ecological implications of his thought, despite its considerable influence, that of contemporary ecological concerns, and the evident contiguity of the two. Nor is it well known how exactly his thought –his later thought especially and, to a lesser extent, his early thought– is panentheistic –or even that it is, perhaps.

Two questions are suggested here. (1) What does Schleiermacher's overall metaphysical-cosmological (MC) account have to say regarding ecology? (2) In what ways, specifically those relevant to ecology, is his MC account panentheistic? In recent years, a handful of sources have explored these questions, directly or indirectly. At least three studies have examined the second question in depth. Still, a question derivative of both questions has yet to be considered in the literature: (3) how the first question is to be answered –what Schleiermacher's (overall) MC account has to offer to ecology– particularly inasmuch as the account is panentheistic and it is supposed that the ecological ontology is monistic. This third question is best answered once Schleiermacher's MC account is understood to be a certain form of panentheistic post-Kantian neo-Spinozism. The answers to all three questions can be drawn out in three steps taken in this chapter.

The first step (Part I) surveys some important observations made in sources addressing the first question. This makes for a literature review regarding the relevance of Schleiermacher's thought to ecology. The second step (Part II) is a delineation of the ways in which Schleiermacher's MC account is panentheistically post-Kantian neo-Spinozistic–addressing the second question. This doubles as an argument for the account being post-Kantian neo-Spinozistic and panentheistic, rather than pantheist, and in a unique way. I rely on helpful points concerning the topic drawn from two groundbreaking studies. The chapter concludes with the third step (Part III), where takeaways from the preceding two parts are summarily connected and the authors' insights are thereby condensed and synthesized in addressing the third question. Through all this I attempt to show how Schleiermacher's MC uniquely panentheistic account has import for, and could and should impact, contemporary ecology.

6.1 Schleiermacher and Ecology

6.1.1 Feeling, Acting, Knowing

In a *Zygon* article, Helen De Cruz brings Schleiermacher's thought and that of certain contemporary philosophers of biology into conversation in order to argue that Schleiermacher predicted insights provided by them regarding the relationship of living things to their environment (De Cruz 2022, p. 162). She points out that, in §3.2–3 of (the 2016 edition) the *Glaubenslehre* (Schleiermacher 2016; hereafter *CF*),

Schleiermacher argues "that self-conscious, living organisms have three properties: they are *feeling*, *doing*, and *knowing*" (De Cruz 2022, p. 164; her italics). Indeed, our self-consciousness and consciousness as agents in relation to the world –what Schleiermacher refers to as "remaining-within-oneself" and "stepping-outside-of-oneself"– are constituted by the three properties (De Cruz 2022, p. 162; quoting *CF* §3.2). Let's examine the three in turn. As we do, many points from De Cruz's analysis of the three will serve well as principal stops on our roadmap, with relevant observations from other scholars interspersed for slight detours where suitable and helpful.

6.1.1.1 Feeling

Schleiermacher takes feeling to be the origin of religion, where feeling is a creature's receptivity to its environment. Through feeling we become aware of how we depend ultimately on God (the feeling of absolute dependence) and how we depend on other creatures in our environment. Let's consider each in turn.

John Crossley, Jr. (2006) explains that the religious ethics implicit in Schleiermacher's dual doctrine of creation/preservation is grounded in the universal feeling of absolute dependence (*das schlechthinnige Abhängigkeitsgefühl*). Opening his article's third section, "The Structure of Religious Ethics of Creation", Crossley, Jr. considers the initial paragraph (par. 36) of Schleiermacher's two-sided doctrine in the *Glaubenslehre* [teaching of faith]:

> The original expression of [the relation between the world and God], i.e., that the world exists only in absolute dependence upon God, is divided in Church doctrine into the two propositions –that the world was created by God, and that God sustains the world (Schleiermacher 1928; quoted by Crossley 2006, p. 600, his bracketed paraphrase).

The doctrine is immediately derived from the feeling of absolute dependence and Schleiermacher directly moves from this feeling to "[t]he proposition that the totality of finite being exists only in dependence upon the Infinite" (Crossley 2006, p. 600; quoting Schleiermacher 1928, p. 142 (par. 36, art. 1)). What extends to the doctrine of the absolute dependence on God of *all* finite things (i.e., the natural world and everything in it), which is objective, starts in the *feeling* of such absolute dependence, which is subjective (Crossley 2006, p. 600).

Whether this absolute dependence is understood in connection with the preservation of the universe as a doctrine of preservation or with the origin of the universe as the doctrine of creation does not make much difference to Schleiermacher, according to Crossley, Jr., since Schleiermacher separates the doctrines for presentational clarity (par. 38, art. 1). Provided that the feeling of absolute dependence is objectively applied to the world, it and all things in it, including *putative* evil (par. 48, art. 1), depend on God for their existence and preservation. *This* makes a difference to Schleiermacher (Crossley 2006, pp. 600–601).

The awareness of our dependence on our environment, including other creatures in it, contributes to the feeling of *relative* dependence: a pre-reflective, immediate self-consciousness experienced by us and other self-conscious creatures when interacting with the world (De Cruz 2022, p. 164). This feeling is divisible into two feelings: (a) a feeling of dependence when "the other parts [of the world] bear an effect on oneself out of their own self-initiated activity", and (b) "a feeling of freedom" when "one […] bears an effect on other parts out of one's own self-initiated activity" (De Cruz 2022, p. 164; quoting from *CF* §32.2). According to Schleiermacher, after all, to produce an effect is to act and for a creature to produce its own effects without coercion is for that creature to be free. The feeling of relative dependence is constituted by the feeling of freedom and the feeling of dependence on the world, the "push and pull" of our acting upon the world and being acted upon by it, as De Cruz (2022, pp. 164–165) helpfully puts it.

Indeed, the feeling of relative dependence is a relation expressed by self-consciousness: "That relation to finite being which can be perceived by the senses, is split into a feeling of partial dependence and a feeling of partial freedom" (De Cruz 2022, p. 168; quoting from *CF* §5.1). In §4.2, Schleiermacher contends that one never can have a feeling of absolute *freedom* due to the constraints imposed on one's action by the world. Whoever thinks themselves absolutely free and self-sufficient are self-deluded. We need our environment, indeed, we are always dependent on it – for air, nourishment, and the other natural necessities for survival. We thus are entangled in "a deep web of interdependence", one constituting the entire universe and grounded by God. This interdependence is both a feeling and "a reality at a deep, metaphysical level" (De Cruz 2022, p. 165). When one becomes aware of God, according to Schleiermacher in §30.1, one becomes aware that God "is designated as the one grounding this interconnected being in all its diverse parts" (De Cruz 2022, p. 165). This indicates that Schleiermacher predicted the view in contemporary philosophy of biology that biological agents are conscious since "they need to be aware of their environment and their place within it in order to act, and consciousness allows for the expression of free agency" (De Cruz 2022, p. 168). As the philosopher of biology Peter Godfrey-Smith puts it, noting how agency and subjectivity are tightly linked in the philosophy of biology: "Nothing is gained biologically from taking in information that is not put to use. The evolution of the mind includes the coupled evolution of agency and subjectivity" (De Cruz 2022, p. 168; quoting Godfrey-Smith 2020, p. 59).

6.1.1.2 Acting

The importance of the agency of organisms in evolution, an insight of the evolutionary biologist Richard Lewontin in his influential article, *The Organism as the Subject and Object of Evolution* (1983), has become commonly accepted in evolutionary theory. Specifically, organisms are agents "who act upon their environment" such that they "exert some influence on long-term evolutionary outcomes" (De Cruz 2022, p. 166). This challenged the notion that organisms are passive toys of the environment such that they are impotent regarding how they are acted upon by natural selection. Lewontin's picture displaced the previously prevailing theory of evolution, according to which natural selection occurs by means of the survival of certain genetic mutations and the dying out of others, due to external forces to which the genes of organisms are subject (e.g., geology, other organisms, weather), also subject to random mutation (De Cruz 2022, p. 166).

Lewontin's now standard picture in evolutionary theory has parallels with Schleiermacher's "threefold distinction of organisms that act [or do], feel, and know" (De Cruz 2022, p. 166). Organisms are influenced by the world and, by influencing it, exercise their *relative* freedom on the world. As agents, they act within the limitations of feeling the push and pull of their environment (De Cruz 2022, p. 166). As James Brandt puts it, "for Schleiermacher the human person or moral agent is the embodied self, shaped by the natural and cultural world that bears down on it [... and] existing in a relation of reciprocity with the 'world' in which it exists" (Brandt 2018, pp. 24–25). Based on his 1843 lectures in *Christian Ethics/Christliche Sittenlehre* (Schleiermacher 2011), the moral agent also is to be understood as "involved in an ongoing process of formation in which the agent's duty is to act in ways aimed at the highest good," an understanding providing "an important warrant that can move toward affirmation of ecological awareness" (Brandt 2018, p. 25). Moreover, the perfection and integration of the goods of human life must be included in the highest good, according to Schleiermacher. Because human life and the natural world are bound up with each other, attentiveness to ecological issues is consonant with Schleiermacher's outlook (Brandt 2018, p. 25).

6.1.1.3 Knowing

Earlier it was noted that, for Schleiermacher, all things in the world depend on God for their subsistence insofar as the feeling of absolute dependence is applied objectively to the world. Schleiermacher does not intend for the doctrine of absolute dependence to undermine "the system of causes and effects operative in the natural world as a relatively self-contained entity" (Crossley 2006, p. 601). Otherwise, Schleiermacher would undermine the entire system of scientific knowledge that he carefully develops in the *Dialektik*. Relatedly, he asserts these two propositions –the first leading to the second– in paragraph 46 of the *Glaubenslehre*:

> Now, as the human soul is just as necessarily disposed towards a knowledge of the world as towards a consciousness of God, it can only be a false wisdom which would put religion aside, and a mis-conceived religion for love of which the progress of knowledge is to be arrested. The religious self-consciousness, by means of which we place all that affects or influences us in absolute dependence on God, coincides entirely with the view that all such things are conditioned and determined by the interdependence of Nature (Schleiermacher 1928, p. 170; quoted by Crossley 2006, p. 601).

When fully extended, knowledge of the world is knowledge of the interdependence of all things in the universe. In article 2 of paragraph 46, Schleiermacher (1928, p. 173) claims both that every finite thing depends on God and that all things are established and grounded in "the universality of the nature-system." Between it and article 1, he contends that the two ideas fully coincide, are not in contradiction, and even reinforce each other (Schleiermacher 1928, pp.172–173). He also says in article 2:

> [D]ivine preservation, as the absolute dependence of all events and changes on God, and natural causation, as the complete determination of all events by the universal nexus, are one and the same thing simply from different points of view, the one being neither separated from the other nor limited by it (Schleiermacher 1928, p. 174; quoted by Crossley 2006, p. 601).

Schleiermacher (par. 47, art. 1) goes on to deprecate the pervasive notion that dependence on God and natural causation are mutually exclusive. Rather, for Schleiermacher they are, as Crossley idiomatically puts it, "two sides of the same coin" (Crossley 2006, pp. 601–602).

Schleiermacher's notion of knowing (*Erkenntnis*) is intimately connected with (the properties of) biological agents acting and feeling in his senses (De Cruz 2022, p. 169). Although he sees feeling as forming the basis of religion (*CF* §3.2), even identifying religion as feeling –to be clear, he does not identify the two in the sense of equating them but, rather, in terms of predication; feeling is predicated of religion but not vice versa– he doesn't take religion to have no connection with acting and knowing. It is connected with both in religious practice and theology, for instance (De Cruz 2022, p. 164). While religion is related to knowing, it is not reducible to *knowledge*. Otherwise, something no one would agree with (he thinks) would follow, namely, that "the best master of Christian faith-doctrine would, at the same time, also unexceptionably be the most pious [or religious] Christian" (De Cruz 2022, p. 169; quoting from *CF* §3.4).

While he is explicit about "lower animals" not having "any complete self-consciousness" and even having no "actual knowledge," the conception of knowledge in contemporary biology and philosophy of biology is quite less restrictive (De Cruz 2022, p. 164; quoting from *CF* §5.1). Philosophers of biology elaborate knowledge, like consciousness and agency, in functional terms, "that is, in order to process perceptions, and to combine their point of view, behavioral flexibility, and goal-directedness, organisms must be able to integrate the information they receive into adaptive decisions" (De Cruz 2022, pp. 169–170; referencing Kaas 2000 and Godfrey-Smith 2020). Organisms having complex behavioral patterns (e.g., feeding, fleeing) require a brain or something analogous (e.g., a nervous system) in order to process information centrally and make decisions. The brain bridges what such organisms feel and do (De Cruz 2022, p. 170).

"Knowledge" refers to information that allows for adaptive action in behavioral biology, and not necessarily to true beliefs (De Cruz 2022, p. 170). Still, the two plausibly are connected, as De Cruz (2022) notes, as in philosophy. The brain enables organisms to flexibly behave in response to their environments. In the case of nonhuman animals, knowledge in this sense would be seen by Schleiermacher, rather than as knowledge as such, to be a form of consciousness. Meanwhile, such mental capacities are seen by contemporary biologists and philosophers of biology to be ways of knowing *per se*, formed by evolutionary forces (De Cruz 2022, p. 170; referencing Holekamp 2007).

6.1.2 *Schleirmachers's Three Features and Ecology*

This only covers a large portion, though not the bulk, of De Cruz's wide-ranging discussion of the striking parallels between Schleiermacher's account of feeling, acting, and knowing as the three fundamental features of the lives of creatures, and work in philosophy of biology. It illustrates the predictive power of the metaphysical-cosmological account of the *Glaubenslehre* with respect to significant discoveries in contemporary biology and insights regarding these from contemporary philosophy. So, firm scientific and philosophical footing is available for connections made between ecology and at least this MC account of Schleiermacher's, if not others. This footing serves as a basis for explicit connections between ecology and the *Glaubenslehre*, in addition to other of his works, to which we now turn.

6.1.2.1 Ecological Economics and the *Naturzusammenhang*

Shelli Poe explains how three main features of ecological economics (ways in which it departs from neoclassical economics) are emphasized by Schleiermacher's mature theology, particularly in the *Glaubenslehre*, even "so robustly that, admitting some anachronism and exaggeration, we might arguably call him an ecological theologian before his time" (Poe 2018, p. 37). We will look at the three features in turn, aided by some insights from a few other authors.

The First Feature

In §38.2 of the *Glaubenslehre*, readers are invited to conceive "the creation of the world and, along with this, the entirety of the interconnectedness of nature to be one divine act" (Poe 2018, p. 37; quoting from *CF*, p. 214), the act of creation/preservation as noted earlier. According to §46.1 (*CF*, p. 248), all things are "determined as one organic, interconnected, natural process" (Poe 2018, p. 37). *Naturzusammenhang* is the term Schleiermacher uses for this "interconnected process of nature" (Poe 2018, p. 36) or "natural nexus" (Waggoner 2018, p. 59).

Indeed, on his conception of the *Naturzusammenhang*, the world is even a living *organism*, a dimension of the view that is important for interpreting his theology, as Waggoner explains (Waggoner 2018, p. 59). In the third *Critique*, Kant makes a distinction between believing the world to be an organism and examining the world as if it is an organism –"a constitutive principle" and "a regulative principle", respectively (Waggoner 2018, p. 59; referencing Kant 2000, pp. 233–234, pp. 250–251, p. 322). Denying this distinction, Schleiermacher writes in the *Dialectic*, "Contrary to Kant, constitutive and regulative principles do not admit of distinction" (Waggoner 2018, p. 59; quoting Schleiermacher 1996a, b, p. 5). Schleiermacher makes explicit that the world is an organism in his *Lectures on Philosophical Ethics*, writing that the world is "the complete unity of finite being as the interrelation of nature and reason within an all-encompassing organism" (Waggoner 2018, p. 59; quoting Schleiermacher 2002b, pp. 148–149).

For Schleiermacher, all parts of the world fit together with each other so that "all could just as well be for the purpose of each part, just as each part could be so for all" (Poe 2018, p. 37; quoting from *CF* §58.2, p. 344). Moreover, the subjective certainty of the Christian that all finite things fully depend on God completely coincides with "the fullest conviction that everything is completely conditioned by and grounded in the totality of the interconnected process of nature [the *Naturzusammenhang*]" (Poe 2018, pp. 37–38; quoting from *CF* §46.2, p. 248).

The Second Feature

Another feature of ecological economics is its emphasis on "the need for each part of the one interdependent organism to properly function within its own limits", an emphasis carried by Schleiermacher's thought (Poe 2018, pp. 37–38). Humanity, for example, must live within specific limitations. The laws of nature governing the lives of all creatures impose some such limitations. Both "objective human consciousness" and human bodies are, as Schleiermacher puts it, "conditioned and determined by the interconnected process of nature [the *Naturzusammenhang*]" (Poe 2018, p. 38; quoting from *CF* §46, p. 248).

The Third Feature

The third feature of ecological economics is a focus on "symbiotic organic relations", specifically "an organic conception of the relation between human persons and the world we call home" (Poe 2018, p. 37, p. 39). This focus is emphasized by Schleiermacher's theology. His presentation of the universe and humanity in it as one organic whole is due to how he understands the relation of Christ to humanity within the natural world. Indeed, as explained in §46 (*CF*, p. 790), the universe and all of humanity with it are determined or shaped as an organic whole by Christ (Poe 2018, p. 38).

Anette Hagan notes that Schleiermacher's view of providence involves both human agency and divine agency, the absolute dependence and the "relative freedom" of finite human agents implied by the *Naturzusammenhang*, and the entire world's redemptive trajectory (Hagan 2018, p. 77). In §164.2 of the *Glaubenslehre* Schleiermacher writes, "The concept of preservation gains its full content only in relation to that element which becomes for us consciousness of grace, leading thereby to the concept of divine causality" (*CF*, 1000; quoted by Hagan 2018, p. 78). We thus can say that "both processes …, the nature of things in their relation to each other and the ordering of their reciprocal influences on each other, subsist in God, just as they subsist in relation to the redemptive revelation of God in Christ or in relation to that revelation of God in Christ which the Spirit is developing toward its consummation" (*CF*, 1000; quoted by Hagan 2018, p. 78). The "single decree to create, sustain, and redeem" includes humans freely acting and reacting in the divinely ordered *Naturzasammenhang*, who therefore are called to advance God's reign "toward its ultimate consummation" (Hagan 2018, p. 78). "Divine providence", Hagan writes, "therefore implies that human beings act toward sustaining the planet and weave ecological, sustainable living into Christian existence" (Hagan 2018, p. 78).

6.1.3 Schleiermacher on Sin and Ecology

It was noted in §6.1.1.1 that everything in the world, including *putative* evil, depends on God for its subsistence (if the objective application to the world of the feeling of absolute dependence is supposed). The cause of genuine evil, however, is to be found within the purview of human beings.

Schleiermacher reverses the customary ordering of the relation between evil and sin, writing in §76.1 of the *Glaubenslehre* that "sin is, above all and overall, the first and original feature, but evil is the derived and secondary feature" (Vander Schel 2018, p. 83; quoting from *CF*, p. 479). Evil, rather than the cause of sin in humans, "names the distortion and deterioration of human living that follows from sin" (Vander Schel 2018, p. 83). Schleiermacher says in §75.1 that "evil is first introduced with the advent of sin, but once sin has appeared it arises inevitably" (*CF*, 475; quoted by Vander Schel 2018, p. 84). Accordingly, the manifestation of sin in humans, the reign of sin in human living, inevitably results in "the spread of evil [...] without fail," "the presence and growth of evil in the world" (Vander Schel 2018, pp. 83–84).

Schleiermacher discusses two classes of evil that affect human life as a result of (human) sin. One, described as a type of "natural evil" (*natürliches Übel*) and independent of human action, arises since "the world with sin appears to be different to human beings than it would have seemed without it" (Vander Schel 2018, p. 84; quoting from *CF* §75.1, p. 475). Natural evil in this sense includes physical inhibitions and bodily afflictions that, although they aren't necessarily incompatible with God-consciousness, through sin are experienced by humans as torments. The other class, "social evil" (*geselliges Übel*), emerges "from the ongoing conflict and opposition between human beings, from oppressive and antagonistic action and the virtually innumerable ways in which persons and communities hinder, injure, and suppress others in pursuit of some private interest or concern" (Vander Schel 2018, p. 84; referencing *CF* §75.2, p. 476). Abuse, alienation, destruction, exploitation, and violence are among the direct effects of social evil in this sense and, so, it can be understood as a form or dimension of "moral evil" as the term is used in the contemporary literature on the problem of evil. Divine punishment for sin is directly exhibited by social evils since the deepening and broadening "deterioration of the world" (*Verschlimmerung der Welt*) results from communal sin (Vander Schel 2018, p. 84; quoting from *CF* §75.3, p. 482).

This deterioration, "a product and consequence of the distortions of collective human action," includes damage and misshaping of the natural world causing destruction and suffering (Vander Schel 2018, p. 85). This destruction and suffering, resulting from "long-standing patterns of domination and exploitation of the natural environment," is a significant and wide-ranging manifestation of the outcome of human sin, social sin (Vander Schel 2018, p. 85). It is *ecological* destruction and suffering: destruction of the environment and its denizens (in addition to the suffering of the latter), organisms not limited to the human species.

Terrence Tice notes that, for Schleiermacher, social sins must be acknowledged, forgiven, redeemed, and lead to purposeful action. Such activity constitutes the ethos of each individual's deeply personal existence in communion with God. Christians receive compassion from and are to return gratitude and tranquil acceptance to God, in addition to actively wishing for others to experience God's love. As they do so, they act in relation to the ecologies constituting the world as well (Tice 2018, p. 94). Tice then adds, "The earth is itself one whole; the universe is one whole; both are created and preserved as one reality by God. Whether individual or in some smaller collectivity, humans are, in fact, inseparable from the earth-world, despite efforts to the contrary. Humanity is part of one indivisible web of being" (Tice 2018, p. 94). Humans compose just one part of the *Naturzusammenhang*.

6.1.4 Schleiermacherean Icoses and Ecology

In the first book-length examination of Schleiermacher's 1813 Academy address (Schleiermacher 2002a), Douglas Robinson discusses Schleiermacher's thought, particularly that set out in the address regarding different modes of translation, with respect to *social* ecologies (what Robinson terms "icoses") in the context of contemporary translation studies (Robinson 2013). What Robinson means by "icosis" is based on and outlined in this extract from the address:

> The need to translate within one's language or dialect, a more or less fleeting emotional need, is too much restricted in its impact to the passing moment to require other guidance than that of gut feelings; if it were submitted to rule, it could only be the kind of rule that impels people to that moral state in which the mind is kept open to that which is more alien, less akin to oneself (Schleiermacher 2002a, p. 226).

In description of an icosis Robinson (2013, p. 10) writes this:

> There is [a] a need [...] which is also [b] [...] a need of the temper or disposition [...] an emotional need, which lasts only a moment, a blink of the eye [...] but [c] it has the force of a rule (Regel) or set of rules, and [d] requires a [...] compliance to [e] a [...] 'moral state' ([... that is in part] a temper [or] more commonly a mood or other affective state).

Robinson considers the interpretation of the need in (a-b) on which it is biologized, since it is a need associated by Schleiermacher with one's mind, which can double as one's nature, even though it is a need to translate in particular and hence a social need. But Robinson finds this interpretation untenable since Schleiermacher specifies in (c-e) that the need complies with some rule(s) "as vague as a moral impulse" (Robinson 2013, p. 10). The need, rather than a biological impulse, is a kind of social order that is felt inwardly, specifically affectively, as a guiding impulse. While not a rule, it has the supervisory force to compel compliance. Yet, it is not something extrinsic like coercion, or conative pressure. Rather, it is felt as a need according to one's own disposition or temper. Robinson describes it as a "socio-affective force" and argues that something like it, an icosis, "is the regulatory impulse that drives, organizes, and 'plausibilizes' *all* of the behavior" that Schleiermacher attempts to define in the address (Robinson 2013, p. 10; original emphasis).

Robinson avows that he derives these icoses from the address "as a speculative clarification of Schleiermacher's feeling-based hermeneutics" (Robinson 2013, p. 11). That his hermeneutics has this basis –specifically in the "situated phenomenology (*Gefühl* or feeling)" of one embodied human being in "a spoken dialogical encounter" with another– should not be in doubt, Robinson thinks (Robinson 2013, p. 11). The importance to Schleiermacher's hermeneutics of *Gefühl* has been known for a long time. Dialogical understanding is made possible by feeling in this sense. It is, per the account of the Schleiermacher of the *Speeches* (*Reden*), at once a "subjective" or inwardly felt (by the individual) self-consciousness, and an "objective" –alternately, "universal"– (i.e., shared at a historical moment by all speakers of a given language) "situated participation in the collective, the communal, communion with other humans and God" (Robinson 2013, p. 12).

I want to consider icoses that are ecological rather than social (ecological), that pertain specifically to, or at least apply to, contemporary ecological concerns. Let's term these "ecological icoses" for ease and clarity of reference. An ecological icosis would be an emotional need felt for a split second (a-b) –whether a social need, a biological need, or *both*. This need could be a social need and a biological need in that the guiding impulse –the inwardly, affectively felt *ecological* order– could be a biological impulse in the sense that it physically and automatically provokes corresponding action. Indeed, it could be that the guiding impulse also is a biological impulse "on the other side" in that it is prompted biologically as well. The momentary emotional need complies with a moral impulse (d) that itself is at least partly an affective state (e) and has the force of some rule(s) (c). An ecological icosis is an ecological-affective force that, following Robinson, is the supervisory impulse motivating, organizing, and "plausibilizing" all ecological behavior of the agent. Ecological icoses exhibit an ecological hermeneutics based on the feeling of an embodied human being in an encounter with other embodied creatures, human and non-human alike. Said feeling expands the collective beyond humans, to other organisms as well.

6.2 Schleiermacherean Panentheism

In this part, I rely on two groundbreaking studies that are highly relevant to the topic of Schleiermacher's panentheistic post-Kantian neo-Spinozism. I survey and synthesize some important observations they make regarding the topic in answer to our second main question: How exactly is Schleiermacher's metaphysical-cosmological account panentheistic, at least in respects relevant to ecology? The takeaways of this part are synthesized with those from the preceding part, in Part III.

6.2.1 Schleiermacher's Post-Kantian Neo-Spinozism

Julia Lamm's *The Living God* (1996) is (self-consciously) the first comprehensive study in English of the relation between Schleiermacher and Spinoza. Specifically, it is the first such study of how Spinoza was appropriated (and understood) by Schleiermacher with respect to the latter's doctrine of God, and of how Schleiermacher's formulations of traditional Christian doctrine may or may not have been influenced by that appropriation. Lamm advises and practices caution in attempting to connect Spinoza's and Schleiermacher's systems. Spinoza's influence on Schleiermacher's doctrine of God was at least not direct in the case of the latter's major theological works. No reference is made to Spinoza's works in the *Reden* and no reference to Spinoza whatsoever is made in the *Glaubenslehre*. Moreover, in his two early philosophical essays on Spinoza, Schleiermacher had only indirect access to the thinker, through Jacobi (Lamm 1996, pp. 4–5).

Given the subtlety of the connection between Schleiermacher's and Spinoza's systems, Lamm uses four terms in her analysis in order to carefully delineate their connection. One is Spinozism in the sense defined in the communications between Lessing and Jacobi (what Lamm refers to simply as "Spinozism"). Another is exemplified by Herder's conversion of Spinoza's substance into substantial force and reflected in an organic *Weltanschauung* [worldview] ("neo-Spinozism"). A third is an adjective indicating specific factual parallels, whether intended or not, between Schleiermacher's thought and that of Spinoza ("Spinozan"). The fourth ("post-Kantian Spinozism") refers to the unique appropriation of Spinoza by Schleiermacher that is characterizable by four themes ("organic monism, ethical determinism, higher realism, and a nonanthropomorphic, nonanthropocentric view of God") and by Schleiermacher's adaption of Spinoza's pronouncement, "One and All" (Lamm 1996, pp. 5–6).

While some themes along the lines of the four terms were inherited by Schleiermacher's mature theological system from his early writings, these lines of influence became increasingly indirect, due to circumstances in which he found himself as his career advanced. Still, Schleiermacher relied on distinctions and clarifications he found in Spinoza in order to surmount difficulties he found in Kant. Indeed, Schleiermacher was the first to try to formulate a Christian doctrine of God within the constraints placed on reason by Kant's critical philosophy. Spinozism and pantheism can be seen not to be attributable to Schleiermacher in the usual ways once it is understood how and why he appropriated Spinoza. Schleiermacher sought a middle path between the anthropomorphized God of prior theology and popular religiosity, on the one hand, and a blind, lifeless necessity in nature, on the other. His alternative is a living God that is not necessarily personal (Lamm 1996, p. 6).

Lamm argues that this conception of God is developed in the *Glaubenslehre* such that the conception is immune to the usual charges of pantheism, even though it is nonetheless influenced by Schleiermacher's appropriation of Spinoza and neo-Spinozism in the sense referred to above. Lamm develops eight claims in support of her thesis but, given its relevance, sufficiency for present purposes, and space constraints, I will briefly focus on her third claim –that certain themes of Schleiermacher's appropriation can be detected from his early essays on Spinoza, the *Reden*, and the *Glaubenslehre*– and, in particular, on one of the four themes. It is "[a] view of nature as interdependent, as a causal nexus, in relation to which humanity is not an exception. Everything has a like cause, which is to say it can be explained in terms of the finite world, and thus the traditional interpretation of miracles is rejected. For Schleiermacher this intimate connection of natural causality and divine causality was expressed in the doctrine of preservation" (Lamm 1996, pp. 6–7). Sound familiar? The view should by now; it is a view, a vision, of none other than the *Naturzasammenhang*.

6.2.2 Schleiermacher and the Panentheistic Array

John Cooper notes that while Baruch Spinoza has been associated with pantheism rather than with panentheism, there is debate regarding the classification of him as a pantheist. Either way, modern panentheism has been shaped significantly by thinkers who have admired and modified Spinoza's views, including Fichte, Hegel, Herder, Lessing, Schelling, and Schleiermacher (Cooper 2007). Indeed, Philip Clayton argues that, when it has been systematically developed in subsequent Western philosophy, Spinoza's pantheism has uniformly morphed into panentheism (Clayton 2000, pp. 215–216; referenced in Cooper 2007, p. 67). Having seen in the preceding section how Schleiermacher's thought is post-Kantian Spinozistic with Lamm's help, we'll now examine how it is panentheistic, with Cooper as our guide and with Forster's help to draw out how exactly his thought is *panentheistically* post-Kantian *neo*-Spinozistic.

Panentheism, rather than a monolithic theology, consists of a group of related positions sharing basic commitments. The term *panentheism*, literally meaning "all-in-God-ism", was coined by Karl Krause, a contemporary of Schleiermacher's (and Hegel and Schelling), although it didn't come into common usage until popularized by Charles Hartshorne in the twentieth century. The commonly accepted, general definition of the term is this: God's Being includes, penetrates, and transcends the entire universe, and every part of the universe exists in God. But there are widely varying ways of construing panentheism and, consequently, much debate among panentheist theologians as to what it is or how it should be understood (Cooper 2007, pp. 26–27).

With this in mind, Cooper outlines five distinctions regarding panentheism that facilitate introduction to the variety and on which he relies throughout his study. One is between explicit and implicit panentheism, between thinkers who explicitly use the term *panentheism* to endorse it and those who do not but whose theologies nonetheless imply panentheism. Thinkers whose theologies qualified as panentheistic before the term was coined are obvious examples of implicit panentheists, and calling them panentheists is anachronistic (Cooper 2007, p. 27), though describing their theologies as panentheistic may be useful, as it is in the present case. Schleiermacher therefore is implicitly panentheistic inasmuch as his theology qualifies as panentheistic, which it does.

Implicit panentheism can be better understood now that the distinction between pantheism and panentheism has been refined. Some theologies that have traditionally been thought to be forms of pantheism are now understood to be forms of panentheism. This applies to Schleiermacher's theology, which he himself referred to as (a form of) pantheism (Cooper 2007, p. 27).

Another distinction is between classical and modern panentheism. The former, what also can be called divine determinist panentheism, affirms divine omnipotence such that creatures exist in, but are unable to affect, God. Cooper lists Schleiermacher as a classical panentheist. The latter, most versions of which can be characterized as cooperative panentheism, implies that creatures have the libertarian freedom not only to act but to influence history and to affect God. Divine-human cooperation is affirmed by nearly all forms of modern panentheism (Cooper 2007, p. 29).

A third distinction is between voluntary and natural panentheism, between views which hold that God was free with respect to the creation of the world and views which hold that God could not exist without the world, that it is necessary for God. As Cooper notes, a compatibilist view of divine freedom can qualify as both voluntary panentheism and natural panentheism. For, it holds that God self-determines and freely creates the world in order to express divine love since this is God's nature; but it would follow that God could not not create the world (Cooper 2007, pp. 28–29). It is likely that Schleiermacher implicitly endorses a compatibilist view of divine freedom –perhaps even consistently (though not uniformly) throughout his career– and thereby might have latitude for qualifying in some way as both a voluntary panentheist and a natural panentheist.

A fourth distinction is between personal and nonpersonal (or Ground of being) panentheism. I will say more about this with respect to Schleiermacher's panentheism below. A fifth is between part-whole and relational panentheism. On part-whole panentheism, the world is viewed to be part of the divine nature in the sense of being an implicit part of the divine mind and explicitly emanating from the divine mind's idea of the (natural) world in creation. (Hereafter I use "the world" in the colloquial sense, i.e., using it and "the universe" interchangeably, and not in the sense of possible worlds theory, i.e., as referring to a maximal state of affairs.) According to relational panentheism, God and the world are completely and intricately bound up with one another historically such that they are ontologically one, analogous to how the (human) mind and body are one according to materialist theories of mind (Cooper 2007, p. 28).

6.2.3 Schleiermacher's Panentheistic Post-Kantian Neo-Spinozism

Need part-whole panentheism and relational panentheism be viewed as mutually exclusive, or need they be understood in ways that are? Consider once more the above analogue of the mind and the body. Now suppose God is a mind and the world constitutes God's ideas –those regarding the world anyway– and that the(se) divine ideas just amount to the divine mind's idea of the world of creation. In that case the world would be implicit in the divine mind and could be said to emanate from its idea of the world of creation. Moreover, God and the world would be completely and intricately, historically bound up with each other, since the "world" here is none other than the history of the universe –and so, here, divine ideas concerning the world would be accounted for by that history. Then God and the world would be ontologically one –although not explanatorily one– just as the human mind and body are ontologically (but not explanatorily) one, or one substance or kind of stuff, on materialist views (both reductive and nonreductive ones). The view we have been considering, then, qualifies as both part-whole panentheist and relational panentheist. It follows that this distinction, like that between voluntary and natural panentheism, is not to be taken to suggest mutual exclusivity between the options.

But inasmuch as a unifying interpretation or view of part-whole panentheism and relational panentheism depends on *substance* monism, neither Schleiermacher's earlier work seems to be amenable to such an interpretation, nor his later work to such a view, for different reasons. As Forster notes, Schleiermacher endorses Spinoza's all-encompassing monistic principle, "One and All", to be sure, in his essays on Spinoza and the *Reden*. But Schleiermacher takes the principle to be a force and a source that unifies multiple earthly forces, due to the influence of Herder (mentioned in the essays), rather than thinking of it as a substance, as Spinoza does (Forster 2022, sec. 11). In this way, Schleiermacher's monism can be seen to be neo-Spinozism in Lamm's sense and, arguably, not pantheistic.

In some of his later work, on the other hand, Schleiermacher disavows Spinoza, initially distancing himself from his own earlier neo-Spinozism and later denying explicitly that he is (or was) a follower of Spinoza. In the *Dialektik*, he accordingly argues for a still higher "transcendental ground" beyond both Spinoza's *natura naturans* ("nature doing what it does") and Herder's highest force (Forster 2022, sec. 11; apparently quoting from Schleiermacher 2001) –described by Lamm as a substantial force, as noted in §6.2.1. It seems that Schleiermacher's primary motivation for the change, as noted by Forster, was an intention of avoiding being accused of Spinozism and *pantheism*. It is for this reason that Schleiermacher develops his conception of God in the *Glaubenslehre* as he does, as Lamm argues (mentioned in §6.2.1). While mention is made there only of the accusation of Spinozism, this was tantamount to the accusation of pantheism (if not vice versa) in the wake of the 1780 Pantheism Controversy. This controversy had a decisive influence on the formation of Schleiermacher's post-Kantian neo-Spinozism, based on his study of Kant's "What Does it Mean to Orient Oneself in Thinking?" essay (Henrich 2003). Still, as Cooper has pointed out, there is debate as to whether Spinoza was in fact a pantheist.

Yet, Schleiermacher's natural substantial force (the *Naturzusammenhang*) indicates his neo-Spinozism in Lamm's sense and, together with his transcendental ground, suggests panentheism. Moreover, as Lamm has pointed out, Schleiermacher's unique appropriation of Spinoza is post-Kantian Spinozism in the sense that the appropriation involves five essential attributes:

(i) organic monism,
(ii) ethical determinism,
(iii) higher realism,
(iv) a non-anthropomorphic and non-anthropocentric conception of God, in addition to
(v) an adaptation of Spinoza's "One and All" which, as both Lamm and Forster note, involves a move from one substance to one world-unifying, if substantial, force (making for neo-Spinozism in Lamm's sense).

Cooper also has noted Schleiermacher's modification of Spinoza's views, and Schleiermacher's modification seems to be a clear example of the morphing of Spinoza's (supposed) pantheism –as it has been developed systematically– into panentheism, which Clayton has argued has uniformly occurred. Suggested by all this is that Schleiermacher's metaphysical-cosmological account implies a panentheistic post-Kantian neo-Spinozism. So, in spite of his protestations to the contrary during the latter part of his career, Schleiermacher's thought remained very much neo-Spinozistic.

Let's focus on (i), (iv), and (v) since these three are borne out in Schleiermacher's *Naturzasammenhang*, as I will explain in a moment, and have bearing for our purposes in this chapter, as we will see in the last part. One of the themes detectable from Schleiermacher's two principal works (the *Reden* and the *Glaubenslehre*) and his essays on Spinoza is his conception of the *Naturzasammenhang* as this is intricately bound up with divine causality via divine creation and preservation (of the *Naturzasammenhang*). Since this natural nexus, which includes all things (all beings), is also to be a living organism, organic monism is affirmed in it. Because the *Naturzasammenhang* is to be a world-unifying force, rather than a single substance, it involves an adaptation of Spinoza's dictum. Since God is the transcendental ground beyond this highest force and of all beings, of which the human species is but one of many parts, Schleiermacher's view that the *Naturzasammenhang* depends on God for its creation and preservation indicates a non-anthropomorphic and non-anthropocentric conception of God. But the divine ground of being may turn out to be a person (the two concepts do not seem to be inherently at odds with each other at least), in which case both personal panentheism and Ground of being panentheism would obtain. (In that case, non-personal panentheism and Ground of being panentheism wouldn't be one and the same; they would come apart at least to some extent.) Were Schleiermacher's God to be a person in addition to the transcendental ground of Nature, he would in effect deny that personal panentheism and Ground of being panentheism are mutually exclusive. So much for the implicit panentheism, the form of panentheistic post-Kantian neo-Spinozism, of Schleiermacher.

6.3 Schleiermacherean Panentheism and Ecology

We're now ready to bring everything together to set out exactly what Schleiermacher's unique form of panentheistic post-Kantian neo-Spinozism has to offer to ecology. This mainly involves recapitulating the high points of the first two parts of the chapter.

In arguing that Schleiermacher predicted insights that have been provided by contemporary philosophers of biology concerning the relationship between living things and their environment, De Cruz (2022) has focused on the three properties attributed to self-conscious organisms by Schleiermacher: feeling, acting, and knowing. For him, feeling is a creature's receptivity to its environment in an exhaustive or global sense (i.e., everything that is outside of the creature). Feeling has two primary forms and epistemic functions.

First, there is the feeling of absolute dependence, through which one becomes aware of how one depends ultimately on God. Crossley, Jr. explains that the universal feeling of absolute dependence grounds the religious ethics implied in Schleiermacher's doctrine of creation/preservation. From the objective application of this feeling it follows that the universe, all finite things (including putative instances of evil), depend on the Infinite, on God, for their subsistence. God is the transcendent(al) ground of the *Naturzusammenhang* comprising all beings, per Schleiermacher's form of panentheistic post-Kantian neo-Spinozism.

Second, there also is the feeling of relative dependence, through which one becomes aware of how one depends on its *natural* environment, including other creatures "in" (partly constituting) it. This feeling, expressed by self-consciousness, comprises the feeling of partial dependence on the universe and the feeling of partial (certainly not absolute) freedom, the "pull" of being acted on by the universe and the "push" of acting on it.

With the feelings of partial dependence and partial freedom, Schleiermacher predicts the view that biological agents are conscious because they require awareness both of their environment and their place in it for action, that consciousness enables free agency to be expressed. This is a prominent view in contemporary philosophy of biology, one premised on an evolution of the mind whereupon subjectivity and agency become inextricably connected.

It also predicts the important role in evolution ascribed to organismic agency, now commonly accepted in evolutionary theory. As Brandt points out, for Schleiermacher the feeling of relative dependence comes into play in moral agency as well. The biological, moral agent has a duty to act in ways aimed at the highest good (including the perfection and synthesis of the goods of the human life), a notion that lends itself to the endorsement of ecological awareness. Schleiermacher's outlook thereby is consonant with attentiveness to ecological issues.

Schleiermacher doesn't mean for the doctrine of absolute dependence to undermine the universe being a (relatively) causally closed system, to put it in modern scientific terms. Rather, the means of religious self-consciousness or knowledge completely coincides with, is not mutually exclusive with, (knowledge of) the interdependently causal network of the natural world (made possible by the feeling of relative dependence). Knowledge of the universe, scientific knowledge, just is the knowledge of this natural nexus, even an interdependent living organism or Schleiermacher's *Naturzusammenhang*. All things both depend on God for their subsistence and are established and conditioned by the universality of this nexus. Both ideas reinforce each other and, like their feeling-relative and epistemic counterparts, coincide. All things are grounded by the non-anthropomorphic God transcending and sustaining the *Naturzasammenhang* unifying them, the One and All that, rather than a (mere) substance, is a substantial force.

Although Schleiermacher sees "lower animals" as having consciousness expressing agency, both of which have a role in the evolution of minds, they have neither "actual knowledge" nor "complete self-consciousness." But while contemporary biology and philosophy of biology employ a less restricted notion of knowledge, it has striking parallels with Schleiermacher's conception. Just as he sees knowledge (*Erkenntnis*) as intricately bound up with feeling (subjectivity) –in self-consciousness– and acting (agency) as properties of biological agents, so do contemporary biologists and philosophers of biology see knowledge, consciousness, and agency as intricately bound up with each other, even understood in functional terms.

We can see that all three features of ecological economics are borne out in Schleiermacher's conception of the *Naturzusammenhang*, even specifically along the lines of the three discussed dimensions of self-conscious organisms, respectively. The first feature of ecological economics –a vision taking into account both the Earth, the whole universe, and not humanity alone– is confirmed particularly by Schleiermacher's view of (human) knowledge of the universe as knowledge of the *Naturzusammenhang* (concerning the knowing property of humans). The importance of the entire creation, and all creatures, in Schleiermacher's thought is collaborative with Schleiermacher's non-anthropocentric view of God.

The second feature –a highlighting of the need for every part of the *Naturzusammenhang* to properly function within its own constraints– is affirmed specifically by the view predicted by Schleiermacher that biological agents are conscious since they need to be aware of their environment and their purview within it in order to act in ways conducive to their survival (pertaining to the acting property). The third feature –a focus on the symbiotic relations between organisms and, specifically, between humans and the universe as an organic whole– is attested especially by the feelings of partial dependence on the universe and partial freedom, exhibiting the reciprocal relation each organism shares both with the universe and with other organisms in it, in constitution of the *Naturzusammenhang* (regarding the feeling property). This living organism-totality vivifies the organic monism of Schleiermacher's panentheistic post-Kantian neo-Spinozism.

This is all well and good as far as how humans *could* live in ecologically economically responsible ways, relating to each other (and other organisms) within the *Naturzusammenhang* in ways directly or indirectly prescribed by Schleiermacher. But as far as how humans do or how they in general tend to live, according to Schleiermacher, they engage in communal sin that causes evil afflicting the *Naturzusammenhang* and other creatures within it (including other humans themselves). This includes a type of natural evil (physical affliction and inhibition experienced by humans as torments) and social evil (deterioration of the world including damage to the natural world causing creaturely suffering and destruction). Social evil in Schleiermacher's sense clearly results in ecological suffering and destruction, as I noted earlier.

The requisite response, for Schleiermacher, is for humans to acknowledge their social sins and to actively wish for others to experience the divine love they themselves have already experienced. This is one way they can act relationally with respect to the ecologies of the world. As far as a concrete way to do so, the divine decree to create, to sustain, and to redeem involves the free action and reaction of humans in the God-ordered *Naturzasammenhang*, people called to promote the kingdom of God in the universe such that they should live sustainably and promote sustainability.

Ecological icoses are a means of sustainably acting in and reacting to one's environment (including the promotion of sustainability itself). An ecological icosis is an affectively, fleetingly felt ecological order complying with a (partly affective) moral impulse having the force of a rule (or a set of rules). An ecological icosis motivates, organizes, and plausibilizes the embodied human agent's (would-be) ecological behavior. It can be felt by her on encountering another embodied creature (or multiple), be it a human or a non-human organism.

But the supervisory ecological order internal to the agent is felt only for a moment, and the instances in which it is felt may be very few and far in between. So, it is incumbent on the agent to endorse such feeling when it emerges within her or him, especially given how infrequently it may emerge. The more agents there are that do so, the more ecological living there will be, and the more the collective *for* those agents will extend beyond them and their kind, to all other natural kinds, as the collective including the agents in fact does, inhabiting and struggling to survive in the *Naturzasammenhang* that is a (temporary) home shared by all.

Such feeling, then, is where ecological living –and ecological mindedness, for that matter– would begin for Schleiermacher, feeling such that it coincides with, and presumably arises from, the feelings of absolute dependence and relative dependence. The intricate connection between the *Naturzasammenhang* and divine causality (by means of creation/preservation), per Schleiermacher's panentheistic post-Kantian neo-Spinozism, is embodied by such sorts of feeling of and by the agent, is represented by the intersection of the three. As one's feelings of absolute dependence and of relative dependence in tandem effervesce into ecological icoses that influence the same in others, and so on such that feedback loops of such feeling are produced among the ever-extending collective, the *communal* interpenetration of the three sorts of feeling will be exhibited, will grow.

Acknowledgements I am grateful to Kevin Jung and David S. Pacini for comments on an earlier draft of this chapter, to Gabriel Vidal for helpful comments on a later draft, to Luca Valera for patiently helping me to prepare the chapter, to Bill Leonard and Hyelin Kim for encouragement regarding the project, and to Elizabeth Corrie for comments on material included in another draft, teaching me much about Schleiermacher's thought, and helping me to appreciate many of its complexities.

References

Brandt JM (2018) Schleiermacher on church and Christian ethics. In: Schleiermacher and sustainability: a theology for ecological living. Westminster John Knox Press, Louisville, pp 17–34

Clayton P (2000) The problem of god in modern thought. Eerdmans, Grand Rapids

Cooper JW (2007) Panentheism: the other god of the philosophers: from Plato to the present. Apollos, Nottingham

Crossley J Jr (2006) The religious ethics implicit in Schleiermacher's doctrine of creation. J Relig Eth 34(4):585–608

De Cruz H (2022) A taste for the infinite: what philosophy of biology can tell us about religious belief. Zygon 57(1):161–180

Forster M (2022) Friedrich Daniel Ernst Schleiermacher. The Stanford encyclopedia of philosophy. Retrieved from: https://plato.stanford.edu/archives/sum2022/entries/schleiermacher/

Godfrey-Smith P (2020) Metazoa: animal life and the birth of the mind. Farrar, Straus, and Girouz, New York

Hagan AI (2018) Divine Providence and human freedom in the quest for ecological living. In: Schleiermacher and sustainability: a theology for ecological living. Westminster John Knox Press, Louisville, pp 64–78

Henrich D (2003) Between Kant and Hegel: lectures on German idealism. Harvard University Press, Cambridge

Holekamp KE (2007) Questioning the social intelligence hypothesis. Trends Cogn Sci 11(2):65–69

Kaas JH (2000) Why is brain size so important: design problems and solutions as neocortex gets bigger or smaller. Brain Mind 1:7–23

Kant I (2000) Critique of the power of judgment. Cambridge University Press, Cambridge

Lamm JA (1996) The living god: Schleiermacher's theological appropriation of Spinoza. The Pennsylvania State University Press, University Park

Lewontin RC (1983) The organism as the subject and object of evolution. Scientia 118:65–82

Poe SM (2018) An ecological Oikos: economics, election, and ecumenism. In: Schleiermacher and sustainability: a theology for ecological living. Westminster John Knox Press, Louisville, pp 35–50

Robinson D (2013) Schleiermacher's Icoses: social ecologies of the different modes of translating. Zeta Books, Bucharest

Schleiermacher F (1843) Die christliche Sitte nach den Grundsätzen der evangelischen Kirche im Zusammenhange dargestellt. G. Reimer, Berlin

Schleiermacher F (1928) The Christian faith. T&T Clark, Edinburgh

Schleiermacher F (1996a) Dialectic; or the art of doing philosophy: a study edition of the 1811 notes. Scholars Press, Atlanta

Schleiermacher F (1996b) On religion: speeches to its cultured despisers. Cambridge University Press, Cambridge

Schleiermacher F (2001) Dialektik. Suhrkamp, Frankfurt

Schleiermacher F (2002a) Ueber die verschiedenen Methoden des Übersetzens. In: Akademievorträge. Berlin-Brandenburg Academy of Sciences and Humanities, Berlin, pp 67–93

Schleiermacher F (2002b) Lectures on philosophical ethics. Cambridge University Press, Cambridge

Schleiermacher F (2011) Selections from Friedrich Schleiermacher's Christian ethics. Westminster John Knox Press, Louisville

Schleiermacher F (2016) Christian faith: a new translation and critical edition, volumes one and two. Westminster John Knox Press, Louisville

Tice TN (2018) Conclusion: Schleiermacher and ecotheology. In: Schleiermacher and sustainability: a theology for ecological living. Westminster John Knox Press, Louisville, pp 93–104

Vander Schel KM (2018) Social sin and the cultivation of nature. In: Schleiermacher and sustainability: a theology for ecological living. Westminster John Knox Press, Louisville, pp 79–92

Waggoner E (2018) Schleiermacher's theological naturalism: critical resources from his views about creation for contemporary ecotheologies. In: Schleiermacher and sustainability: a theology for ecological living. Westminster John Knox Press, Louisville, pp 51–63

Chapter 7
Rumi and Tagore on Being-With-Nature

Abubakr Khan and Zahra Rashid

Abstract The relationship between humans and nature has undergone much desecration in modern times. Nature is objectified, used and abused, without any moral limits. To counter such a devaluation of the living earth, we propose the concept of 'being-with-nature' by building on Heidegger's work, and then deepen it through the philosophy and poetry of Rumi and Tagore. We emphasize the fundamental ontological connection between human beings and nature. The value of the latter also depends on how human beings understand the divine, and how much and in what way nature is connected to the divine. Thus, we highlight some traditions and perspectives — rooted in pantheism or panentheism — that uphold the sacredness of nature. Rumi seeks and sees the Divine Beloved everywhere in nature. Through exceptionally vibrant imagery, he evokes the energy and presence of the living natural world. Tagore reminds us of the kinship we share with nature. For him, transcendence is to be found in immanence, and in this connection. Both Rumi and Tagore emphasize the dynamism of nature, and they see the sacred in it. There is a power and beauty to nature that cannot be reduced to its material or useful dimensions. Such intellectual traditions hold a lot of promise for environmental ethics and praxis. They can counter objectifying and extractive attitudes that stem from anthropocentric en-framings of the earth, and therefore share an important link with contemporary movements that seek to restore the sanctity of the natural world.

Keywords Environmental ethics · Heidegger · South Asian literature · Sufism · Immanence

A. Khan (✉) · Z. Rashid
Faculty of Humanities and Social Sciences, Information Technology University of the Punjab, Lahore, Pakistan
e-mail: abubakr.khan@itu.edu.pk

L. Valera (ed.), *Pantheism and Ecology*, Ecology and Ethics 6,
https://doi.org/10.1007/978-3-031-40040-7_7

7.1 Introduction

How human beings conceive of divine reality inevitably affects not only how they understand their own humanness and humanity, but also their relation to the non-human natural world. To believe in deities that are transcendent (in the sense of being wholly removed from the world) would necessarily have consequences for the worth that is given to the immanent surrounding world. And the situation would probably be very different in the case of alternative conceptions of the divine that allow for a deeper connection with nature, such as those rooted in or inspired by pantheism (or panentheism). For instance, there has been a lot of interesting scholarship on the reverence that Native American tribes have for the natural world, which for them is a manifestation of the Great Spirit (Baudot 1999, p. 20). The crux of this chapter lies in the fundamental relationship between divinity and nature, and how varying interpretations of the divine can affect human attitudes and comportment towards nature.

The link — or lack thereof — between divinity and nature surely has some influence on how the latter is valued. An overlapping of the two would endow the natural world with at least bits of divine prominence. In contrast, an unbridgeable dichotomy between the two may have potentially dire consequences for nature, so that it may be completely divested of any special characteristics, or of any mystery or force within, and given the status of a mere object (or a collection or storehouse of objects). This could very well result in nature being treated in a purely technical or extractive manner. Okocho Ryogi expresses this in the following words: "the objective and objectivized conception of nature as determined by the natural-scientific way of thinking is in our time the dominant one" (1991, p. 201).

Nature may also have to suffer on account of those theological perspectives that cater to anthropocentric positions whereby "humans inhabit a special position at the very top of the natural order" (Keller 2010, p. 59). If nature is conceived as that which is outside us, and — on top of that — perhaps also as inferior to us, then this would obviously contaminate the moral and practical dimensions of our relationship to nature.

Our contemporary lives are subject to the capitalist pressures of endless production and accumulation, which ultimately rely on a relationship with the natural world that is premised upon the nefarious principle of *extraction without any moral limits*. From such a perspective, nature appears to humans only as an exploitable phenomenon. This is what Martin Heidegger meant when he spoke out against nature being viewed as merely a "stockpile" (*Bestand*) for human beings. He presented the argument that nature has — in the age of science and technology — come to be objectified because of machination (*Machenschaft*), i.e., the "dominance of subjectivity", whereby everything in nature becomes an object for human agency. The earth becomes "a region submitted to the scientific gaze and the willful power of man" (de Beistegui 1998, p. 77). The implication is that nature comes to be understood and approached only in terms of *natural resources*. This means that it can be nothing other than a repository of disposable reserves. It is only a resource, and it is to be viewed and worked upon in this manner — that is, merely as a source of raw materials or (at best) tools. Nature has no purpose other than being of use to man. Might this explain why, in so many contexts, humans are not humane to nature.

Philosophical and cultural traditions that do not create strict oppositions between immanence and transcendence hold immense potential for more humane contributions to environmental ethics (Rozzi 2013, 2015). With a focus on perspectives and interpretations that regard the sacred as immersed in the natural world, we wish to deepen the possibility of a fundamental ontological connection between human beings and nature. The purpose is to find new grounds for the in-itself value of the nonhuman natural world. In this chapter, we build on Heidegger's framework of being-with [*Mitsein*] to propose the concept of "being-with-nature", which will be presented and enriched via the philosophical and poetic contributions of Jelal-ud-din Rumi and Rabindranath Tagore.

7.2 The Primordial Concept of "Being-With"

The belief that human agency overrides all that is nature goes back to some of the fathers of Western modernity, particularly Descartes, who took great pains to posit that the fundament of everything is the human consciousness (Gare 1995, pp. 52–53). Not only did they introduce and exaggerate a strict opposition between humans and world but they also put the latter in the throes of mere objecthood. Over the last century or a bit more, these views have been called into question. Heidegger has been an important voice in this regard. He criticizes Descartes for endorsing and upholding a "bare subject without a world" (Heidegger in Ha 2010, p. 355). In his view, the subject-object dichotomy must be interrogated and done away with. Among other things, this would also provide us with some very important tools to rethink human-nature relations. A crucial argument that comes up again and again throughout his oeuvre is about how whatever exists cannot exist without a world. The basic argument is that no entity can exist outside of an environment, detached and isolated from its surroundings. To human beings, nature is what surrounds us. However: Does this mean that they are wholly detached from it? Is it possible to conceive of or make sense of a human existence without nature? And how are we to, then, conceive of and value the earth on which we live, the earth that lives with us? To examine and dive into this series of questions, we will bring in the work of Rumi and Tagore. However, first, let us consider Heidegger's framing of 'being-with', which is crucial to the later explorations and arguments of this chapter.

At this point, it is important to emphasize some fundamental principles that permeate Heidegger's overall philosophy and are crucial to his study of human existence and its relation to the world, nature, and the beyond. Heidegger understands human existence in terms of a "being-here" that is not divorced from the 'there' of the world, so that we may be able to conceive of a fundamental connection between the 'here' of humanness and the 'there' of nature. This points to an existential openness, i.e., the sense of any existing entity cannot be like that of an enclosed monad.

In other words, existence cannot and must not be understood as something that is locked in within itself.

Heidegger often uses the term *gleichursprünglich*, which can be translated as "equioriginally" or "equiprimordially". This basically refers to the co-equal and simultaneous origin of two phenomena (Sheehan 2011, p. 45). This implies that relationality is embedded in the very foundations of Heidegger's philosophical framework. Another term often used by Heidegger is *Zusammenhang*, which means interconnection, though a more precise way of explaining this in the context of Heidegger's overall philosophy would be to put it in the following words: *an always already belonging together*. These are important for understanding the various relations that are in play in Heidegger's philosophy, and this will help us uncover the possibility of a more connected bond between humans and nature.

Heidegger's term for human existence, when one has taken into account its ontological relation with being on the whole (all that is the world in which one exists), is Dasein. The "Da" is meant to emphasize the situatedness of an existing being, and it also points to the *ekstatic* possibilities and intertwinings woven into the existential grounds of this being that lives and exists. Moreover, for Heidegger, Dasein is also, at once, *Mitsein* and *Miteinander-sein*, which might be translated as *being-with* and *being-with-one-another*, respectively. He also relies on the notion of a with-world (*Mitwelt*). In short, Dasein exists only in terms of a "withness."

Dasein is not at all "a windowless capsule" (*ein fensterloses Gehäuse*). It is in its very essence being-with-others; its being-toward-itself is inextricably tied to this being-with. And Heidegger traces the unity of these relations to *being-in-the-world*. In his 1925 lectures, Heidegger explains: "Dasein as being-with is lived by the Dasein-with (*Mitdasein*) of others and by the world that concerns it thus and such" (Heidegger 1979, p. 337). To a large extent, therefore, Heidegger's work is "a theory of interaction with the things we encounter in the world around us" (Sembera 2007, p. 35). However, it is important to detach it from the notion of a "human" or "subjective" comportment towards something else, and this is to be achieved by clarifying being-in-the-world as the primordial "truth" of Dasein — meanings emerge only in terms of a primordial withness.

7.3 Being-With-Earth, or Lords of the Earth?

In the context of our discussion, the primordial withness that is couched in the concept of being-in-the-world concerns the earth. Scholars such as Kenneth Maly (2009) have focused on the importance of the notion of *earth* in terms of Heidegger's overall philosophy, which guides us to new ways of thinking about the earth and our relationship to it. The emphasis is on an unbidden and profound connectedness. In short, our existence corresponds to being of and with the earth (Maly 2009, p. 53). In *Der Ursprung des Kuntswerkes*, Heidegger writes: "It is one thing to just use the earth: it is quite another thing to receive the blessing of the earth and to become at home in the laws of this reception, in order to shepherd the mystery of being and to pay attention to the inviolability of the possible" (Heidegger in Maly 2009, p. 47).

In the later *Contributions to Philosophy*, Heidegger refers to the Da as the Between [*Zwischen*] of earth and world, which is also a reference to the imbrication of immanence and transcendence when it comes to existence (Heidegger 1999, p. 198). In that particular context, it seems like he has used the term "earth" (*Erde*) in terraneanean, physical terms, whereas the term "world" (*Welt*) relates more to the domains of sense and meaningfulness. In any authentic consideration of existence, however, earth is worldly and worlds emerge with the earth. The facticity of existence must concern both. The overall point seems to be that human existence is always between earth and world, or rather, *always in the Between of the earth and the world*. It is through such a lens then that Heidegger critiques the modern culture of domination and control over earth, because according to his framework, the earth exists *with* us, it is not something that comes *under* our possession. A longer description of this critique is found in Deluca (2005, p. 75), who also quotes directly from Heidegger (1999, pp. 86–87):

> Machination is unconditional controllability, the domination of all beings, the world, and earth through calculation, acceleration, technicity, and giganticism. Calculation represents a reduction of knowing to mathematics and science and a reduction of the world and earth to what is calculable, a step taken decisively by Descartes (1999, pp. 84–96). The unrestrained domination of machination produces a totalizing worldview that enchants … by virtue of which everything is [pressed forth] into calculation, usage, breeding, manageability, and regulation.

With regard to our relationship with the earth, Heidegger links machination to the technological enframing that is primarily concerned with subordinating and appropriating the earth. So, machination is the process, or schema, according to which the earth is reduced to a region of planning, calculation, arranging, and exploitation. As a consequence, everything on earth is "objectified." With machination, truth itself comes to be about the "the securing of beings in their perfectly accessible disposability" (Heidegger 1987, p. 174). It is in a similar vein that Heidegger critiques the ideological task of science, as elucidated by Daniel Ray White: "How convenient for entrepreneurial power that physics defines nature as transformations of energy, an idea that is purely quantitative and not subject to ethical consideration" (1997, p. 84). It is precisely on this point that he wrote about "the distress of the technological enframing of the earth" (Deluca 2005, p. 68). Heidegger ties modern technology to this notion of perfect disposability. To be clear, by this term "technology" he is not referring to specific technologies, or the machines that we use but rather a perspective on life and the living world and earth, a view from a distance, a view from up above: an "en-framing". According to Heidegger, "the revealing that rules throughout modern technology has the character of a setting-upon" (Deluca 2005, p. 79). Nature is approached with a specific purpose in mind: it is to be set upon, it is to be taken from. Man bears down upon nature. This setting-upon is "in the sense of a challenging-forth … which puts to nature the unreasonable demand that it supply energy which can be extracted and stored as such" (Schimmel 1993, pp. 320–321). Nature, then, is reduced to a "standing-reserve … a calculable coherence of forces" (Schimmel 1993, p. 322, p. 326), so that "nature reports itself in some way or other that is identifiable through calculation and that it remains orderable as a system of information" (Schimmel 1993, p. 328; in Deluca 2005, p. 79).

The relationship between man and nature is a broken one, and not only that, but it is also suspicious and corrupt. The Romantic poets — Wordsworth, Keats, Coleridge and so on — did not take kindly to such a view of nature. No wonder, then, that they are widely regarded as wellsprings for modern-day environmental thought (Hinchman and Hinchman 2007, p. 333). Romanticism's hostility to science was tied to the severe existential, aesthetic, and ethical consequences of modernity's technological conquest of nature (Economides 2016, p. 36). Seyyed Hossein Nasr, one of the most prominent scholars of Rumi and Islamic philosophy, has a similar contention with the "modern scientific view" that renders nature as lifeless and inert, compressing it into the form of "a machine to be dominated and manipulated by a purely earthly man" (1996, p. 4). We have, then, a specific way of dealing with nature which stands in stark contrast to any divine or living possibilities that one may recognize or seek in nature. The earth is, when framed in this manner, simply a means to a higher end: utility for humanity. It must be no more than a commodity. To this end, the scientific gaze reduces the earth and all that is nature to mere object, to be studied or to be used, or rather, to be studied so that it may be of use. Nature has been framed — so that it may be captured, and made available, made disposable. Modern technological thinking makes it appear that "the Earth waits on us" (Hiltner 2003, p. 118). Nature just lies in wait, waiting to be used.

This deranged sense of nature is tied to a narcissistic delusion on the part of humans, that they are the lord of the earth, or rather, an overlord: a master that bears down on the earth from above.

At the same time, however, we are well aware that nature is much more than this, and so, we must not only critique anthropocentric en-framings that manacle the earth but also look towards other ways of being with nature.

7.4 "Living in the Heart of Nature": Tagore on Divine Immanence, Harmony, and Union

Rabindranath Tagore was a Bengali poet and philosopher who was born in 1861 and died in 1941. This was the British colonial era, one of the peaks of technological and economic exploitation in the history of the world. Tagore wrote both poetry and prose, and his intricate writings (in Bengali, Hindi, and English) create a sense of nature which stands in sharp contrast to the sorry state that nature finds itself in when enframed in the manner discussed in the previous section, i.e., as something to be dominated and objectified. Many of Tagore's ideas therefore align with Heidegger's criticism of Western metaphysics and Enlightenment ideals. Scholars have thus found interesting affinities in the work of Tagore and Heidegger, especially regarding the former's reflections on nature and technology (Gupta 2005). Tagore mirrored Heidegger in bemoaning "the Europeanization of the world" (Banerji 2015, p. 6). He was very critical of "the imperialistic demand that nature should yield itself up in order to satisfy human needs" (Gupta 2005, p. 65). A very important philosophical ideal for Tagore was to make oneself at home in the world (Bandyopadhyay 2013). Thus, both Heidegger and Tagore share a concern for the homelessness that may have unfortunately set in as "the destiny of the world" in the technological era (Gupta 2005, p. 65). Homelessness here implies an alienation from the earth, the soil, and everything else that makes up nature. The tragedy can be summed up in this manner: To not be at home in a world in which one lives. As we discussed above, for Heidegger, this *not-at-home-ness* owes to a warped conceptualization of nature, whereby it comes to be seen as 'a standing reserve'. The earth is no longer a place that one can call home. One can only work on it. Nature is out there, and one must only ever take from it. It is reduced to "stockpiles of material being ordered about by humans who hubristically posit themselves as 'lords over the earth'" (Davis 2010, p. 126). Thus, the earth reveals itself as nothing but "a mining district, the soil as a mineral deposit" (Gupta 2005, p. 65).

This *technological manner of enframing and revealing* may indeed be the reality of our times. The purpose here is not to deny this, or to uncover the truth of its falsity, or to expose it as a Platonic shadow, but to emphasize that it is nevertheless "monstrous," i.e., that it has severe consequences for almost all dimensions of human and natural existence. For Heidegger and Tagore both, the problem is that it "drives out every other possibility of revealing": it does not, that is, allow nature to speak to us in its rich variety of aspects, and thereby precludes any realization of our deeper alliance with it. Instead we are constantly set over against nature, as it were in a master–slave relation forcing it to our advantage. For example, "the Rhine itself appears to be something at our command … a water power supplier," and no longer the river which once gave to the peoples living on its banks a sense of their community (Gupta 2005, p. 65).

Our *fundamental* connection to nature is of crucial importance to Tagore. In the modern world, our relationship to nature is dictated by "the power of possession," whereas Tagore's hope is to reintroduce us to the "the power of union" with the natural world (Gupta 2005, p. 66). This is the crux of Tagore's critical humanism, which therefore resonates with Heidegger's anti-humanism (Banerji 2015). Kalyan Sen Gupta (2005, p. 63) explains Tagore's approach succinctly:

> Scientific knowledge of nature is indeed a considerable achievement of the human mind. There is no doubt, then, that when we attain such knowledge, there is much that we gain. What Tagore insists upon, however, is that what we attain and gain in this way cannot be sufficient. To think otherwise is to have succumbed to a scientistic obsession with the establishment of causal laws. Undoubtedly, there are such laws to be discovered, not only in nature but also in the social and cultural spheres. But the discovery of laws, Rabindranath holds, cannot be a final aim.

This is reminiscent of Heidegger's critique of calculative laws which, according to him, do not and cannot encapsulate all that life has to offer. Scientific knowledge is imperative and useful but it cannot exhaust all the possibilities of life, knowledge, and so on. Hence, Tagore presents us with a conception of nature that is beyond the dominion of human subjectivity. He harbored a deep longing to nurture the delicate relationship between humans and nature (Bandyopadhyay 2019), which, for him, exist together. In *Creative Unity*, written in 1922, Tagore (1962, p. 312) writes:

> For us the highest purpose of this world is not merely living in it, knowing it and making use of it, but realizing our own selves in it through expansion of sympathy, not alienating ourselves from it, and dominating it, but comprehending and uniting it with ourselves in perfect union.

For Tagore, the relationship between humans and nature should be seen in this manner: "The earth does not merely hold their body, but it gladdens their minds, for its contact is more than physical contact –it is living presence" (Tagore 1915, p. 9). To bring into the awareness of human beings the reality and force of this "living presence," he writes about nature using vivid imagery and words that flow forth in a manner that reverberates with the dynamism of nature itself.

> Before entering on a journey on the Padma, I feel nervous lest she on account of constant company, look unattractive to me. But the moment I float on the river, all my apprehensions vanish into nothingness. The kulkul noise of the ripples, the gentle tremor of the boat, the light-bathed sky, the vast expanse of soft blue water, the fresh foliage of the trees — an ensemble of colour, music, dance and beauty lend radiance to the superb melody of nature (Tagore in Roy 1915, p. 106).

The excessive dynamism of this *natural* existence is important because it signifies, as Tagore (2013, p. 16) writes, "my intimacy with Nature … that [Nature] which satisfies our personality with manifestations that make our life rich and stimulate our imagination in their harmony of forms, colours, sounds and movements."

This dynamic living presence of nature can be understood more clearly through Heidegger's idea of *equiprimordiality*, because it allows one to conceive of non-human natural phenomena as imbricated with the phenomenon of human existence. The point is that what is out there and what is the human being must originate co-equally. Time and again, Tagore returns to themes that evoke this co-originality and co-inherence, reminding us of the marvelous bond that exists between nature and human beings: "There is a bond of harmony between our two eyes, which makes them act in unison. Likewise there is an unbreakable continuity of relation in the physical world between heat and cold, light and darkness, motion and rest, as between the bass and treble notes of a piano" (Tagore 1915, p. 53).

This *unbreakable continuity* runs through all of the cosmos and all of existence. It speaks to a natural bond that beckons a healthier ethical and practical relationship between humanity and nature — by virtue of which humans can also create and cultivate more open ways of existence (reminiscent of a Heideggerian existential open-ness). With his poems and other writings, Tagore wishes to constantly remind us of the sense of such an authentic relationship:

Let me dance all day long, having kissed
each flower bud, having hugged
the satin-soft green corn fields.
Let me swing on each of the waves all day long
On the hammock of joy (Tagore 1969, p. 50).

Ah my heart dances like a peacock
the rain patters on the new leaves of summer,
the tremor of the cricket's chirp troubles the shade of the tree,
the river overflows its bank washing the village meadows.
My heart dances (Tagore 2004, p. 334).

For Tagore, "people's harmony with nature" was especially important because, as Gupta explains, he considered this to be "an essential aspect of their transcending self-centered existence" (2005, p. 58). An openness to nature assists in breaking the fetters imprisoning one's heart. An ego decentering is required. Connectivity with nature requires *ekstasis*. This is indeed resonant in Tagore's ideas of spirituality and divinity, which center on "the realization and extension of one's being in the open panorama of nature" (Gupta 2005, p. 59). Thus, Tagore talks about nature as *the most sacred place for pilgrimage* (Gupta 2005, p. 59). Since his ideas regarding spirituality are situated within the immanence of nature, he vehemently protests "against the philosophy which demands retirement from active life and sees the purpose of life in hazy mysticism" (Lesny and McKeever Phillips 1939, p. 98). Tagore is very critical of approaches that see this world as a mere illusion or a dream or a play. The 'here' of the world is crucial for him, and thus he prefers "living in the heart of nature against the lofty philosophizing of the ascetic" (Gupta 2005, p. 60).

Tagore's elegant expressions of being-with-nature represent a remarkable attempt at imbuing natural phenomena with divine *potentia*. Transcendence is to be found in immanence, and all the gods are rooted to the earth. With regard to the sense and force of existence, Tagore values nature as a partner for humans. This holds the promise of new paths for ecological ethics and praxis, and we find similar or resonant voices in other Eastern traditions, such as the Sufi poetry of Jelal-ud-din Rumi.

7.5 Rumi on Being-with-Nature: The Immanent Scent of the Divine

Rumi, the thirteenth century Persian Sufi poet and scholar, wove immaculate rhythmic patterns to portray the marvels of nature. His poetry is replete with references to nature as animate. Nature lives and walks and talks. Whether it is the cypress tree walking towards the garden of truth or the lily flower speaking to all lovers, nature appears to be alive, expressive, thriving, and dynamic.

> Touched by a spring breeze, the twigs and branches become intoxicated and are caught up in the great dance that permeates all levels of Creation. Do the little twigs not look as if they were merrily stamping their feet on cruel January's tomb and clapping their hands in mystical dance, led by the plane tree, whose leaves have always been compared to human hands? *'The twigs start dancing like repenters [who have recently entered the mystical path], the leaves clap their hands like minstrels* (M IV 3264)[1] (Schimmel 1992, p. 60).

[1] Rumi's poetry is cited as either: (M, volume #, verse #) or (D, poem #). The former represents his Mathnawi Ma'anvi, followed by the volume number and verse number, whereas the latter represents the Diwan-e-Shams, followed by the poem number.

Rumi's poetic genius brings forth vibrant imagery and metaphors to represent the "living presence" of the natural world. It is also worth bearing in mind that such descriptions of nature have been a familiar feature of the Sufi poetic heritage that Rumi inherited. It has been written about Goethe that he really appreciated the "sacred flame" of veneration for nature that was persistently present in Persian poetry (Mommsen 2014, p. 153). In one place, Rumi compares the trees that he sees to Virgin Mary's immaculate conception: "Do not the trees in their youthful beauty and innocence surrender to the Divine Breath as manifested in the breeze, which fertilizes them so that they bring forth beautiful flowers and delicious fruits? The comparison itself is not new, but Maulana vividly reworks the traditional image" (Schimmel 1992, pp. 62–63). With great care, Rumi wishes to emphasize that *all* nature is living: "Everything has Soul…" (Clarke 2002, p. 175). Such is the liveliness of all forms in nature, whether it is the breeze or the birds or even "the silly gourd, which does not understand why the gardener binds its throat — can it not learn to dance along that very rope by which it is bound?" (Schimmel 1992, p. 64). No one and nothing is exempt, and this sense of living presence is extended to even the minutest or least relevant (from a narrow humanistic or subjectivist standpoint) of organisms. For instance, the dance of the *zarre-ha* (i.e., particles that can be seen drifting in the beams of sunlight) is a regular feature of his poetic imagery, and this points towards a dynamism inherent in nature, even at this miniscule level.

Once the participation of nature in all forms of vitality is established, it becomes easier to see how it does not stand apart from human beings but rather coexists with them. There is a striking affinity between nature's movement and human energy, such that they seem to be partners in this co-emergent landscape around them. In the following verses (that we have translated from the original Persian), Rumi tells us that the lover's weeping eyes are always connected to the raining clouds outside in nature.

Without the eyes—two clouds—the lightning of the heart:
The fire of God's threat, how could it be allayed?
How would the herbage grow of union, sweet to taste?
How would the fountains all gush forth with water pure?
How would the rosebed tell its secret to the meadow?
How would the violet make contracts with jasmine?
How would the plane tree lift its hands in prayer, say?
How would the trees' heads toss free in the air of Love?
How would the blossoms shake their sleeves in days of spring
To shed their lovely coins about the garden wide?
How would the tulip's cheek be red like flames and blood?
How would the rose draw out its gold now from its purse?
How would the nightingale follow the rose's scent?
How would the ringdoves call like seekers, "Where, oh where?"
How would the stork repeat his laklak from his soul,
To say: "O Helper high, Thine is the kingdom, Thine!"
How would the dust reveal the secrets of its heart? How would the sky become a garden full of light? (M II 1655–1664).

According to Schimmel, these various expressions and movements and the consequent rhythm of nature correspond "exactly to human behavior" (1992, p. 59). There is a sense of partnership or *correspondence* between the human and non-human world through resonance and mirroring: "Just as the raindrops can help to produce the garden's beauty, the lover's tears will eventually result in a manifestation of Divine Loving-kindness' (Schimmel 1992, p. 59). This also relates intimately to, as Clarke has termed it, "an intuitive, ecstatic mode of *experiencing* nature" (Clarke 2002, p. 160). One that does not simply place nature as outside of human existence, but within it, or more appropriately, *with* it.

This brings to the fore the same sense of *equiprimordiality* that Heidegger wanted to establish, and it is also reminiscent of the harmonies of union that echo across the entire oeuvre of Rabindranath Tagore. Consequently, as Heidegger was led towards reevaluating the status accorded to nature, and Tagore had taken up sacralizing nature to revere it as a 'sacred site of pilgrimage,' the aliveness of nature for Rumi is divine. After all, the natural world for the Sufis is a manifestation of the Divine Beloved. And thus Rumi sees the Beloved in the turning of the seasons: "*A thousand new springs would arrive if the Beloved scatters the pearls of compassion and kindness, Would it be a loss if their anger shreds this spring into pieces?*" (D, 57).

The Sufis find the Beloved even in the tulip, as its color is taken from the cheeks of the Beloved, and also in the cooing of the bird that repetitively utters *ku-ku* (in Persian, *ku* means 'where?', thereby implying that the bird is also seeking the Beloved). Everything in nature, with its beauty and expressions, is a mirroring of God. Nature is immanently divine because, for Rumi, "*Everywhere the scent of God wafts*" (D, 837).

In the immanent midst of a heavenly fragrance, Rumi posits nature as sacred and thereby as essentially related to and intertwined with the Divine Beloved. Let us bear in mind that this is not a matter of "bestowing sacredness upon nature, which is beyond the power of man," as noted by Seyyed Hossein Nasr. Rather, the actual point is about "lifting aside the veils of ignorance and pride" that constitute the reductionist view of nature ("as a lifeless mass") and blind humans from the inherent sacredness of nature (Nasr 1996, pp. 4–7). For Rumi, beholding natural beauty is not different from the ultimate witness of Divine beauty:

In the garden are hundreds of charming beloveds
And roses and tulips dancing around
And limpid water running in the brook.
All this is a pretext –it is He alone! (Schimmel 1992, p. 70).

Here, Rumi is not dismissing natural beauty or reducing it to a bad imitation of some perfect Form of the divine (in the Platonic sense) but rather upholding it as sacred in its own right — as the Beloved manifest. That which must otherwise remain hidden is nevertheless apparent in nature, and not in a distant or removed manner. And thus we can witness its divine potential and appreciate the spiritually immanent divine coursing through it. Therefore, in Rumi's philosophy, seeking the divine through nature is much more valuable than retiring to an asceticism which shuns the world, as is summarized in the following verse:

"Thanks to the gaze of the sun, the soil became a tulip bed–
To sit at home is now a plague, a plague!" (D, 1349).

Rumi inculcates in his readers the spiritual significance of Nature because, as Schimmel (1992, p. 71) writes, "the subtlety of spirit needs matter to become visible". Rumi's words, poetry, and teachings therefore clearly brings out a perspective that values nature not just as bare material but spiritually, as something with force, a force that overlaps with divinity. There is a power and beauty to nature that cannot be reduced to its material or useful dimensions. This enables us to witness "the glory of the Divine Beloved everywhere, especially in spring, when the divine breeze, the rain of Grace, and the Sun of reality work together to melt the snow of coarse matter and call every atom to the intoxicated dance of spiritual resurrection" (Schimmel 1992, pp. 71–72). Similar to Tagore, Rumi's perspective on nature initiates us towards an open-ness towards nature, or in other words, a 'being-with-nature' that echoes throughout this chapter. Both thinkers ground this open-ness in spirituality – a spirituality that is animated by an engagement with nature. This spirituality, as we have seen, can enable and provide tools for contemporary movements that want to overturn the domination-centric modes of relations to nature and restore the sanctity of the natural world.

At this point, it also becomes relevant to investigate whether the open and *ekstatic* spirituality that is inherent in our being-with-nature is, for Rumi and Tagore, pantheistic or panentheistic.

7.6 Conclusion

In current debates on environmental ethics and praxis, the academic distinction between pantheism and panentheism has become quite relevant. This is essentially a question of the relationship that God has to the finite ways in which Godself is manifested (Valera and Vidal 2022). Contrasting interpretations in this field tie into the various ways in which we may understand the connection between human beings, divinity and the nonhuman natural world. While there exist many variations of pantheistic philosophies across the world, the notable similarity amongst them is the principle that "everything that exists constitutes a divine Unity" (Levine 1994, p. 26). Even though there may be differences between what each philosophy construes as *divinity* or *unity*, but, on the whole, as Levine argues, this basic ontological principle prevails (Levine 1994, p. 25).

Panentheism is seen as a middle ground between theism (belief in the Divine as a separate and distinct being) and pantheism because it acknowledges the immanence of the Divine yet leaves room for its transcendence as well. This is "the belief that the Being of God includes and penetrates the whole universe, so that every part of it exists in him, but (against pantheism) that his being is more than, and is not exhausted by the universe" (Owen in Levine 1994, p. 20). Both terms come from the Greek and share the initial πᾶν (*pan-*) and the θεός (*theos*) at the end, that is, "all" and "God", respectively. The etymological difference relies on the extra ἐν (*en*), that is, the 'in' in the middle. For panentheists, all is *in* God. Whereas for pantheists, all is God, and there is no more: the cosmos is God and God is the cosmos. Panentheists do not think this suffices and want to allow the divine a little more room. For them, "God is more than the cosmos" (Brierley 2008, pp. 639–640). Both frameworks offer interesting repercussions for our investigation into the conception and valuation of nature. The former emphasizes nature *as* Divine, while the latter focuses on nature *as an expression* of the Divine. Berkes intriguingly argues that the decline in our relationship with nature seems to be related to the decline in traditional pantheistic cultures (Berkes 2012, p. 267). It becomes important then to turn to the traditions that view nature differently than the technical and scientific models that modernity presents. In this chapter, we tried to bring the poetic words of Tagore and Rumi to the fore as a stream of thought relevant for contemporary ecological ethics. As Berkes quotes Paul Shepard: "Although ecology is a science, its greater and overriding wisdom is universal. That wisdom can be approached mathematically, experimentally, or it can be danced or told as myth. It is in Australian aboriginal people's 'dreamtime' and in Gary Snyder's poetry" (Shepard in Berkes 2012, p. xvii).

Tagore is usually understood as a pantheist, especially since his ideas are believed to stem from Indian mysticism (Gide 1913; Awad 2014; Rahman 2014). Consider the following lines from Gitanjali: "Thy sunbeam comes upon this earth of mine with arms outstretched and stands at my door the livelong day to carry back to thy feet clouds made of my tears and sighs and songs" (Tagore in Paul 2006, p. 280). With sublime exuberance, he unravels the harmonies at play in the cosmos. Tagore writes:

> The same stream of life that runs through my veins night and day runs through the world and dances in rhythmic measures. It is the same life that shoots in joy through the dust of the earth in numberless blades of grass and breaks into tumultuous waves of leaves and flowers (Tagore in Paul 2006, p. 283).

There is a beautiful sentence in Julie Morley's *Future Sacred* that carries this Tagorean impulse: "*nature consists of interactive entelechies that create the warp and weft of the cosmos*" (Morley 2019, p. 118). As we have seen, Tagore is very clear about how he views the being that corresponds to our existence as fused with the rest of the universe; the concept of 'union' is indeed an important part of his philosophical modus operandi. There is nevertheless the possibility that a certain sense of panentheism may also lay hidden in his words, for example, when he speaks of that "realization of this Greater-than-all in the heart of the all" (Tagore in Shrivastava 2020, p. 25), which throws open the possibility of an interpretation that leans towards a divine entity or being immersed in but nevertheless beyond the immanent realities.

As for Rumi, it is probably pertinent to place his poetry within the broader context and prominent debates in Sufi mysticism. In terms of pantheistic ideas, Ibn Arabi is popularly known as the greatest and foremost proponent of *wahdat-ul-wujud* (Chittick 1994, p. 72) and his sophisticated writings emphasize the unity of existence: "All of *wujud* is one on reality, there is nothing along with it" (Ibn Arabi in Chittick 1994, p. 72). *Wujud* refers to existence, a term which was initially used in the Islamic tradition for the development of *falsafa* or philosophy (Chittick 1994, p. 70). Even though this concept has been interpreted in various forms, Ibn Arabi and *wahdat-ul-wujud* remain within the Islamicate world and Western Philosophical tradition to be broadly construed as pantheistic.[2]

Given this background, it becomes easier to understand certain interpretations of Rumi as a pantheist or monist, since his poetry frequently speaks of the immanently Divine. However, many scholars have criticized such simplistic interpretations (Bausani 1968; Nicholson 1923; Schimmel 1993). The emphasis remains on the separate existence of the Divine even if their immanent presence is felt in the world around us. Given this background, it is important to remember that it remains impossible to classify Rumi's exposition of unity with the Divine into neat categorizations of pantheism, panentheism, existential monism, monism, or others. This is so because the formulations of Divine Existence in mystic traditions escape the captures of any formulaic discourse and also due to the fact that mystic poets such as Rumi were not writing as prosaic philosophers insofar as their aim was not to arrive at definitive conclusions but rather to simply attempt at explaining their ineffable spiritual experiences.

No matter how the words of these literary geniuses are read and understood across diverse streams of intellectual thought, the fundamental intuition of a 'being-with-nature' remains steady throughout. As nature is imbricated with the divine — aesthetically, ontologically, and existentially — it demands a respect and appreciation that modern life has, for the most part, blinded itself to. The point is to question subjectivist approaches that rely on separating, reducing, objectifying, and appropriating the natural world only in terms of its being-of-use-to-humanity. As Borgmann writes: "For most of human time, reality/nature were divine and one thing" (Borgmann 1995, p. viii). Hence, one must be wary of human/nature dualism (Hart 2022) insofar as it is premised in a strictly dichotomous subject/object bifurcation. Related to this is the need to critique the Anthropocene epoch that denies nature a status in its own right. We simply cannot accept nature as "a victim of human hubris" (Merchant, 2013, p. 161). And we must also bear in mind that the sense of our own existence also suffers as a result of world-dismissing and nature-objectifying attitudes, perspectives and practices. As Michel de Beistegui (1998, p. 77) rightly points out: "Not only the earth but man himself … becomes subjected to this process of machination: a commodity like any other commodity, an instrument of global planning, disposable waste (industrial, biological, political)."

The intention of this project was to look into philosophical, literary, and cultural perspectives that emphasize the "coinherence and co-dependence of God and the world" (Main 2017, p. 1104). The ultimate purpose is to seek a deeper connection

[2]As a reaction, the proponents of *wahdat-ush-shuhud* (unity in perception), such as Sirhindi, argued for a unity in perception to separate Islamic mysticism from such influences. The argument is that the Sufis who achieve union with the Divine in their ecstatic states mistake this experience for a unity in their overwhelm, and therefore that the unity does not exist in existence but rather in witnessing or perception (Khan and Tantray 2021).

with the world around us — a connection based on a primary withness, and a respect and reverence for what we otherwise, for the most part, objectify: nature, others, and ourselves. With Heidegger, Rumi, and Tagore, we carry forth the promise of a fundamental 'being-with' that brings together human beings, the giving and living earth, and the beyond.

References

Awad AR (2014) Arab countries. In: Kämpchen M, Bangha I (eds) Rabindranath Tagore: one hundred years of global reception. Orient Blackswan, Hyderabad, pp 117–142

Bandyopadhyay D (2013) Tagore, environment and ecology: a place-space dynamics. In: Dasgupta S, Guha C (eds) Tagore — At home in the world. Sage, New Delhi, pp 305–316

Bandyopadhyay D (2019) Rabindranath Tagore: a life of intimacy with nature. Rupa, New Delhi

Banerji D (2015) Rabindranath Tagore in the 21st century: theoretical renewals. Springer, New Delhi

Baudot BS (1999) Introduction. In: Baudot BS, Moomaw W (eds) People and their planet: searching for balance. St. Martin's Press, New York, pp 1–28

Bausani A (1968) Theism and pantheism in Rumi. Iran Stud 1(1):8–24

Berkes F (2012) Sacred ecology. Taylor and Francis, Philadelphia

Borgmann A (1995) The nature of reality and the reality of nature. In: Soulé ME, Lease G (eds) Reinventing nature? Responses to postmodern deconstruction. Island Press, Washington, DC, pp 31–46

Brierley MW (2008) The potential of Panentheism for dialogue between science and religion. In: Clayton P, Simpson Z (eds) The Oxford handbook of religion and science. Oxford University Press, Oxford, pp 635–651

Chittick WC (1994) Rumi and Wahdat al-Wujud. In: Banani A, Hovannisian R, Sabagh G (eds) Poetry and mysticism in Islam: the heritage of Rumi. Cambridge University Press, Cambridge, pp 70–111

Clarke L (2002) The universe alive: nature in the Masnavi of Jalal al-din Rumi. In: Robinson TM, Westra L (eds) Thinking about the environment: our debt to the classical and medieval past. Lexington Books, New York, pp 159–178

Davis BW (2010) The global displacement of Western modernity. In: Jones D, Wirth JM, Schwartz M (eds) The gift of logos: essays in continental philosophy. Cambridge Scholars Press, Newcastle upon Tyne, pp 115–132

de Beistegui M (1998) Heidegger and the political: dystopias. Routledge, New York

Deluca KM (2005) Thinking with Heidegger: rethinking environmental theory and practice. Eth Environ 10(1):67–87

Economides L (2016) The ecology of wonder in romantic and postmodern literature. Palgrave Macmillan, London

Gare A (1995) Postmodernism and the environmental crisis. Routledge, New York

Gide A (1913) L'Offrande Lyrique. Éditions de la Nouvelle Revue Française 60:833–851

Gupta KS (2005) The philosophy of Rabindranath Tagore. Ashgate, Farnham

Ha P (2010) Heidegger's concept of solus ipse and the problem of intersubjectivity. In: Lau KY, Cheung CF, Kwan TW (eds) Identity and alterity: phenomenology and cultural traditions. Konigshausen and Neumann, Würzburg, pp 351–366

Hart WD (2022) Nature and human nature. In: Schweiker W (ed) Encyclopedia of religious ethics. Wiley-Blackwell, Hoboken, pp 453–460

Heidegger M (1979) Prolegomena zur Geschichte des Zeitbegriffs. Klostermann, Frankfurt

Heidegger M (1987) Nietzsche, Vol. III: the will to power as knowledge and metaphysics (trans. Krell DF). Harper and Row, New York

Heidegger M (1999) Contributions to philosophy (from Enowning) (trans. Emad P, Maly K). Indiana University Press, Bloomington
Hiltner K (2003) Milton and Ecology. Cambridge University Press, Cambridge
Hinchman LP, Hinchman SK (2007) What we owe the romantics. Environ Values 16(3):333–354
Keller DR (2010) Environmental ethics: the big questions. Wiley-Blackwell, Hoboken
Khan TR, Tantray MA (2021) Concepts in Muslim philosophy. Rudra Publications, Bilaspur
Lesny V, McKeever Phillips G (1939) Rabindranath Tagore: his personality and work. George Allen and Unwin, Sydney
Levine MP (1994) Pantheism: a non-theistic concept of deity. Routledge, London
Main R (2017) Panentheism and the undoing of disenchantment. Zygon 52(4):1098–1122
Maly K (2009) Earth-thinking and transformation. In: Mcwhorter L, Stenstad G (eds) Heidegger and the earth: essays in environmental philosophy. University of Toronto Press, Toronto, pp 45–61
Merchant C (2013) Reinventing Eden: the fate of nature in Western culture. Routledge, London
Mommsen K (2014) Goethe and the poets of Arabia (trans. Metzger MM). Camden House, Rochester
Morley JJ (2019) Future sacred: the connected creativity of nature. Park Street Press, Windsor
Nasr SH (1996) Religion and the order of nature. Oxford University Press, Oxford
Nicholson RA (1923) The idea of personality in Sufism. Cambridge University Press, Cambridge
Paul SK (2006) The complete poems of Rabindranath Tagore's Gitanjali: texts and critical evaluation. Sarup & Sons, New Delhi
Rahman B (2014) Egypt. In: Kämpchen M, Tagore IBR (eds) One hundred years of global reception. Orient Blackswan, Hyderabad, pp 143–161
Roy BK (1915) Rabindranath Tagore: the man and his poetry. Dodd, Mead & Company, New York
Rozzi R (2013) Biocultural ethics: from biocultural homogenization toward biocultural conservation. In: Rozzi R, Pickett STA, Palmer C, Armesto JJ, Callicott JB (eds) Linking ecology and ethics for a changing world: values, philosophy, and action. Springer, Dordrecht, pp 9–32
Rozzi R (2015) Earth stewardship and the biocultural ethic: Latin American perspectives. In: Rozzi R, Chapin FS, Callicott JB, Pickett STA, Power ME, Armesto JJ, May RH Jr (eds) Earth stewardship: linking ecology and ethics in theory and practice. Springer, Dordrecht, pp 87–112
Ryogi O (1991) Nietzsche's conception of nature from an East-Asian point of view. In: Parkes G (ed) Nietzsche and Asian thought. Chicago University Press, Chicago, pp 200–213
Schimmel A (1992) Rumi's world: the life and work of the great Sufi poet. Shambhala, Boston
Schimmel A (1993) The triumphal sun: a study of the works of Jalāloddin Rumi. SUNY Press, Albany
Sembera R (2007) Rephrasing Heidegger: a companion to being and time. University of Ottawa Press, Ottawa
Sheehan T (2011) Facticity and Ereignis. In: Dahlstrom DO (ed) Interpreting Heidegger: critical essays. Cambridge University Press, Cambridge, pp 42–68
Shrivastava A (2020) An ecology of the Spirit: Rabindranath's experience of nature. In: Chaudhuri S (ed) The Cambridge companion to Rabindranath Tagore. Cambridge University Press, Cambridge, pp 323–336
Tagore R (1915) Sadhana: the realization of life. Macmillan, New York
Tagore R (1962) Creative Unity. Macmillan, New York
Tagore R (1969) Fifteen longer poems of Rabindranath Tagore (trans. Choudhury RN). Oxford Book & Stationery Company, Oxford
Tagore R (2004) The English writings of Rabindranath Tagore: poems. Sahitya Akademi, New Delhi
Tagore R (2013) The religion of man. Martino Publishing, Windham
Valera L, Vidal G (2022) Pantheism, Panentheism, and Ecosophy: getting Back to Spinoza? Zygon 57(3):545–563
White DR (1997) Postmodern ecology: communication, evolution and play. SUNY Press, Albany

Chapter 8
The Withdrawal of God and Humanity as Co-creator in Hans Jonas' Cosmogonic Conjecture

Juan Jesús Gutierro Carrasco

Abstract Interiority, reason, freedom, and the transcendent are cosmic figures that are studied from cosmology and about whose first cause Hans Jonas formulated cosmogonic conjectures. Accordingly, he proposes a cosmogonic eros that accompanies the emergence of the spirit, which grows at the same time as the freedom of the living being and is fulfilled in the human being. Human beings, with their transcendental freedom (which is rooted in the spirit), in the passage from being to duty, hold in their hands not only their own future, but also that of God, who, with breath held, places in the hands of humans their responsibility, the fruits of a radical sacrifice.

Keywords Cosmogony · Hans Jonas · God · Self-alienation · Anthropology

> *I think that the human spirit proves that immanence is present in transcendence* (Jonas 2005, p. 375)

In Hans Jonas' view, value, will, and power are closely linked. The will that posits values and has the power to turn them into laws is rooted in its connection to divine wisdom and a will revealed to faith. The problem arises when faith disappears, and the only ways left to ground the moral rule are human will and power. In the Socratic-Platonic proposal, the law survives even without God because there are intangible essences. In Scholastic thought, however, these essences collapse in the absence of God, unless someone protects them. And this guardian and creator of values are human beings. Without a vision of eternal wisdom to appeal to, they will have to exercise their power without guidance. When the Divine falls or disappears, only humanity and their law are left, immersed in the flow of their being. This will be the framework of our contribution.

J. J. Gutierro Carrasco (✉)
ESCUNI, University Centre for Education, Madrid, Spain

Comillas Pontifical University, Madrid, Spain

L. Valera (ed.), *Pantheism and Ecology*, Ecology and Ethics 6,
https://doi.org/10.1007/978-3-031-40040-7_8

If there is one issue in which Hans Jonas has been taken as a specialist, it is ecology. He set the groundwork for the well-known ethics of responsibility. Having said that, our aim here is to delve into the metaphysical foundations of what cannot be understood in our author without going into the theological issue: the place of God in the creation of the world and the spirit within it. Thus, our description wishes to accompany the creative work of a God that –withholding divine breath– leaves responsibility in human hands as the result of a radical renunciation.

8.1 A Cosmogonic Eros

For Hans Jonas, the eternity of the world and, within it, of the spirit, discussed in the medieval debate between Aristotelians and creationists about eternity or the temporal beginning of the world (who eventually adopted the latter and from which Spinoza distanced himself), fails in view of our current scientific knowledge as well as theologically, since a purely immanent pantheism or pan-psychism might as well be a pan-demonism or pan-devilism.

In fact, our cosmologic knowledge about the beginning of the world and the late and infrequent appearance of the spirit, leads us to reframe the cosmogonic question:

> Modern knowledge of nature has refuted both the biblical belief in a simultaneous and sudden Creation and the Aristotelian issue about the eternal existence of multiple species, complete and different from each other. And this adds an equally relevant second issue to our cosmologic objection. Our knowledge about the original bang is not as certain as our knowledge about the late, precarious, and cosmically very isolated arrival of the spirit from a gradual and meandering evolving of life. In fact, it is a local exception in an immense space-temporal universe of matter and emptiness, lacking life and spirit (Jonas 2012, p. 264).

In Jonas' view, rejecting as he does a cosmogonic logo guiding evolution and causing the appearance of subjectivity, there is a cosmogonic eros that instils a tendency or desire in matter. This matter is subjectivity in a dormant state. In fact, according to Jonas, the aim of the empty cosmos was life insofar as it is an end in itself that is wished and aimed at (purpose in the cosmos). Jonas moves in the immanent level, the result of the phenomenological witness forwarded by life, with no need to appeal explicitly to transcendence to explain the presence of subjectivity: "We must consider the interior dimension as such, from the most imprecise feelings down to the most clear perception and the most acute pleasure and pain, as a result of the merits of the general universal substance, though depending on special external conditions" (Jonas 2012, p. 245).

In fact, says Jonas, if life-forms were mere mechanical automatons, a cosmologic explanation referring to matter would suffice. But the enigma is the subjective dimension, an inner experience "about whose existence no physical model reveals the slightest notion […]. Not even the most exhaustive register of the exterior state of a brain even in its finest functional structures and mechanisms would allow anyone to suspect the participation of conscience, should we not know it on account of our own inner experience, in other words, conscience itself" (Jonas 2012, p. 238).

In Jonas' view, subjectivity or interiority is enigmatic, trans-physical and immaterial. It is what makes life interesting because subjectivity provides an aim, an aspiration, a target (teleology). In fact, life acknowledges such interiority as ontologically essential since lacking it, being would be incomplete. The interior horizon that is present in organisms is surprising because it was neither "previously included in the data from which it is possible to build the genesis of organic systems" nor "can it be added *a posteriori* to organic matter, as a complement" (Jonas 2012, p. 239).

In fact, we should inquire if this potential end –Jonas calls it "teleological end"– limited itself to waiting for its heteronomous appearance or if it contributed to creating the circumstances that would allow its appearance: "We might think of a 'desire' to see that happening, that *might* have been *active* with causal effects and that was intensified as from the first material opportunities (in other words, with the exponential accumulation of these opportunities) to work towards its fulfilment" (Jonas 2012, p. 245).

That is why Jonas will inquire precisely about that first cause (Böttigheimer 2015), about yearning, some "secret teleology" or a "material disposition":

> As life occurred gifted with interiority, interest and finalistic will from the universal matter, these traits cannot be totally strange to its essence and, if not to its essence, then (and at this point the argument becomes cosmogonic) neither to its origins: the matter that was constituted with the original bang must already have had the possibility of subjectivity imbedded in it; a dimension of interiority in a latent state, that was expecting the cosmic external opportunity to manifest itself (Jonas 2012, p. 254).

Who granted matter that possibility? Was there an original creating will? And, if so, what was its next action? It is difficult –a tough demand to our thinking– our author will say, to suggest the possibility that will and values appeared from an unconcerned matter. Jonas will reject a cosmogonic logos guiding evolution and causing the advent of subjectivity (external teleology). But he does suggest, albeit cautiously, something along the line of a "cosmogonic eros" that instils a tendency, a desire in matter.

This *eros*, if not *logos*, takes up the opportunity offered by chance to appear as subjectivity. "From the beginning, matter is latent subjectivity, although for this potential to become actual millions of years are needed compounded by an exceptional happy chance" (Jonas 2012, p. 244). According to Jonas, the secret "desired" aim of the empty cosmos was life, for it is an end in itself, wished for and pursued. For this end to appear, an end "that in its unflinching saying yes to itself is infinitely superior to all that is indifferent and lacks ends," is necessary and nature cannot be alien to it. The aim, he remarks, must be "according to nature, determined by nature and autonomously produced by nature" (Jonas 2012, p. 244). Having said that, "we must not presume any 'foreseeing' intelligence in the beginning, no providence of what was eventually produced. An unconscious tendency to understand what we observe in every living being is sufficient" (Jonas 2012, p. 255).

Jonas will hold that, as a secondary cause, there is a material disposition that unwittingly holds the information for the development process and does its work unknowingly; a process where there is no spirit. In fact, Jonas states in a footnote that the spirit comes "only and exclusively from communication." This would produce an ontogenesis of spiritual information which comes from an external spirit that contrives to enact the potential produced by the blind material inner information. These are mere hypotheses because the path we might follow, Jonas states, from eukaryotes down to human beings would distance us even further from the spirit, even if the spirit is the one who walked that path.

Nevertheless, our author holds that "matter is sleeping spirit from the beginning" (Jonas 2012, p. 256). Therefore, its creating cause "can only be an awoken spirit, the cause of the sole potential, just an actual cause" (Jonas 2012, p. 256). And we arrive at this testimony of the spirit through human self-experience. This allows us to presume that humans see something else besides the merely physical precisely because they can appeal to the spirit that is beyond life and subjectivity. Jonas "reaches transcendence through something that is immanent to us, through our self-experience of the spirit. And he starts not from a postulate of reason but through empirical evidence" (Godoy-Fonseca 2019, p. NN). Therefore, "the anthropic testimony [...] leads us to the postulate of something spiritual, transcendent, and supratemporal in the origin of things: that must be the first cause if there is only one, or a co-participant cause, should there be more than one" (Jonas 2012, p. 257).

Let us focus for a moment in the anthropic. Jonas may be blamed for anthropomorphism, and whoever judged him so would have good reasons. In fact, he admits it himself, although he tries to walk away from the human vanity implicit in it. Jonas states that when we talk about the divine, we create it according to our own image. Therefore, the only way to start opening up the idea of the divine is to do it from the spirit within us, since it is the highest being we know in the whole universe. That leads him to an inevitable anthropomorphising of the idea of God, stressing that the awareness of inadequacy should never be lost (Torres Queiruga 2011, p. 180; Spaemann 2002).

8.2 Self-Alienation of God

If matter yet lacking spirit was gifted with its possibility, and if its cause cannot be but itself a spirit that will not intervene in the course of the world, how was matter granted the spirit? (Moltmann 1987). Jonas insists on the original gift of matter, on that cosmogonic *eros* focused on the spirit, albeit unplanned, even when all the rest depended on the dynamics of matter: but "what could these dynamics be, he asks himself, which, despite its unplanned style, allowed the plan of the universal logos to unfurl?" (Jonas 2012, p. 265). Jonas refers here to Hegel's proposal about the alienation of the spirit in the beginning of the world, a spirit that gradually recovers itself in the sequence of the world (Polkinghorne 2008). Thus, the universal process of thesis-antithesis-synthesis in its process toward the reign of reason, begins with the alienation of the creative spirit in the beginning of all things (Duch 2007). Jonas might share this, but its sequence, that "majestic Hegelian development of all becoming, in dialectic steps in our direction and through us toward perfect termination and, in general, all this edifying idea of intelligible legality of *one sole* global process whose success is guaranteed from the start is something that we, as a most disenchanted audience of the theatre of the world in its large and small scale –of nature and history– are forced to deny" (Jonas 2012, p. 266).

Quantitively, the immensity of the universe turns the manifestation of the spirit in its midst into a small and solitary flame lost in the dark night of the universe. But further still, qualitatively, Jonas asks himself if, should that step or regal course of Hegelian reason be taken, would we be the climax of that universal spirit? Would we ourselves be the definite form of its truth?

Therefore, let us start by describing Jonas' proposal according to whom, following Hegel, the spirit decided to offer itself and its destiny to the flow of the world, to space-time possibilities: "In the beginning, for unknowable reasons, the ground of being, or the Divine, chose to give itself over to the chance and risk and endless variety of becoming. And wholly so: entering into the adventure of space and time, the deity held back nothing of itself: no uncommitted or unimpaired part remained to direct, correct, and ultimately guarantee the devious working-out of its destiny in creation" (Jonas 2001, p. 275).

God chooses to be-in-the world, to fully offer Godself to the future, giving up God's own power, why? "Because it is only in the infinite play of the finite, in inexhaustible chance, in the surprises of what has not been planned, *compounded* by the urgency of mortality, where the spirit may experience itself in the multiplicity of its possibilities, and the divine too" (Jonas 2012, p. 269).

For the world to exist and exist by itself, God relinquishes self-being, self-abandons divine integrity, and renounces divinity, "to receive it back from the Odyssey of time weighted with the chance harvest of unforeseeable temporal experience" (Jonas 2001, p. 275). In this relinquishing, Jonas notes (2012, p. 215), there is no previous knowledge "except that connected to the *possibilities* offered by the cosmic being in view of its own conditions" and those to which the divinity rendered its cause.

Over eons God's interests were in the hands of cosmic chance, living with patient expectations or with patient memory. Thus, creation went on its way along time and space with no transcendent intervention, just the transformations of natural law, "the exterior chance within that law and its own inner gift" (Jonas 2012, p. 269). And only that kind of universe could contrive to offer the opportunity for the spirit to come, "and if the purpose of the Creator was the appearance and self-experience of the spirit, he had to create an immense universe and leave its free course toward the finite to it" (Jonas 2012, p. 270).

The first breath of life is the cosmic chance that *ongoing divinity* was expecting. It introduced a new language that would start filling eternity with self-assurance through feeling, perception, inclination, or actions that would become stronger with time. It is then, when that God that awakens may say that all creation "was very good" (*Gn* 1: 31) (Cobb 1996).

With life, death came; life is an adventure of mortality. But it is in the finite individuals' feeling, living, and suffering that the divine scenery displays all its hues, and divinity begins experiencing itself. Each new species produced by evolution adds further possibilities to feeling and doing, even with divine innocence due to the absence of knowledge: "Note also that nothing with life's innocence, before the advent of knowledge God's cause, cannot go wrong. [...] Every new dimension of world-response opened up in its course means another modality for God's trying out his hidden essence and discovering himself through the surprises of the world-adventure. And all its harvest of anxious toil, whether bright or dark, swells the transcendent treasure of temporally lived eternity" (Jonas 2001, p. 276).

The divine subject adds and improves its treasure of eternity that dwells in time and that is even more exposed with movement and animal perception contributed by instinct, fear, pleasure, pain, triumph, and shortage. Experiencing them so often (innumerably) does not weaken the experience; rather, it provides the purified essence from which divinity constitutes and re-constitutes itself: "The deep character of its intensity, of all experimenting in general, is an asset of the divine subject" (Jonas 2000, p. 319).

Referring to a Jewish prayer, Jonas states that God as the original creating basis, is not only a God that is alive but the God that loves life, "and both for the sake of life itself, through the soul, and as the cradle of the spirit. Therefore, this allows us to speak, up to a certain point, about the sanctity of life, even if dreadful things may occur in it, as well as in the spirit" (Jonas 2012, p. 270).

8.3 The Spirit in Humanity: Created Creator

The spirit, which is perceived in animals as well, with whom humanity shares subjectivity and interior life, flourishes and is sustained in organic life. Now, it is the human being as bearer of knowledge based on thinking insofar as it is spirit, who may act based on free will. It is with the coming of humanity, namely the coming of knowledge and freedom, when life leaves innocence behind and God's wager is exposed to the criteria of success and failure: "With this supremely double-edged gift the innocence of the mere subject of self-fulfilling life has given way to the charge of responsibility under the disjunction of good and evil" (Jonas 2001, p. 277).

That will make humanity responsible for *being*. Now, *must be* is not derived from *being*. It is necessary to know and listen to that *being* though testimonies of life and spirit: "our *capacity* to see and listen turns us into the recipients of the call of his precept to acknowledge *being*, thus becoming subjects of a *duty* to him" (Jonas 2012, p. 217). This *being* must be protected from ourselves, we must rescue the powerless divinity from our actions, from ourselves. Jonas says, "it is a cosmic duty, because whatever we may cause to fail and destroy with and in ourselves is a component of a cosmic experiment" (Jonas 2012, 272).

In Jonas' opinion, in human beings the divine experiments itself fully, albeit depending on human acts. It remains quiet, undecided over eons but with humanity, with his responsible choices, says Jonas, the image of God may be fulfilled, saved, or ruined.

> The image of God started hesitantly in the physical universe and over the years it has been working –undecided– in the wide spirals of pre-human life, but their reach has gradually shrunk, so that with this latest move and with a dramatic acceleration of movement, the image of God moves under the questionable care of man to be fulfilled, saved or ruined by what man does with himself and with the world. And this appalling impact of man's actions over the divine destiny, particularly its effects on the situation of the eternal being, constitutes human immortality (Jonas 2000, p. 319).

Here we can glimpse Jonas' notion of divinity, a transcendence that accompanies human actions "and henceforth accompanies his doings with the bated breath of suspense, hoping and beckoning, rejoicing and grieving, approving and frowning – and, I daresay, making itself felt to him even while not intervening in the dynamics of his worldly scene" (Jonas 2001, p. 277). This is Jonas' proposal: a God that is close by from eternity, being present at every perceptive moment; God may be perceived but God does not interfere: "By foregoing its own inviolateness the eternal ground allowed the world to be. To this self-denial all creature owes its existence, and with it has received all there is to receive from beyond. Having given himself whole to the becoming world, God has no more to give: it is man's now to give to him" (Jonas 2001, p. 279).

Therefore, in his passage to *must be* the human should make sure that God will not have to repent from having allowed the world to be. It is precisely in this passage from contemplation of value to complying with it, from *being* to must be, where moral freedom is born in human beings. Nevertheless, here that *eros* that walks along with full awareness by the spirit of humanity "does not yet provide any guarantee as a *guide*" (Jonas 2012, p. 247) because this freedom, the most transcendent and dangerous, Jonas will say, "is also the freedom to refuse, of wilful deafness, and even the opposite option that may reach up to radical evil and which, additionally, may appear as the supreme being" (Jonas 2012, p. 247).

Thus, the double responsibility of human beings. On the one hand, for his actions which have future unpredictable effects depending on chance and luck. On the other, on account of the effects of our actions on the eternal sphere. The latter is what Jonas calls "instant metaphysics." The moment is the key that unites us with eternity and places whoever acts responsibly between eternity and time. Each person's possibility of a new beginning and their history flows now and again every minute. In every instant of decision making, humans must think as if they were in the presence of eternity, thus open to the call of their full truth. "To act as if in the face of the end is to act as if in the face of eternity, if either is taken as a summons to unedging truth of selfhood" (Jonas 2001, p. 269). That is how the absolute comes into us: in each decision. Because the divine face will depend on the divine face, leaving its mark in eternity.

From the perspective of anthropology, the human, morally free, knows good and evil and must choose between them "here, where the deepest abysses of corruption and perversion of sight and will are opened, we also find the place where the highest peaks of holiness of the will and surrender of life to the imperative of life also rise.

Projecting its supernatural light over the earthly chaos, illuminations of time by an instant of eternity" (Jonas 2012, p. 248).

Now, humanity enjoys this moral freedom insofar as they also enjoy freedom for self-reflection. Although this subjectivity is also present in plants and animals, it takes a qualitative leap in human beings, since it can go deep into the subject, into the self. Human being is both valuer and valued because "reflection submits him to the judgement of moral conscience" (Jonas 2012, p. 249). Humans worry, and is thus responsible for the world outside, as well as for the inner world, "for the possibility and the duty to be good as a person" (Jonas 2012, p. 249).

The leap from the morals of behavior to personal ethics is the fruit of humanity's concern for themselves. Furthermore, this concern turns into something infinite and absolute when it is submitted to transcendent criteria, in other words, with a view to eternity, "that is how the unexpected, daring and, at the same time, repentant attitude, of great explorers into the entrails of the soul, that burn themselves in their ardent love of the supreme good and that endure the suffering caused by their passionate self-analysis, comes to be" (Jonas 2012, p. 250). From a cosmogonical stance, the fabulation proposed by Jonas responds to his defence of metaphysics, despite the likely objections of analytic philosophers. A metaphysics that, in the face of menaces about the future of humanity, should go beyond optimisms and leave some room for blind, unplanned, sheer chance… briefly, "for the immeasurable *adventure* embarked in by the first principle, presuming the spirit took part in it when launching the Creation" (Jonas 2012, p. 268).

To date, the most urgent and pressing transcendent duty, says Jonas, is preventive judiciousness. Because now the "minute" of decision is in the hands of humans in their global actions. Humanity can destroy itself –thus following the evolutionary cycle– but the interests of eternity are in humanity's hands too: "We literally hold in our faltering hands the future of the divine adventure and must not fail Him, even if we would fail ourselves" (Jonas 2001, p. 281).

Human beings, the "crown of Creation, summit of the universal spirit, must assume as a task, as a precept, to be up to the condition of being the image of God" (Jonas 2012, p. 259). Despite that, they betray it:

> We must feel grateful for the rare confirmations that glitter once and again, and sometimes amid the deepest darkness: lacking that and in view of the historical universal list of proofs to the contrary, the combination of atrocities and apathy, we should surrender completely to doubts about the meaning of the human adventure. The example of the just rescues us from that and rescues itself again and again (Jonas 2012, p. 259).

The task of humanity, created in the image of God, has among their work the building of vital ways that will eventually turn into traits of the divine being. A transcendent divinity that grows with the horribly ambiguous harvest of our works. Consequently, we are the mortal trustees of an immortal deposit because each personal being, unrepeatable and fleeting, is a wager of eternity.

If we were ordered and fabricated, we would be finite in the sense that we would be finished. And we would not have the chance of *becoming (gignomai)* children of God. This is where the *genethetho* of creation returns! We were not born children of God, as through generation. We are children through adoption and grace, this requires our consent, our freedom and, daring to say it, our own adoption of God ("choice"). And that is because creation it not birth (only the Son was born from God), not an emanation, where our origins would come "by nature" and would be totally complete. As the Son, who is only truly so when he has also been adopted by his father, "we risk" our filiation. But in the words of the Book of Wisdom: on God's knees. For us too, the articulation of creation will be an articulation of trust. A trust completely surrounded and confirmed by a Cosmos that the Father has created precisely "waiting with eagerness" (*Rom 8: 19*) that we will accept to become sons and daughters (Gesché 2005, p. 153).

8.4 Conclusion

Hans Jonas embarked in these issues because he was convinced that eventually we must learn how to swim and risk jumping into deep waters, even if it is impossible to go forward with positive guarantees (Jonas 2012, p. 275).

The origins of Jonas' myth lie in the question about all those who could not write their name in the Book of Life, because their own lives were interrupted. The point of his answer was "to clarify the feeling of eternity through the notion of a deity whose luck is placed in man's hands" (Bouretz 2012, p. 866). In this way, it goes through the three steps we developed in previous pages: the self-alienation of God, spirit and life, and humanity as subject to duty insofar as they are created beings.

Inner life, reason, freedom, and the transcendent are cosmic data studied by cosmology. Jonas stated the cosmogonic conjectures mentioned above. Conjectures, because knowledge will give way to faith which, though based on reason and not on revelation (though the latter must also be heard), unable to provide a demonstration, strives to give an answer about the first cause. In Jonas' view, the key lies in the first cause, in yearning.

The presence of the spirit in creation opens a multitude of issues about its origin and teleology. Something divine has been seen in the spirit since ancient times:

> When Pythagoras shuddered upon grasping the atemporal truth of his theorem, when the prophets in Israel first perceived in the word of God the unconditional character of the moral mandate and in similar instances in other cultures, a whole horizon of transcendence within immanence was opened. Beyond what was directly stated, it had something to say about the characteristic of being *within which* that opening happened; and this being is both that of the perceiver and of what is perceived (Jonas 2012, p. 252).

Humans, who with transcendent freedom that lies in the spirit, in their passage from being to must be, holds in their hands not only their own future but that of God, a God who, as we said above, awaits while withholding divine breath the fruits of God's action leaving humans responsibility for the fruits derived from a radical renunciation.

References

Böttigheimer C (2015) ¿Cómo actúa Dios en el mundo? Sígueme, Salamanca
Bouretz P (2012) Testigos del futuro. Filosofía y mesianismo. Trotta, Madrid
Cobb J (1996) Hans Jonas as process theologian. In: Lubarsky SB, Griffin DR (eds) Jewish theology and process thought. SUNY Press, New York, pp 159–162
Duch L (2007) Un extraño en nuestra casa. Herder, Barcelona
Gesché A (2005) El cosmos. Sígueme, Salamanca
Gesché A (2010) El hombre. Dios para pensar II. Sígueme, Salamanca
Godoy Fonseca LS (2019) Hans Jonas, cosmologia e criaçao. In: Vieria Da Costa AH (ed) Heidegger, Jonas e Levinas. ANPOF, Brasilia, pp 160–169
Jonas H (1995) El principio de responsabilidad. Ensayo para una ética de la civilización tecnológica (trans. Fernández Retenaga J). Herder, Barcelona
Jonas H (2000) El principio vida. Hacia una biología filosófica (trans. Sierra JM). Trotta, Madrid
Jonas H (2001) The Phenomenon of Life. Northwestern University Press, Evanston
Jonas H (2005) Memorias (trans. Comín IG). Losada, Buenos Aires
Jonas H (2006) Conoscere Dio. Una sfida al pensiero. AlboVersorio, Milán
Jonas H (2010) Philosophical essays: from ancient creed to technological man. Antropos Press, New York/Dresden
Jonas H (2012) Pensar sobre Dios y otros ensayos (trans. Ackermann A). Herder, Barcelona
Moltmann J (1987) Dios en la creación. Sígueme
Polkinghorne J (2008) La obra del amor. La creación como kénosis. Editorial del Verbo Divino, Navarra
Spaemann R (2002) Realidad como antropomorfismo. Anuario filosófico 35:713–730
Torres Queiruga A (2011) Repensar el mal. De la ponerología a la teodicea. Trotta, Madrid

Chapter 9
Hans Jonas and Pantheism: On Ecology and the Problematic Relationship Between God, World, and Humanity

Jelson R. de Oliveira

Abstract The problem of the relationship between God, the world, and the human being presents itself as one of the most central themes of Hans Jonas' philosophy and can be understood from three complementary perspectives. First, the analysis developed in his theses on Gnostic nihilism that, according to him, would have replaced old pantheistic ideas by a radical rupture, according to which the world would have been created by evil forces as a prison for humans, who must, therefore, fight against the world (hostility); in this perspective, the "gnostic principle" would be anti-cosmic and demonizing and would unfold in existentialist philosophy through an indifference towards the world. Second, the attempt to recover a concept of God as separate from the world (and therefore not identical with it) in the form of an "immanent transcendence" (a God who cancels Godself so that the world can exist, along the lines of the doctrine of *Tzimtzum*), in which it is possible to think of a reciprocal participation of the creator in the creature without falling into a "romantic" pantheism. Third, the challenge of thinking about human responsibility for life (understood through an integral monism that results from its phenomenological project) as a new type of bond between humans and the world. Jonas, in this case, distances himself both from the transcendent and anti-cosmic Gnostic split and from radical pantheism (God is everything), to guarantee the possibility of philosophical reflection on the concept of God and, at the same time, to safeguard a non-religious foundation for ecological responsibility, facing the environmental challenges that put the existence of humanity and the world at risk.

Keywords Pantheism · Nihilism · Indifference · Ecological responsibility · Hans Jonas

---

J. R. de Oliveira (✉)
Department of Philosophy, Pontificia Universidade Catolica de Paraná, Curitiba, Brazil

L. Valera (ed.), *Pantheism and Ecology*, Ecology and Ethics 6,
https://doi.org/10.1007/978-3-031-40040-7_9

9.1 Introduction

The problem of the relationship between God, the world, and humanity presents itself as one of the most central themes of Hans Jonas' philosophy and can be understood from three complementary perspectives. First, the analysis developed in his theses on Gnostic nihilism that, according to him, would have replaced old pantheistic ideas by a radical rupture, according to which the world would have been created by evil forces as a prison for human beings, who must fight against the world (hostility); in this perspective, the "gnostic principle" would be anti-cosmic and demonizing and would unfold, in existentialist philosophy, through an indifference towards the world. Second, the attempt to recover a concept of God as separate from the world (and therefore not identical with it) in the form of an "immanent transcendence" (a God who cancels Godself so that the world can exist, along the lines of the doctrine of *Tzimtzum*), in which it is possible to think of a reciprocal participation of the creator in the creature without falling into a "romantic" pantheism. Third, the challenge of thinking about human responsibility for life (understood through an integral monism that results from its phenomenological project) as a new type of bond between humans and the world.

Jonas, in this case, distances himself both from the transcendent and anti-cosmic Gnostic split and from radical pantheism (God is everything), to guarantee the possibility of philosophical reflection on the concept of God and, at the same time, to safeguard a non-religious foundation for responsibility, facing the environmental challenges that put the existence of humanity and the world at risk.

9.2 Pantheism and Gnostic Anti-cosmism

In the summer of 1921, Hans Jonas, aged 18, moved from his hometown of Mönchengladbach to Freiburg, where he studied with Husserl and Heidegger, whom he met again in Marburg in 1924, when he met Rudolf Bultmann. It was precisely in this intellectual environment that he encountered the theme of gnosis, which he addressed in his thesis that he defended in 1928 under the guidance of Heidegger and strongly influenced by his existential analytics. The work, published in two parts (the first in 1934,[1] and the second, twenty years later, in 1954[2]), is a central study for understanding a practically ignored period in the history of philosophy. Jonas analyzes these nearly five centuries (III B.C. to A.D. III) both from the point of view of mythology (theme of the first part of his thesis[3]) and of its influences on

[1] *Gnosis und spätantiker Geist. Erster Teil: Die mythologische Gnosis.*

[2] *Gnosis und spätantiker Geist. Teil II, 1: Von der Mythologie zur mysti schen Philosophie.*

[3] The first part of the work deals with *gnosis in its mystical form* (which contains an introduction to *History and research methodology* followed by four chapters: *The Logos of Gnosis*; *The Existential Attitude of Gnosis*; *Gnostic Mythology and Speculation*; and *New Gnostic Texts*), and the second,

the nascent Christianity (which he calls, in the second part of the work, mystical philosophy, especially considering Philo of Alexandria, Origen, and Plotinus[4]). Jonas' objective is to understand what the Gnostic movements are and to analyze how their mystical speculation, their soteriological culture, and their eschatology influenced early Christianity, and to identify their "fundamental motive": "a clear God-world dualism, that is, an ultramundane idea of God and an anti-divine character of the world as such" (Jonas 2010, p. 22). This dualism is associated with an idea of the cosmos as a "kingdom of darkness," of creation as a result of a "depravity" or "partial fall" of a violent nature caused by the anti-divine action that makes primordial humans trapped in the world as foreigners, although a supra-cosmic soul (pneuma, the inner life of humans, the pneumatic) also resides in them as an inheritance of their old belonging (Jonas 2010).

Following the philosophical intuitions of the so-called "first Heidegger," developed in *Being and Time*, Jonas philosophically approaches this theme by considering gnosis an objectification of the spirit of an era. It is the discovery of an "*a priori* binding horizon" (Jonas 2010, p. 22) between the events of a time, objectified by the subjects present in that time. As an existential core, gnosis would be characterized by a principle of de-worldliness, understood as a tendency to abandon earthly reality, in radical opposition to divine reality, and, in this sense, its first impact would be the denial of Greek or Roman pantheisms[5]: Gnosticism, for maintaining a relationship of hostility with a world, turns against the idea of identification between God and the world, typical of pantheism. On the contrary, the Gnostic anti-cosmic dualism would transform the world into a prison for humanity, a space of ignorance and evil to be abandoned, against which they must oppose and fight for their liberation.

The Gnostic principle identified by Jonas describes a radical opposition between God, the world, and human beings themselves. For the philosopher, such a principle could be considered as a "transcendental 'foundation'" of historicity, which must be sought "in that original stratum of the imaginative apperception of the being from which it is generated as the complex composition of the 'world' and the relationship of existence with the world in an entire era" (Jonas 2010, p. 33).

Jonas identified the origin of this de-mundane principle in Gnosticism:

---

that is called *From mythology to mystical philosophy* (which begins with an introduction *on the problem of objectification and its change of form* and is followed by six chapters: *On the dissolution of the ancient concept of Arete in the scope of gnosis*; *Anticipation of the eschaton and formation of a gnostic concept of Arete*; *Knowledge of God, vision and realization in Philo of Alexandria*; *From the second to the third century: from mythological gnosis to mystical-philosophical gnosis; third century systems: Origen*; *Fragments on Plotinus*; *Parerga on gnosis*).

[4] The texts about Plotinus, announced by Jonas since the beginning as part of his project, were finally added to the work only in 1993. They are texts and notes accumulated over several years, which did not reach a finished version, even in the posthumous publication.

[5] In the stoic pantheism by Cicero, for example, the association between the Cosmos and God remains, as the Cosmos is the All, and this is God. Even the Aristotelian association between Nature and the Divine Nous can lead to an understanding of the world as a manifestation of the divine; not to mention Plotinus' transcendentalism, primarily inherited from a particular interpretation of Platonic texts.

> In the thesis according to which the 'pressure of the terrible times' (*Reitzenstein*) would have produced fatalism, pessimism, and, subsequently, also the reaction to them. The catastrophe and the diversities of that time [Jonas refers to late Ancient times], added to the political impotence of man, would have generated a feeling of helpless abandonment to an evil dominion of the world and the awareness of one's very insufficiency that, falling into interiority, was transformed into a deep sense of sin and guilt, in a universal desire for salvation, and, in consequence of all this, in a disposition of spirit to escape from the world (Jonas 2010, p. 91).

We are facing the concept of pseudomorphosis: when humans formulate an image of the world, they also project an idea of themselves or, in Jonas' words, "an aspect of man gains expression" (Jonas 2010, p. 602), and, with it, a way of relating to the world comes to light. The radical solitude of the ego expresses the rupture that could only be corrected through gnosis, that is, through knowledge, and this, in turn, can only be achieved through asceticism and escape from the world since the cosmos appears as a radically negative concept, opposed to the divine transmundane reality.

For Jonas, "the rupture between man and total reality is the foundation of nihilism" (Jonas 2001a, b, p. 440). This rupture would also be a mark of Western history in at least three aspects: in Ancient times through Gnosticism; in Modernity through modern science characterized by the ontology of death; in Contemporaneity through the rise of philosophical existentialism (especially that of Heidegger).[6] With this, Jonas demonstrates how the hostility against the world, typical of the Gnostic movements, extended to modernity, affecting not only religion but science and philosophy itself, either because teleology was denied to nature or because life was interpreted only as matter (closing the eyes to the event of interiority). The result, thus, is that the world has been made devoid of God (by Gnosticism, which broke the pantheistic paradigm), nature has been made devoid of teleology (by philosophy and modern science), and life has been made devoid of spirit (by scientific biology). Religion, science, and philosophy contributed, consequently, a worldview that justified the exploitation of nature (as *res Extensa*) by technological activity insofar as nature did not carry any value. Thus, from Gnostic *hostility*, Western civilization went to a modern *indifference*, whose paradigm is Heidegger, but whose turning point is Pascal, considered by Jonas as the last Gnostic and the first existentialist, as he expressed the "solitude of man in the physical universe of modern cosmology" (Jonas 2001a, b, p. 339).

[6] Jonas realizes that Heideggerian philosophy would assume the idea of "being thrown" (*Geworfenheit*) in the world and unconscious of one's own finitude, which he characterized as the bearer of an inauthentic life. The philosophy produced by Heidegger in the 1930s, thus, would precisely evoke the fall and decadence (*Verfallenheit*), anguish (*Angst*), dissatisfaction, the sense of crisis and estrangement (*Unheimlichkeit*) as constitutive elements of the spirit of that time, made real by the advent of Nazism and technology.

9.3 Modern Nihilism: From Hostility to Indifference

In his 1952 text, *Gnosticism and Modern Nihilism*,[7] Jonas articulates the historical moments of gnosticism and existentialism under the guiding thread of nihilism, as a way "to draw a comparison between two movements, or positions, or systems of thought widely separated in time and space" (Jonas 2001a, b, p. 320). For Jonas, following the Cartesian tradition of rationality, modernity accentuated humanity's estrangement from the world as it understood humans as the only being capable of thinking about themselves in this universe, something that made them "stranged from the community of being in one whole" and, therefore, "his consciousness only makes him a foreigner in the world, and in every act of true reflection tells of this stark foreignness" (Jonas 2001a, b, p. 323). In this way, the modern condition of humans before the *cosmos* is marked, therefore, by foreignness because the *cosmos*, previously ordered by *logos*, is now marked by the infinite opening of modern physics, in which it is no longer possible to justify the "here" of humanity nor understand the world as their home.

The result of this is indifference towards nature: "With the ejection of teleology from the system of natural causes, nature, itself purposeless, ceased to provide any sanction to possible human purposes" (Jonas 2001a, b, p. 323). Without purpose, goal, or meaning, the universe is "without [an] intrinsic hierarchy of being" and no longer provides any ontological foundation for values. For Nathalie Frogneux, this gives human beings absolute freedom (associated with the lack of norms or rules), but it also throws them into a total indifference to actions that end up making humans "deprived of the effectiveness of his acts and closed in on himself and in a narrow subjectivity" (Jonas 2001a, b, p. 64). Such a model of subjectivity, closed in its solipsism, grants unique values and meanings from itself: we enter the time of the crisis of supreme values and the "Nietzschean phase, of the situation in which European nihilism breaks the surface" (Jonas 2001a, b, p. 324), that of the homeless human being living in solitude, whose "way in which his existence exists" (Frogneux 2001, p. 67) occurs as indifference. The world would be, after all, reduced to its magnitude and quantity and, as such, a mere scenario or occasion for the exercise of human power. For Jonas, interpreting Pascal, in modernity "the universe does not reveal the creator's purpose by the pattern of its order, nor his goodness by the abundance of created things, nor his wisdom by their fitness, nor his perfection by the beauty of the whole - but reveals solely his power by its magnitude, its spatial and temporal immensity" (Jonas 2001a, b, p. 324). This is because, after all, "what magnitude can tell of its power" (Jonas 2001a, b, p. 324) and what human beings can do

[7]Originally published in "Social Research," XIX, 430-52; translated into German as "Gnosis, Existentialismus und Nihilismus" and published in Organismus und Freiheit: Ansätze zu einer philosophischen Biologie (Göttingen: Vandenhoeck & Ruprecht, 1973) and later as Das Prinzip Leben: Ansätze zu einer philosophischen Biologie (Frankfurt, Insel, 1994). This text has been revised by Jonas at least seven other times. We will use here the version that appears as an epilogue to *Gnostic Religion* (2001).

before the world is to exercise their "will to unlimited power" [*Willens zu grenzenloser Macht*] (Jonas 1985, p. 34).

For Jonas, the Gnostic dualism, which separates humanity and world, in terms of experience has as "its logical ground" (Jonas 2001a, b, p. 326) another one, which separates God and world, awakening in humans a feeling of "an absolute rift" (Jonas 2001a, b, p. 327) between them and "that in which he finds himself lodged – the world" (Jonas 2001a, b, p. 327). The same sense is found in Heidegger's existentialist theses because, for Jonas, the "existentialist depreciation of the concept of nature" identified by humanity, belatedly, in the thought of their former master "obviously reflects its spiritual denudation at the hands of physical science, and it has something in common with the gnostic contempt for nature" (Jonas 2001a, b, p. 337). This makes him address Heidegger's philosophy with new colors: "no philosophy has ever been less concerned about nature than Existentialism, for which it has no dignity left" (Jonas 2001a, b, p. 337). Indebted to nihilism, the Heideggerian method allowed itself to be filled with the same content that dealt with nature according to "a mere onlooking curiosity" (Jonas 2001a, b, p. 337).

In this sense, "modern nihilism infinitely more radical and more desperate than gnostic nihilism ever could be for all its panic terror of the world and its defiant contempt of its laws" (Jonas 2001a, b, p. 339). That loss of eternity described by Nietzsche and interpreted by Heidegger from the announcement of the death of God would have led to the inauthenticity of the present time, now trapped in a constant crisis between the past and the future, in which the human project lacks a *nomos*, since, if before the Gnostic God had "more of the *nihil* than the *ens* in his concept" (Jonas 2001a, b, p. 332), now "no *nomos* emanates from him, no law for nature and thus none for human action as part of the natural order" (Jonas 2001a, b, p. 332). It is up to the human being, then, to be just "nothing but his own project" and, therefore, "all is permitted to him" (Jonas 2001a, b, p. 332).

9.4 From Indifference to Ecological Responsibility

Such reflection explains how, for Jonas, it would be necessary to think about the concept of God and the concept of responsibility without falling into immanent pantheism (which would dispense with human duty) or Gnostic anti-cosmism (which would lead to hostility or indifference). His theoretical alternative evokes a new way of understanding God and thinking about human responsibility before the world. Thus, if the Gnostic principle was anti-cosmic, the responsibility principle understands the world as an authentic place for the realization of human freedom and responsibility; if indifference was the product of foreignness and, in this case, one of the causes of the ecological crisis, then responsibility must be understood as its ethical response.

Jonas considered his text *The concept of God after Auschwitz, a Jewish voice*, as "a piece of frankly speculative theology" (Jonas 1996a, b, p. 131), claiming the philosopher's legitimacy to deal theoretically with the concept of God, especially when opposed to the events of Auschwitz: which God, after all, would have allowed that to happen to God's chosen people? Jonas highlights that such a question is challenging for Jews, as they, unlike Christians, understand that "God is eminently the Lord of history" (Jonas 1996a, b, p. 133) because salvation and justice are in the hereafter. For this, he makes use of a myth (described as "a vehicle of imaginative but credible conjecture") (Jonas 1996a, b, p. 134). inspired by the *Tzimtzum* tradition: God, having created the world, would have given up omnipotence and retreated before becoming. *Tzimtzum* is, indeed, the center of Yitzchaq Luria's thought and means "concentration" or even "contraction," although it is also possible to translate it as "retreat" or even "return." The doctrine refers to the idea that God would have withdrawn from everywhere so that the universe could exist. This is precisely what Jonas refers to at the end of his text: "*Tzimtzum* means contraction, withdrawal, self-limitation. To make room for the world, the En-Sof (Infinite; literally, No-End) of the beginning had to contract himself so that, vacated by him, empty space could expand outside of him: the 'Nothing' in which and from which God could then create the world. Without this retreat into himself, there could be no 'other' outside God, and only his continued holding-himself-in preserves the finite things from losing their separate being again into the divine 'all in all'" (Jonas 2001a, b, p. 142). It is precisely at this point that this idea cannot be confused with pantheism, insofar as the retreat in itself is what makes possible the existence of a reality outside of God; without this, according to the "all in all" logic, nothing would be possible outside of God, not even the world.

Thus, Jonas does not think of God as identical to the world but as correlated with it and depending on it for God's own regeneration. For him, the modern temper is a result of this "unconditional immanence" and "it is its courage or despair, in any case, its bitter honesty, to take our being-in-the-world seriously: to view the world as left to itself, its laws as brooking no interference, and the rigor of our belonging to it as not softened by extramundane providence" (Jonas 1996a, b, p. 134). Therefore, according to Jonas, "the same our myth postulates for God's being in the world" but "not, however, in the sense of pantheistic immanence: if world and God are simply the same, the world at each moment and in each state represents his fullness, and God can neither lose nor gain" (Jonas 1996a, b, p. 134). In this way, the myth forged by Jonas wants to demonstrate that God gives up being and renounces power, and depends on humans to rescue themselves in their own history. In the experience of life, thus, says Jonas, in its revocability and finitude, "the divine landscape bursts into color and the deity comes to experience itself" (Jonas 1996a, b, p. 135). Each form of life, each species, and each one of the beings would be, in this sense, a form of enrichment of this experience that God has of Godself. In the adventure of the world, God recognizes God's own essence. Jonas even states (making a direct connection with his ontology of life) that the experience of animal life in its perception and movement (the degrees of freedom that are typical of animal life in its process of opening to the world and differentiating from the plants[8]), with the refinement of his instincts "supplies the tempered essence from which the

[8] See especially the Fourth Essay of *The phenomenon of life*: "To Move and to Feel: On the Animal Soul" (Jonas 1966, pp. 99–108).

Godhead reconstitutes itself" (Jonas 1996a, b, p. 134). This means that God updates Godself through living in the countless and repeated lives of God's own creatures, so that evolution is understood as a process of self-recognition of divinity: "evolution provides in the mere lavishness of its play and the sternness of its spur" (Jonas 1996a, b, p. 135).

The evolution of life appears as a deepening of the experience of God's selfhood, whose peak is the emergence of the human being: "the advent of man means the advent of knowledge and freedom, and with this supremely double-edged gift the innocence of the mere subject of self-fulfilling life has given way to the charge of responsibility under the disjunction of good and evil" (Jonas 1996a, b, p. 135). In humanity, therefore, God reaches God's highest potential and God's highest risk, since it is on humanity, now, that the divine cause depends: "the image of God," writes Jonas, "haltingly begun by the universe, for so long worked upon-and left undecided-in the wide and then narrowing spirals of prehuman life, passes man's precarious trust, to be completed, saved, or spoiled by what he will do to himself and the world" (Jonas 2001a, b, p. 136). For the philosopher, in this sense, "this awesome impact of his deeds on God's destiny, on the very complexion of eternal being, lies the immortality of man" (Jonas 2001a, b, p. 136).

The concept of God mythically described by Jonas, therefore, is that of a "suffering God" but also a "becoming God" (Jonas 1996a, b, p. 137), a God who, not being anti-cosmic and not even being, on the contrary, the world itself, is a deity that "emerging in time instead of possessing a completed being that remains identical with itself throughout eternity" (Jonas 2001a, b, p. 137). To this extent, the new concept of God forged by Jonas denies the supratemporality, impassibility, and immutability of God and, at the same time, overcomes the ontological antithesis between being and becoming, in order to recognize that God is affected ("altered, made different") by the events of the world: "if God is in any relation to the world-which is the cardinal assumption of religion- then by that token alone the Eternal has 'temporalized' himself and progressively becomes different through the actualizations of the world process" (Jonas 2001a, b, p. 137).

Another characteristic of God is caring for and caring about the world and is not, therefore, a distant God closed in on Godself or a wizard who acts directly on the world: God is, instead, a God who depends on the human being, since having given the power of intervention to human beings, God also made divine concern depend on them. Thus, risk and constant danger demonstrate that the world is not in absolute perfection but depends on human responsibility for its care. And this is because God withdraws from God's own power: dependent on human freedom and responsibility, God can no longer be recognized as an all-powerful being who acts in the world to guide its destiny. At this point, Jonas demonstrates the logical falsity of the idea of an all-powerful power, since all power is always a relation, which would imply the existence of another instance that would also have some kind of power: "power is exercised only in relation to something that itself has power" (Jonas 2001a, b, p. 139). Creation, in this case, would be the logical proof of God's retreat from God's own almighty power and, at the same time, the moral proof of the existence of evil, against which God, being absolute Goodness, can do nothing but be dependent on human action. This explains why, in the face of the horrors of Auschwitz,[9] God remained silent: God cannot even revoke the decision God made and, with that, God also suffers with those victims: "God was silent. And there I say, or my myth says: Not because he chose not to, but because he *could* not intervene did he fail to intervene" (Jonas 2001a, b, p. 140). This is because, by granting freedom to humanity, God renounces power and thus opens the way for the demand for responsibility. All evil, in this case, is nothing but the product of human action and not of an action external to the world – Jonas expresses himself oracularly on this point, stating that Auschwitz is not Lisbon: "Auschwitz rather than the earthquake of Lisbon" (Jonas 1996a, b, p. 141) – that is, an evil produced by the human being and not by an action of nature itself, as was the case of the Lisbon earthquake of 1755.

Note, in this case, how the mythical-speculative hypothesis of Hans Jonas is supported both in his ontology of life (from 1966) and in his ethics of responsibility (from 1979): these are the two bases on which the author thinks his concept of a God who, surrendered to the future of the world but not confused with it, maintains the call for human beings to assume their responsibility for the care of "creation." Only then will God have no reason to regret God's actions. Believing in the superiority of good over evil, Jonas professes his faith in human virtue, whose apex is precisely the ability to authentically fulfill oneself through responsibility.

This type of spiritual progress manifests, as we have seen, the divine dynamics itself and how God relates to the created world, with which Jonas distances himself from both a dualistic (world and God are disconnected) and a monistic (world and God are the same) point of view. For Jonas, neither God is entirely exiled (gnosticism and existentialism) nor is God fully "at home" in the world (pantheism). The new concept of God, mythically supported, is a third way: "Jonas's speculations aimed at *amending* modernity by interweaving dualism and monism together, thereby integrating contesting world-views into one theological master theory" (Hotam 2010, p. 482) that represents "a third road ... one by which the dualistic rift can be avoided and yet enough of the dualistic insight retained to uphold the humanity of man" (Jonas 2001a, b, p. 340).

Hotam emphasizes that to correct the dualism, "Jonas defends a particular presence of God in the world," which, as we have seen, "is not characterized by direct dominion over the world (monotheism), or (ontic) unity with him (pantheism); nor does he reveal himself in the form of a hidden spark exclusive to the human being (Gnosis)," but, on the contrary, such presence "is patently manifested in the message of God encoded in the world; that is, in nature as such," whose laws translate a "Cosmogonic Eros"[10] (Jonas 1996a, b, p. 172, author's translation), according to

[9] Or, if we can actualize evil, the same question – which God, after all, allows this to happen to his creatures? – could be made in the face of the horrors imposed on living beings (human and non-human) today, through environmental catastrophes that also, metaphorically as in Auschwitz, lead to the death and extinction of species.

[10] An expression that Jonas would have inherited from Ludwig Klages, who was considered a "philosopher of life" (cf. Hotam 2010, p. 482, author's translation).

which natural laws echo a kind of divine desire. For Hotam, the Jonasian alternative presents itself as a new concept of God who is now "at home in exile" (Hotam 2010, p. 482), as a transcendence within immanence, whose meaning could only be understood through the anti-utopian effort that mobilizes Jonas' thought as an effort to understand the human duty of responsibility: as it happens when he analyzes the "utopia of technical progress" in *The Imperative of Responsibility*, to reaffirm the urgency of ethics and oppose the "heuristic of fear"[11] (Jonas 1984, p. 27) to the irresponsible tendency of utopia that closes its eyes to the catastrophe, blinded by the optimism it maintains concerning its own potential, here also the refusal of the idea of a God who acts directly in the world (or even of a world that constantly adapts to the will of God) translates the ethical obligation that humans assume their responsibility in the face of the facts engendered by them. Therefore, only the rejection of pantheism gives rise to responsibility, which means that responsibility cannot be religiously founded on the idea of God, but assumed as an ethical principle of humanity who, independent of divinity, achieves their own authenticity through responsibility.

9.5 Final Considerations

Contrary to what Hotam concludes, for whom "Jonas' theological speculations provided the foundations for his philosophy of the organic" (2010, p. 483), we insist on the fact that it is the philosophy of the organic (and, we should add, the ethics of responsibility) that provide the basis for the author's philosophical speculations. Thus, in addition to a new ontology and ethics, Jonas formulates a new theology (in the sense of a theological philosophy), which reaches a full meaning when understood in the light of his previous texts. However, such speculative theology is nothing more than a product of his ontology and ethics, evidenced by the idea that Jonas refuses a religious foundation for his ethics. One of the texts in which the philosopher is most explicit in this regard is called *To be and ought to be, how we can establish our duty to future generations and the Earth regardless of Faith*, an unprecedented conference[12] contained in Konstanz's university archives (HJ 5-9-5). In this text, although he considers that faith can offer some foundation for ethics, Jonas states that "a foundation independent of faith is necessary precisely because, as it is known, many people today no longer believe in the Judeo-Christian God and in the creation of the world by his work" and everyone must be able to "justify what is today the most urgent of duties: common human responsibility (a "*summa contra gentiles*," so to speak)" (Jonas 2003, p. 43, author's translation).

[11] About this concept, see Tibaldeo (2015).

[12] It is not possible to specify the exact date of the conference, but according to Paolo Becchi (Jonas 2003), translator of the text into Italian, it must have been after 1979, in Germany and Switzerland.

In this way, the foundation of responsibility is not religious but ontological, given that the self-transcendence of life is nothing but the effort for its own continuity, and that is where Jonas founds the idea of the intrinsic value of life in general. Having a value and being an end in itself, all forms of life deserve to live and depend, for this, on the human duty to take responsibility: as they are the only beings capable of doing so, humanity must then assume such responsibility and ensure that future generations can do the same. As a critic of technology, Jonas believes it poses a risk precisely because it does not allow itself to be guided ethically. That is why Jonas speaks of a "power over power," that is, of an ethical capacity to guide the use of technical powers – and this is, in the limit, what responsibility means. If technological civilization was built on indifference, responsibility insists on a new type of bond between humans and nature. In ethical terms, the topic ends here. But if one wanted to go a step further, in theological terms, this would require a reassessment of the concept of God – and this is what Jonas does in his "piece of frankly speculative theology" (Jonas 1996a, b, p. 131).

This reflection establishes a direct bridge among philosophy, ethics, theology, and ecology, insofar as it requires a vision of nature that is not that modern one, marked by materialism that opens everything to exploitation and depredation. As Leonardo Boff pointed out: "Ecology, more than any other science, confronts us with nature as an organic, differentiated, and unique entity. It helps us understand the theological concept of creation, through which God and the universe are differentiated and at the same time united" (Boff 1994, p. 74, author's translation). On this point, we also disagree with Charmenant, for whom Boff, with such a statement, "invited us to develop a Christian panentheism according to which God is present in everything and everything is present in God, and to distinguish it clearly from pantheism in which everything is God" (2018, p. 115). In fact, Boff does not suggest any type of pantheism, but, on the contrary, approaches the non-dualistic Jonasian proposal, precisely emphasizing the fact that God and nature maintain an ambiguous relationship with each other – mediated by the existence of humanity.

References

Boff L (1994) La Terre en devenir. Une nouvelle théologie de la liberation. Albin Michel, Paris

Charmetant E (2018) Écologie profonde et spiritualité: un lien si fort. Revue d'éthique et de théologie morale SI:103–115

Frogneux N (2001) Hans Jonas on la vie dans le monde. De Boeck Université, Bruxelles

Hotam Y (2010) Ecology and pedagogy: on the educational implications of postwar environmental philosophy. Policy Futur Educ 8(3–4):478–487

Jonas H (1966) The phenomenon of life: toward a philosophical biology. Northwestern University Press, Evanston

Jonas H (1984) The imperative of responsibility: in search of ethics for the technological age. University of Chicago Press, Chicago

Jonas H (1985) Technik, Medizin und Ethik: Zur Praxis des Prinzips Verantwortung. Suhrkamp, Frankfurt am Main

Jonas H (1996a) The concept of god after Auschwitz: a Jewish voice. In: Vogel L (ed) Mortality and morality: a search for the good after Auschwitz. Northwestern University Press, Evanston, pp 131–143

Jonas H (1996b) Matter, mind and creation: cosmological evidence and cosmogonic speculation. In: Vogel L (ed) Mortality and morality: a search for the good after Auschwitz. Northwestern University Press, Evanston, pp 165–197

Jonas H (2001a) La domanda senza risposta. Alcune riflessioni su scienza, ateismo e la nozione di Dio (trans. Spinelli E). Il Nuovo Melangolo, Genova

Jonas H (2001b) The gnostic religion: the message of the alien god and the beginnings of Christianity. Beacon Press, Boston

Jonas H (2003) Essere e dover essere. Come possiamo fondare indipendentemente dalla fede il nostro dovere nei confronti delle generazioni future e della terra? Micro-mega 5:39–54

Jonas H (2010) Gnosi e spirit tardoantico (trans. Bonaldi C). Bompiani, Milan

Tibaldeo R (2015) The heuristics of fear: can the ambivalence of fear teach us anything in the technological age? Eth Prog 6(1):225–238

Chapter 10
The Evolutionary Process Leading Up to the Anthropocene as Seen Through Pierre Teilhard de Chardin's Cosmic Christology

Lucio Florio

Abstract In the past few centuries, we have become increasingly aware that both the planet we inhabit and the life it contains share a long history. Due to the intervention of one of Earth's species, *Homo sapiens*, the natural flow of the biosphere has been altered. Today's climate crisis is so significant that a specific name has been coined in order to describe a whole new era: the Anthropocene. This state of affairs allows us to take a fresh look at the history of the Cosmos and its meaning. Here we dwell on a few key insights developed in the last century by Pierre Teilhard de Chardin, a Christian paleontologist, thinker, and Roman Catholic priest. This starting point allows us to study the Anthropocene from a perspective in which science, philosophy, and theology converge. Our goal is to highlight Chardin's evolutionary views in order to study the underlying ontological issue in depth.

Keywords Cosmogenesis · Biogenesis · Noosphere · Anthropocene · Omega point

10.1 Introduction

The evolutionary history of life is widely accepted as a fact by the scientific, philosophic, and theological community. Indeed, it may be stated that the evolutionary viewpoint is a keystone of most twenty-first century people's thinking. We ourselves share the certainty that life has both unity and a history of its own. Furthermore, we hasten to add that species are indeed modified by genetic and environmental factors. Human beings themselves are involved in the evolutionary process by being part and consequence of it. Therefore, this worldview does not only convey the mere perception of a fact; it is widely accepted via various theories.

L. Florio (✉)
Facultad Filosofía y Letras, Pontificia Universidad Católica Argentina, Buenos Aires, Argentina

L. Valera (ed.), *Pantheism and Ecology*, Ecology and Ethics 6,
https://doi.org/10.1007/978-3-031-40040-7_10

According to the terms stated above, the geological notion of "Anthropocene" constitutes an attempt to explain the planetary situation as far as Earth and the biosphere are concerned. This notion was developed within the field of geological science, and its scope being to identify a new era in the history of Earth and the life it harbors. This phase of planetary history is characterized by the overwhelming action of one of its species, *Homo sapiens*, which has affected the planetary system as a whole (that is, physically, chemically, and biologically).

In this chapter, I bring to light a few insights reached by the twentieth century paleontologist and Christian thinker Pierre Teilhard de Chardin. Our aim is to reconsider the Anthropocene from a scientific, philosophical, and theological perspective. Even though certain limitations pertaining to Chardin's time must be allowed for, we believe it is worthwhile to reconsider a few of his insights in order to build a comprehensive thinking system which will enable us to approach the Anthropocene in a fruitful way. We will also attempt to underscore the implications of the evolutionary viewpoint as shown in Teilhard's writings in order to address the underlying ontological question. More specifically, we will identify the variant of theism that the French scientist implicitly puts forward as he approaches the mystery of the Cosmos from a dual (evolutionary/theological) perspective.

10.2 Evolution and Christian Faith: Teilhard de Chardin's Contribution

The history of evolution has been slowly acknowledged by Christian theological thinkers. As a matter of fact, it has often lacked historiographical recognition. Pierre Teilhard de Chardin (1881–1955), a Jesuit priest and paleontologist, has played a paramount role to understand the evolution. Even though his main efforts were focused on field research, his theories remain important in order to address the history of the link between evolution and faith (Artigas et al. 2010). However, such attempts did not fully take shape in the theological arena, mostly due to problems related to a literal interpretation of Biblical texts which deal with Creation. After Theilhard de Chardin's death, however, a growing interest in his work has developed and continues to do so (Giustozzi 2015, 2016, 2021).[1]

We should therefore rediscover some of his ways of thinking, which are relevant for those who address two issues. On the one hand, his way of interrelating a cosmic, biological perspective with Biblically-rooted theology. On the other, the planetary stage we have referred to as "Anthropocene," which Teilhard calls "noosphere," following the Russian evolutionist Validimir Vernadsky's idea of the "biosphere"

[1] Along with the Centre Teilhard de Chardin (https://www.centreteilharddechardin.fr), his work has given rise to several societies. Moreover, it should be stressed that even though we cannot refer to a "Teilhardian system" *stricto sensu*, numerous theologians and biologists/theologians find inspiration in his great insights (Schmitz-Moorman 2005; Edwards 2006; Martelet 2007; Galleni 2010; Haught 2003), etc.

(Vernadsky 1929). In both cases, we will attempt to present his theoretical framework from the perspective of an evolutionary, finalistic Christology (that is, a Christology in which Christ Himself is part of the evolutionary process while being its "Omega Point"). Additionally, we believe it is worthwhile to underscore Teilhard's wholistic calling, which led him to sympathize to some extent with pantheistic insights.

10.3 The Noosphere and the Anthropocene

The term "Anthropocene" currently describes a new geological era characterized by human interference with the lithosphere, the hydrosphere, the biosphere, and the atmosphere. In other words, it refers to a generalized kind of planetary change. We are actually the witnesses of a new era in planetary history.

As a label, "Anthropocene" highlights the way in which the planet has been altered during the past few centuries due to *Homo sapiens*' growing interference. One example may suffice to prove our point: human activity has triggered a sixth mass extinction of species –both faster and more powerful than ever. At the very least, five mass extinctions have taken place during the last 600 million years. They were all due to natural causes. The most recent one was the Cretaceous-Paleogene extinction event, which happened 66 million years ago. It is reckoned that almost 70% of species disappeared within a few thousand years, including most dinosaurs –indeed, only the dinosaurs that are the evolutionary antecedents of today's birds managed to survive. We are currently undergoing a sixth mass extinction. Unlike those which preceded it, it is caused by human action. It is also a lot faster due to human population growth (2000 million in 1930 as opposed to 7800 million today) and technology. It might destroy up to 70% of all living species, including *Homo sapiens*, just like the Cretaceous-Paleogene extinction event did (Ceballos et al. 2021, p. 33).

Let us now consider geologist and paleontologist Teilhard de Chardin's view on the stages of our planet's history and life itself. His approach to the Cosmos is complex. The term "cosmogenesis" describes the growing complexity of the Universe – stretching back to the beginning and open to a distant future. "Cosmogenesis" is the foundation of "biogenesis," that is, the origin and diversification of life (Nuñez de Castro 2008). Life, which presents itself as the biosphere and follows an evolutionary path, has undergone various changes during around 3500 million years. As a paleontologist, Teilhard is convinced of the scientific validity of the evolutionary process: he is an "average" Darwinian of his time. However, his interpretation of the mechanics of evolution differs from Darwin's (Galleni 2010). Unlike most Darwinists, Teilhard asserts on the existence of a direction underpinning the whole process, which would therefore not be restricted to mere chance.

Besides, the French scientist-philosopher states the existence of "noogenesis," a process which has led to *Homo sapiens* through the means of biogenesis. A new dimension of reality, the "noosphere," has come into existence – the spiritual sphere.

This (relatively recent) realm of the spirit (*nous*), which is unique to the human species, is characterized by consciousness and reflexivity.

Teilhard maintains that evolution happens as the result of a unifying creative activity. Even though reality presents itself as a complex unity, it is intrinsically coherent (Plašienková 2017). Spiritually speaking, humanity is at the very nucleus of the Universe in terms of biological evolution. Each human person may be considered as a center which finds its reflection in the Universe and renews the Earth by adding to it a new, spiritual dimension: the noosphere.

Moreover, Teilhard goes on to add yet another dimension to the universal evolutionary process: "Christogenesis," which is born of his Christian faith. Christogenesis involves the presence of the Son, made flesh within the cosmic, biological, and human process of evolution. Relying on Eph. 1: 3–14, the Biblical text which places Christ at the center of the mystery of Creation, Teilhard goes on to envisage a dynamic, cosmic process of growing complexity focused on the Incarnation of the Son and therefore leading to a hypothetically final "Omega Point." This pole of attraction, which can be empirically and philosophically established, coincides with Christ himself due to the fact that, according to the New Testament, *is* present at the very end of human and cosmic history. The whole process of cosmic evolution, animated and shaped by love, leads to Christ.[2]

As far as the biosphere is concerned, *Homo sapiens* was relatively unimportant for a long time. From the evolutionary perspective, it just meant that yet another species had come into existence (Martelet 2007). From the ontological point of view, however, the arrival of *Homo sapiens* implied the emergence of an entirely new dimension of the biosphere, that is, the noosphere – the human world, which includes human culture.

In a nutshell, Teilhard de Chardin defines the "noosphere" as the new human space which came into being during the last phase of evolutionary history. The term "Anthropocene" was coined more recently in order to convey the situation which human intervention has on a planetary scale. Both terms define the same reality, even though each implies its own nuances (Plašienková and Florio 2022). On the one hand, the term "noosphere" underscores the positive connotations of human existence – that is, a deeper level of consciousness and reflexivity, socialization, communication, the refinement of nature through human means, etc. Despite being a geological/descriptive term, "Anthropocene" appears to highlight the disturbing side of anthropic domination. Indeed, it stresses the threat which human activity poses to the biosphere as a whole (Monastersky 2015). It is clear that both terms are not semantically the same: "noosphere" connotes optimism, whereas "Anthropocene" is concerning, even socking (Ceballos et al. 2021, pp. 32–41). As a matter of fact, the latter term implies that globalization is synonymous with deep damage inflicted on all living beings. Some highlight that the most fortunate living beings under Anthropic domination are domestic animals –i.e., those which are rewarded for

[2]An overview of Teilhard's ideas on Christology and evolution may be found in a long-unpublished article which bears the Spanish title of "Cristología y evolución" (Teilhard de Chardin 2005, pp. 69–83). By way of commentary, cfr. (Plašienková and Florio 2017).

their relationship with *Homo Sapiens*. Paradoxically, these animals could now be the at the cutting edge of evolution itself (Francis 2016).

The Teilhardian noosphere is characterized by the fact that evolution builds a growing complexity of functions, organs, and relationships along the lines of biogenesis. Therefore, human beings are able to create societies bound together by an ever-growing network of deep, intense links. The whole process seems positive, since greater perfection is its goal. Needless to say, such neat lineaments do not apply to human history. The sphere of evil and suffering cannot be denied. However, Teilhard de Chardin's views rather minimize the place of evil in the great Christ-oriented cosmogenesis. If anything, evil appears as a derivative of the general process that leads to fulfillment in the Omega Point. Some Teilhardian scholars do, however, limit this boundless optimism by acknowledging the dark side of evolution. For instance, some hold that we should compare the "noosphere" with the ambivalent effects of technology and science on life in general. Along these lines, it might be worthwhile to dwell on some observations regarding the ambiguity of the online world (Plašienková et al. 2020) – which is, perhaps, an almost perfect example of integration and noogenesis.

Let us not forget that "Anthropocene" is a geological term. As such, it is merely descriptive, since it aims to describe the state of the planet during a given period of its history. It does, however, provide a good starting point for reflection. To begin with, science shows that our planet is undergoing significant disturbances from an evolutionary point of view. It is undeniable that the biosphere's evolution is unstable: great weather upheavals, mass extinctions, and even involution have all happened before. Within this context, human activity appears to be just another action or interference carried out by living beings. However, according to the Teilhardian mindset, the noosphere is not just a derivative of cosmic causes: it actually involves the emergence of a new dimension which implies a jump ahead in terms of biogenesis. What is more, Teilhard states that the evolutionary process has reached the noosphere due to a pole of attraction which leads to growing complexity – even after consciousness is achieved. This pole of attraction is the Omega Point.

10.4 What Is the Omega Point?

In order to understand the meaning of the Omega Point, we must bear in mind that Teilhard de Chardin did not regard himself just as a philosopher or even a theologian, but rather as a researcher of phenomena – what the Greeks used to call a physicist. In the preface to *Le Phénomène Humain*, he wrote that his work must not be regarded as metaphysical, and even less as a sort of theological essay. His aim was to write science. In his view, his scientific understanding of the world should be termed "hyperphysical" instead of "metaphysical." He was, of course, aware of the fact that metaphysics is more than just an extension of science or a way of classifying its findings. However, he wanted to pursue what he called "the strictly experimental film" (De Lubac 1967).

According to the French scientist, therefore, all things converge in a hypothetical Omega Point which constitutes the end and keystone of evolution. Scientifically speaking, this could be defined as the evolutionary force, which grows ever more complex as time goes by. However, this pole of attraction can be philosophically and theologically grasped as a transcendental cause which underpins the process of evolution. From Teilhard's holistic perspective, theological and metaphysical reflections on the Omega Point are also scientifically valid. He maintains that the world is growing through the stages of biosphere and noosphere towards its supreme destiny, a supremely independent, self-subsistent, and final unity (Teilhard de Chardin 1965, pp. 321–326).[3]

10.5 Teilhard's View: A Theology which Embraces Science While Also Having an Ontological Impact

The evolutionary view of the Cosmos and life itself is currently accepted both by specialists and well-informed laypeople. It is therefore impossible to avoid it if we wish to approach reality from a metaphysical perspective. Regardless of its interpretation, the long history of the cosmos and the biosphere is necessarily our starting point in order to examine reality as such. The universe has a long history of its own; life itself, such as we know it on Earth, has existed and evolved for at least 3500 million years. Both cosmology and evolutionary biology need to transcend their own methods. On the other hand, ontology and theology can go further in the meta-empirical field by using their own devices.[4]

Indeed, theology constitutes a discipline which is structured around a community of people who believe that the so-called Biblical Revelation is inherently rational. Through the centuries, it has incorporated pre-existing world views. Lately, in the twentieth century, it has even absorbed (in a slow but definite way) the evolutionary view of the Universe.[5] Teilhard de Chardin's approach helps us glimpse a type of theology which embraces evolutionary data, especially as far as the history of the Universe and life are concerned.

This intertwining of science and theology may also be of use to study other aspects of metaphysical reality; even non-believers may benefit from it. As it stands, the Teilhardian system allows those who welcome it to rethink universal ontology in terms of the dynamic movement of finished beings, which are never dissolved into a mere process of transformation. The Teilhardian system is provisional for two reasons: in the first place, due to the intrinsic provisionality of scientific theories (as

[3] In order to find out more about the interrelation between the Omega Point (regarded as a metaphysical and theological principle) and evolutionary history, see Florio (2016).

[4] As far as biology is concerned, an interesting update regarding its relationship with religious thinking may be found in Gregersen (2020).

[5] An example of how evolutionary thinking was received by Catholics may be found in Auletta et al. (2011).

pointed out by epistemologists Popper and Kuhn, among others). This does not make it untrue. It merely states the fact that it remains open and alert in order to absorb new data. The historicity of the Cosmos and life seems, however, to have become a pillar of certainty in the scientific community. Teilhard de Chardin's oeuvre is provisional for biographical reasons too, since he was not allowed to publish texts dealing with religious issues. Therefore, his works remain an unfinished corpus to this day (see Prats 2022). De Chardin's system, thus, allows us to accept transcendence as an integral part of cosmic immanence without disappearing into it. The French philosopher stayed carefully distant from a form of theism in which God does not get involved in Creation –not only as far as deism is concerned, but staying away also from late scholastic philosophy, in which divine intervention is reduced to a fleeting instant at the beginning of time, followed by a sort of accompaniment (through the partaking of being) in a vague *creatio continua*. He did not fall into the trap of pantheism, either. Indeed, pure, monist pantheism would not be compatible with Christian theology. The latter is based on the historical reality of Jesus of Nazareth, who lived, died, and resurrected. The Christological axis, which is paramount in the Teilhardian system, prevents him from overfocusing on mere evolutionary creation. Behind these ideas lies the certainty that Christianity, if deprived from its historical dimension, turns into philosophy (Giustozzi 2016, p. 279).

Teilhard's system has been labeled as "panentheism" by some authors, while others refer to it as "pan-Christism" (Giustozzi 2021, pp. 132–135) due to its Christological nature. Either way, "panentheism" seems more apt for philosophical debate than "pan-Christism" (which is strictly theological). According to some of his commentators, the Jesuit priest was the first theologian ever to attempt a synthesis between theism and pantheism, that is, an attempt to transcend both Descartes and the Enlightenment (the idea of a God who moves the wheel of the world from the outside in order to avoid being polluted by matter) and the ancient Pagan antithesis (later recovered by Spinoza) of a God who is identical with the Universe, Creator and Creation all in one –and ultimately, neither of them, since all Being itself is eternal and therefore lacks a beginning (Campa 2021, pp. 17–18).

Teilhard's view is made possible by the fact that he does not separate ontology from reason (enlightened by Revelation), unlike late Scholasticism (with which he disagreed on this point). His world view is illuminated by actual contact between human and divine reality, as revealed in the Epiphany of Jesus.[6] The world appears to him as a sacred place where God and Humanity, Humanity and World, have come together in the Incarnation, which intertwines Spinoza's God with metaphysical theism (Giustozzi 2021, p. 143). We may sum up his thinking by saying that: "As far as Teilhard is concerned, indeed, God is not identical to the world, but not completely different from it either - he cannot be regarded as a parcel of reality standing outside the world" (Giustozzi 2021, p. 143).

[6] Teilhard dwells on this spiritual vision in Teilhard de Chardin (1966).

10.6 Conclusion

Handled with care (that is, being respectful of each other's objects and methods), scientific and religious ways of thinking can be extremely fruitful when combined. We have attempted to show in which ways an evolutionary worldview, in conjunction with theological thinking, can influence our understanding of the ecological crisis of the Anthropocene, as well as attempting an ontology which includes the evolutionary nature of the Universe. It can even bring us to reconsider the ancient conflict between theism and pantheism. In order to achieve such tasks, we have found some of Teilhard de Chardin's insights and concepts to be useful –especially as far as his clear understanding of the dynamic dimension of the Cosmos and the biosphere is concerned. His theological work breathes new life into cosmic Christology, which has existed since the very beginning of Biblical speculative thinking.

Let us point out a final consideration regarding this combined path: there appears to be a way out of the paradigmatic "Science versus Religion" conflict. Barbour's well-known typology, which exemplifies the ways in which science and religion can relate to each other (conflict, independence, dialogue, and integration) deserves a mention here (Barbour 1997). The history of human thought proves the existence of an often-fruitful interrelation between science and religion. The very origin of ideas about the beginning and expansion of the Universe (or even biological evolution) shows how relevant that relationship has always been, even though it has undergone temporary conflict. A carefully-handled "theological science" may help us delve further into ontological and ethical thinking, as shown above. Since this dual discipline is based on mutual respect, constant effort is needed to achieve dialogue (Florio and Oviedo 2020; Galleni 2016). This approach may lead us to a deeper understanding of phenomena than the one we currently have (which is based on the work of isolated disciplines). After all, reality has proven to be far more complex than we could ever have expected.

References

Artigas M, Glick TF, Martínez RA (2010) Seis católicos evolucionistas. BAC, Madrid

Auletta G, Leclerc M, Martínez RA (2011) Biological evolution: facts and theories. A critical appraisal 150 years after "the origin of species". Gregorian Biblical Press, Rome

Barbour IG (1997) Religion and science: historical and contemporary issues. HarperOne, New York

Campa R (2021) Prefazione. In: Giustozzi G (ed) Pierre Teilhard de Chardin. La "reinvenzione" dell'esperienza religiosa. Edizione Studium, Rome, pp 11–25

Ceballos G, Ehrlich AH, Ehrlich P (2021) La aniquilación de la naturaleza. La extinción de aves y mamíferos por el ser humano. Océano de México, Tlalnepantla

De Lubac H (1967) El pensamiento religioso de Teilhard de Chardin. Taurus, Madrid

Edwards D (2006) El Dios de la evolución. Una teología trinitaria. Maliaño, Sal Terrae

Florio L (2016) The omega point revisited from the new tree of life and the ecological crisis. Studia Aloisiana 3:31–40

Florio L, Oviedo L (2020) Hermeneutic and pedagogical perspectives of the relationships between science and religion. Eur J Sci Theol 16(5):1–11
Francis R (2016) Addomesticati. L'insolita evoluzione degli animali che vivono accanto all'uomo. Bollati Boringhieri, Turin
Galleni L (2010) Darwin, Teilhard de Chardin y los otros. Epifanía, Buenos Aires
Galleni L (2016) Una proposta: il programma di un corso su Teologia e Scienza. Quaerentibus 4(6):82–116
Giustozzi G (2015) Leggere Teilhard senza encomi, senza deprecazioni, senza annessioni. Per una normalizzazione dell'ermeneutica degli scritti teilhardiani. Quaerentibus 5:27–74
Giustozzi G (2016) Pierre Teilhard de Chardin. Geobiologia/Geotecnica/Neo-cristianesimo. Edizione Studium, Rome
Giustozzi G (2021) Pierre Teilhard de Chardin. La "reinvenzione" dell'esperienza religiosa. Edizione Studium, Rome
Gregersen NH (2020) Religious implications of multilevel systems biology. In: Reiss MJ, Watts F, Wiseman H (eds) Rethinking biology. Public understandings. World Scientific Publishing, London, pp 233–263
Haught J (2003) Deeper than Darwin. The prospect for religion in the age of evolution. Westview Press, Boulder
Martelet G (2007) E se Teilhard dicesse il vero...? Jaca Book, Milan
Monastersky R (2015) Anthropocene: the human age. Nature 519:144–147
Nuñez de Castro I (2008) The bio-philosophy of Teilhard de Chardin. In: Heller del Riego C (ed) God seen by science: anthropic evolution of the universe. Universidad Pontificia de Comillas, Madrid, pp 99–126
Plašienková Z (2017) Evolution-science-religion: Teilhard de Chardin's inspirations in the contemporary world. Comenius University, Bratislava
Plašienková Z, Florio L (2017) Ideas inspiradoras de Teilhard de Chardin para la educación en el tiempo de crisis ecológica. Quaerentibus 8:100–112
Plašienková Z, Florio L (2022) El antropoceno en perspectiva teilhardiana: visión comparativa con el concepto de noósfera. En: Florio L (ed) X Congreso Latinoamericano de Ciencia y Religión. La Originalidad y la Fragilidad de la Vida en el Planeta Tierra (págs. 187–200). City Bell, La Plata: DECYR., 16.2. https://repositorio.uca.edu.ar/bitstream/123456789/13850/1/antropoceno-perspectiva-teilhardiana.pdf
Plašienková Z, Sámelová A, Vertranová S (2020) La noosfera en el pensamiento de Teilhard de Chardin y la ambivalencia de la producción mediática online. Quaerentibus 8(14):3–14
Prats M (2022) Une parole attendue. La circulation des polycopiés de Teilhard. Salvator, Paris
Schmitz-Moorman K (2005) Teología de la creación de un mundo en evolución. Verbo Divino, Navarra
Teilhard de Chardin P (1965) El fenómeno humano. Taurus, Madrid
Teilhard de Chardin P (1966) El medio divino. Taurus, Madrid
Teilhard de Chardin P (2005) Lo que yo creo. Trotta, Madrid
Vernadsky VI (1929) La Biosphére. Alcan, Paris

Chapter 11
Influences of the Spinozian Philosophy in the Environmental Activism of Arne Næss

Alicia Irene Bugallo

Abstract Næss was significantly influenced by Spinoza's relational ontology, both for the promotion of his personal ecosophy and for the proposal of the deep ecology movement; for example, the idea that everything is connected with every other form of life. For him, the value dualism spirit-matter does not hold in Spinoza nor is it of any use in field ecology and ecosophies. The realization of the union with the whole of nature is made, in Spinoza, through the understanding of particular things as a manifold of expressions or manifestations of nature (God). Næss highlights the panentheistic interpretation of Spinoza, accentuating that nature or God is nothing apart from the manifestations. These ideas are significant for the promotion of care and respect for all forms of life.

Keywords Ecosophy T · Gestalt ontology · Panentheism · Rhetoric

11.1 Introduction

Norwegian philosopher Arne Næss (1912–2009) was well known for his studies on logic, empirical semantics, and the theory of communicative action, among other contributions. In 1941, he published in Norwegian *Communication and argument. Elements of applied semantics (*translated into English in 1966 as *Communication and Argument*). He set certain grounds for a dialogic theory of conversational games which would be recovered by dialogue formalists and pragma-dialecticians. The relevance of these contributions by Næss lies in his early attention to the first rules for a public and critical discussion within contexts of use of natural language, anticipating later proposals by researchers on the subject (Santibáñez 2012).

Other works followed, such as *Democracy in a World of Tensions* (1951), *Interpretation and Preciseness* (1953), and *Democracy, Ideology and Objectivity*

A. I. Bugallo (✉)
Department of Posgrade Studies in Philosophy, Universidad Nacional de San Juan ,
San Juan, Argentina

L. Valera (ed.), *Pantheism and Ecology*, Ecology and Ethics 6,
https://doi.org/10.1007/978-3-031-40040-7_11

(1956). A non-minor fact is that he goes from his concern for logic or empirical semantics to ecologism. And he became the early founder of the Deep Ecology Movement in 1972. This would not be "either mere chance or a personal whim: Næss was one of those philosophers and citizens who turned their thinking and work into a consistent pair in response to the contextual demand" (Santibáñez 2012, p. 23).

The decision to get actively involved in mitigating an expanding environmental crisis and working on his personal *ecosophy*, allowed him to grant a novel expression and cultural hues to practically all the dreams and reflections of his youth: "The global ecological movement touched off by Rachel Carson gave me an opportunity to gather together most of the themes of my philosophical works in systematic form under the heading ecosophy" (Næss 2005a, p. 315).

Næss acknowledged four main philosophical or ideological sources that signaled his personal journey of reflection and also served as a guide in that stage of his life when he excelled as an environmental activist: Baruch Spinoza, William James, George Herbert Mead, and Mahatma Gandhi (Bugallo 2015). We will follow the track of Spinoza's influence which certainly implies focusing only on part of the rich world of his critical environmental thought.

The context where the *deep ecology movement* and his personal *ecosophy T,*[1] witnesses to a highly worrying environmental and existential situation. Philosophy would be an adequate tool to discuss environmental matters and come down to the ground of the problem, though it should be approached as wisdom, as *sophia.*

According to Næss (1995a, p. 8), ecology as a science does not necessarily provide guidelines to monitor human actions: "A philosophy as a kind of sofía or wisdom, is openly normative, it contains both norms, rules, postulates, value priority announcements and hypotheses concerning the state of affairs in our universe. Wisdom is policy wisdom, prescription, not only scientific description and prediction".

As we said before, among the many sources of inspiration to launch and continue environmental activism, Næss finds most appropriate guidelines in Spinoza, particularly in the combination of ontology, ethics, and politics (Næss 1992, p. 8).

For Spinoza, all reality, and thinking in particular, is power of action. Næss picks this pro-active disposition. Interaction with things and knowledge of things must not be analysed separately. According to Spinoza, greater progress of the body is paired by greater progress of the soul: "I say conception rather than perception, because the word perception seems to imply that the mind is passive in respect to the object; whereas conception seems to express an activity of the mind" (1941, EIIDef 3).

Næss stresses the active and dynamic character of "concept" as "conception;" as a noun for "conceive." Units of comprehension are not "propositions" but "actions." Comprehension is more connected to cause, set free, apprehend, grasp knowledge rather than simply possess it. On this subject, he says that "I have elected to use the

[1] The name alludes to the philosophy of life or total vision that, for the most part, was elaborated in his mountain hut in Tvergastein [crossed stones], in the foothills of the Hallingskarvet massif (NW of Oslo).

more dynamic term 'lambanological,' which comes from the Greek verb λαμβάνω (to grasp), to the static and platonic 'epistemological.' To be active or to act and to understand cannot be systematically distinguished in the Ethics" (Næss 1975, p. 12).

In the face of a worsening environmental crisis, it may prove demoralizing to see millions of humans criticizing, albeit passively, the state of things. An increase in rationality and freedom is in proportion with an increase in activity (as opposed to inaction).

11.2 Rhetoric Mediations for Eco-activism

From Næss' point of view, the search for knowledge implied having developed a life's philosophy, a total philosophical view, acknowledging the values from which our own actions are prioritized and fully living in accordance with them.

At the same time, in some special circumstances and with special aims, it may be convenient to propose an eco-politics move –as organising and defending a grass-roots social movement (as the *deep ecology movement*). This presumes an exercise of argumentation, of narrative, as a rhetoric tool. Næss was familiar with this practice because of his interventions during the Nazi occupation in Norway and as the leader of the UNESCO project, "East-West Controversy" at the beginning of the Cold War (Bugallo 2018, p. 18).

In his *Art of Rhetoric*, Aristotle suggested we should understand rhetoric as the capacity to know what may persuade, in each case (Aristotle, I2 1355b). On the rhetorical character of ethics as from Aristotle, and particularly its usefulness in environmental ethics, persuading is, in a certain sense, stirring. A stirring that is encouraged though advice, since *suadeo (suadere)* means precisely to advise, encourage. Therefore, the function of rhetoric is to advise with the purpose of spurring the receiver of the message to move toward a suggested cause (Buganza 2011, p. 128).

As opposed to sophists, Aristotle limits the persuasion factor to its fair terms and rationalizes it by providing it with arguments, since he understands that "persuasion is a kind of demonstration" (I, 1, 1354b). That is why he sometimes talks about arguments of persuasion. Along the same line, in Næss' opinion the eco-sophist is responsible for suggesting, under the various circumstances, a set –neither too wide nor too limited– of postulates for actions, shared by the widest possible group of supporters.

These ideas are not suggested as *a priori* recipes to be applied and merely solve problems. Rather, they are brought forward as a program to work on and particularly as an indication on how existent realities might change. Both in 1973 versions as in 1984 of the *deep ecology movement* the idea of fostering diversity and encourage symbiosis impels guaranteeing the continued existence of a world rich and varied in ways of life, where the preference for living and letting live establishes an ethics of tolerance (Bugallo 2015).

11.2.1 Biosphere Egalitarianism –in Principle

The postulates of the first version of the *deep ecology movement,* published in 1973,[2] are seven: "1. Rejection of the human-in-environment image in favour of the relational, total-field image; 2. Biospherical egalitarianism –in principle; 3. Principles of diversity and of symbiosis: 4. Anti-class posture; 5. Fight against pollution and resource depletion; 6. Complexity, not complication; 7. Local autonomy and decentralization (Næss 1995a, pp. 3–7).

Although it shows the internalization of concepts from the field of ecology studies, such as *complexity, diversity,* and *symbiosis*, Næss adds a proviso. The norms and tendencies of the deep ecology movement are not derived from ecology logically nor by induction: "Ecological knowledge and life-style of the ecological field-worker have suggested, inspired and fortified the perspective of the Deep Ecology Movement" (Næss 1995a, p. 7).

As to his notion of "Biospherical egalitarianism –in principle" in Postulate 2, Næss (1995a, p. 4) comments: "The 'in principle' clause is inserted because any realistic praxis necessitates some killing, exploitation and suppression".

Among other grounds, he refers to the Spinozian concept of a natural right of all things. According to Spinoza: "Whatever exists expresses God's nature or essence in a definite and determinate way, that is, whatever exists expresses God's power" (EIP36Dem).

Every being has the right to do whatever his power or potency allows. Potency indicates each one's willingness to display his own nature and it is not identical with coercing others. Both in our author and in many eco-philosophers, ecologists, and field biologists, there is an intuitive or direct understanding of nature as perfect in the Spinozian sense: as complete in its fulness and reality. Spinoza understands *perfection* as reality ("by reality and perfection I mean the same thing" –EIVPref.), namely, the essence of anything insofar as it exists and works in a certain way, without considering its duration.

For Næss, the fact that I am part of nature is always amazing. The ideas of diversity and symbiosis in Postulate 3 must be interpreted "in the sense of ability to coexist and cooperate in complex relationships, rather than ability to kill, exploit, and suppress" (Næss 1995a, p. 4).

11.2.2 The Value of Biodiversity

The first three hypotheses of the "Platform Principles of the Deep Ecology", 1984, designed jointly with philosopher George Sessions, propose: "1. The well-being and flourishing of human and nonhuman Life on Earth have value in themselves (synonyms: intrinsic value, inherent value). These values are independent of the

[2] Printed for the first time in Næss (1973).

usefulness of the nonhuman world for human purposes; 2. Richness and diversity of life forms contribute to the realization of these values and are also values in themselves; 3. Humans have no right to reduce this richness and diversity except to satisfy vital needs" (Næss and Sessions 1985, p. 70).

Within the context of environmental ethics, Næss stands out because he does not agree with granting moral rights to all living beings, as stated by other environmentalists. He prefers –under the influence of Spinoza– to limit himself to the idea that "animals and humans may be said to have at least *one kind of right* in common, namely the right to live and blossom" (Næss 1992, p. 14).

It is true that every being has an intrinsic value –subjective and objective– based on its *conatus* (its innate inclination to continue to exist and flourish), but for both Spinoza and Næss the actual moral qualification corresponds to the socio-cultural civil sphere. In Spinoza there is no pre-established moral world order; in other words, human justice, civil status, are not laws of nature. According to Spinoza, in the state of nature, no one is by common consent master of anything, nor is there anything in nature, which can be said to belong to one person rather than another: all things are common to all. "Hence, in the state of nature, we can conceive no wish to render to every man his own, or to deprive a man of that which belongs to him; in other words, there is nothing in the state of nature answering to justice and injustice" (1941, EIVP37Esc 2).

Everything acts in a neutral playing field, and, in its own way, each one is responsible for its development; the conclusion is that all values are conventional. Nevertheless, the "amoral" view of nature, its need, does not exclude the further practice of each part which, in the case of humanity, is valued according to its utility. This implies an ethical and political effort in order to benefit from what each one has received.

Furthermore, and at the same time, Næss holds that no natural law prevents or limits the task of widening the field of justice. There is no natural law preventing the task of outspreading our notions of pity and justice beyond those established by civil society in relationships among humans. Consequently, besides the Spinozian inspiration, an environmentally sustainable future society would imply, in Næss' (1978, p. 420) view, "the balance between a submissive, amoral attitude towards all kinds of life struggle, and a shallow moralistic and antagonistic attitude. Future societies in ecological equilibrium presuppose such a 'third way'."

11.2.3 Relational Ontology

Næss finds in Spinoza's *Ethics* a relational, structural, or systemic character in the notions of body, singular object, and individual, that appear significant for his approaches of environmental philosophy.

The form of an individual (his objective reality, according to Spinoza) is a conjunction of bodies, a union that is the connection between those component bodies. And the (objective reality) is constituted by those internal connections between the parts and not by the mere sum of them. At the same time, and beyond this individual composed by bodies, the Jewish philosopher conceives another individual of a second gender, conformed by composite individuals and to which the same properties applied to individuals of the first gender are applied. And then, a third gender of individuals formed by composite individuals of the second gender, *ad infinitum* (1941, EIIDef 7 EIIDef) (1941, EEIIP 13 Def) (1941, EEIIL 4). "We may easily proceed thus to infinity, and conceive the whole of nature as one individual, whose parts, that is, all bodies, vary in infinite ways, without any change in the individual as a whole" (1941, EIIL 7).

Up to a certain point, the notion of a traditional substantial identity gives way to a composed identity which remains at the level of the internal connections of the composite. According to Spinoza, what we identify as a body is a temporally stable relationship. In Næss' (1995b, p. 242) opinion, "It is better to talk about *subordinate* and *superordinate* gestalts."

In the face of the motto that "the whole is more than the sum of the parts" or that it is "less than the sum of the parts," according to each case, he suggests the alternative phrase "A 'part' of a gestalt is more than a part" since any entity is itself relational, it is *relata* (Næss 2005b, p. 119; Bugallo 2011, pp. 17–19).

One of the first steps to be taken is the correction –an *emendatio*– of that predominant distorting tendency: the belief that humanity is something placed *in* the environment, instead of seeing itself *with* the environment. Thus, as from the first postulate of the *deep ecology movement,* Næss (1995a, pp. 3-4) establishes the necessary "rejection of the human-in-environment image in favor of the relational, total-field- image. [...] The total-field model dissolves not only the human-in-environment concept, but every compact thing-in-milieu concept –except when talking at a superficial or preliminary level of communication".

Næss (1992, p. 5) stresses that "Ethics furnishes no basis for assuming that the immanent God expresses its nature, essence or power, in any other way then through each existent being." In fact, the Dutch philosopher holds that the better we know singular things, the more we know God (Spinoza 1941, EVP 24).

Næss holds a tendency toward the identification of God with Nature, of *Natura naturans* with *Natura naturata* though at the level of denotation, extension, or reference, albeit not at a conceptual or connotative level. This would imply revaluing nature with perfection, value, and sacredness. For Næss (2005b, p. 384) "particularity and divinity may perhaps be said to be equally basic aspects of 'The Whole'."

It is not some kind of mystical union *(unio mystica)* whose result would mean turning his back on singular things and nature. This implies, in Spinoza as well as in Næss, a rejection of ascetic life: the more active the body, the more deeply subtle the mind.

At the same time, Næss acknowledges (as do other scholars) that Spinoza never managed to suggest a system where God is a real, consistent, immanent presence to everyday objects. The existence of God follows from his essence, but it is only an *existence as essence.* So, Næss (1992, p. 4) persuades us: "I have a suspicion that he never completely gave up his Jewish faith, the trascendent God he loved in his youth." Thus, Næss attributes a panentheistic stance to Spinoza. *Panentheism* consists in affirming that God is both immanent and transcendent to the Universe; it encompasses the Universe, but is not limited or reduced to it, as opposed to *pantheism*, which affirms the identity between God and the Universe.

11.3 The Narrative Artifice of *Ecosophy T*

The *deep ecology movement* must not be confused with Næss' personal philosophy, his *ecosophy T*. For our thinker, in times of environmental crises we ought to be activists. If one has a philosophy, a religion, a total view of oneself, he must act from that perspective. As well as we can recognize the "ontology-ethics-politics" connection that permeates Spinoza's work, in Næss we may underline an integration of "ontology-lifestyles-social activism" quite analogous to the Spinozian approach. (Bugallo 2015).

The narrative of *ecosophy T* emphasizes basic norms and hypotheses concerning humanity's relationship with a more encompassing reality, with the basic intention of promoting self-realization for all living beings.

Næss deems it impossible –and even senseless– to try to formulate a complete philosophy of life, an ecosophy, in view of its complexity or flexibility. From a strictly logical point of view, it is equally absurd to try to fully posit a cosmovision or total vision.

With no dependence from the deductive or reasoning model typical of the seventeenth century, Næss assumes that Spinoza's system, with its details and complex connections, may prove scarcely comprehensible for many in these times. And it has not much chance of being simply acknowledged by most people, "We must not succumb to any irrational reverance for the formal priority. [...] I believe there is need for deeply different articulations, including the poetic" (Næss 1992, p. 2).

Notwithstanding that, Næss deems it legitimate to suggest an argumentative model in which the place of hypotheses and premisses is occupied by an interplay of norms and hypotheses, expressed in many ways, more or less complex. Aristotle warned about persuasive arguments: "Its task [the task of rhetoric] is not to persuade, but to recognise the most relevant means of persuasion in each case" (I1, 1355b).

From the basic norm of "Self-realization!" jointly with hypotheses about the world, Næss derives a series of principles for green politics. The latter, on their part, connected to diagnostics or hypotheses about life in communities, will lead to upholding other derived norms. Lastly, adding reflections on class and social justice, the practical personal system of the Næssian vision is gradually completed (Næss 1989; Bugallo 2015).

One of the varied formulations of his *ecosophy T* would be, schematically, the following (Næss 1989, pp. 197–207):

N1: Self-realization!

H1: The higher the Self-realization attained by anyone, the broader and deeper the identification with others.

H2: The higher the level of Self-realization attained by anyone, the more its further increase depends upon the Self-realization of others.
H3: Complete Self-realization of anyone depends on that of all.
N2: Self-realization for all living beings!

H4: Diversity of life increases Self-realization potentials.
N3: Diversity of life!
H5: Complexity of life increases Self-realization potentials.
N4: Complexity!
H6: Life resources of the Earth are limited.
H7: Symbiosis maximizes Self-realization potentials under conditions of limited resources.
N5: Symbiosis!

H8: Local self-sufficiency and cooperation favors increase of Self-realization.
H9: Local autonomy increases the chances of maintaining local self-sufficiency.
H10: Centralization decreases local self-sufficiency and autonomy.
N6: Local self-sufficiency and cooperation!
N7: Local autonomy!
N8: No centralization!

H11: Self-realization requires realization of all potentials.
H12: Exploitation reduces or eliminates all potentials.
N9: No exploitation!
H13: Subjection reduces potentials.
N10: No subjection!
N11: All have equal rights to Self-realization!
H14: Class societies deny equal rights to Self-realization!
N12: No class societies!
H15: Self-determination favors Self-realization.
N13: Self-determination!

In this way, abstract problems of philosophy are connected to concrete issues of the contemporary political conflict.

In the case of the human being, Self-realization implies going further from ego or the *self in small* letters toward the realization of a *Self*, in capitals, ecological or systemic. According to Spinozian inspiration, humans may discover how their particular potency is merely a singular expression of the infinite potency of God. We are not mere creatures in nature, but authentic creative poles.

Excess of dominant anthropocentrism diminishes the pleasure and satisfaction of being in contact with other ways of life. Næss and his followers' great challenge is to persuade us that bio-spherical degradation not only harms human and non-human interests, but it also implies a diminishment of a real joyful life for everyone.

From the perspective of rhetoric argumentation, and about the use of exclamation marks, Næss admits that there is someone who releases the idea, but no special receiver is identified. Nor is the use of exclamation marks linked exclusively to the imperative mode; they may indicate a deep experience (surprise, admiration, for example).

In both the *deep ecology movement* and in *ecosophy T* some stylistic *tropes* of rhetoric are eventually acknowledged. As an example of synecdoche, in the expression "deep ecology," "ecology" is used as "environmentalism" or "ecosystem ontology." It can also be interpreted as a metonymy, since "ecology" may include the ecosystem approximation and some de-centring of strong anthropocentrism. On the other hand, the expression "deep ecology" is a metaphor of deep relational, critical thinking (and its opposite would be "superficial ecology"). Such narrative mediations are not mere formal resources: they appear as a social way of communication, with the purpose of establishing agreements in daily life.

11.4 Deep Compatible Compositions

Since Spinoza, the analysis of bodies allows us to start a practical project: the acknowledgement of similar compositions or relationships between bodies provides criteria to favor compatible encounters (joyful passions) and avoid incompatible encounters (sad passions). This fosters ethical, socio-political projects and, ultimately, a practical philosophy of joy since it favors active affections and discourages passive ones.

But as Spinoza said in his *Tractatus de intellectus emendatione*, reflection by itself is not enough for widening joy. Its fulfilment and preservation require appropriate civil institutions, in an adequate, democratic society. Along the line of these considerations, the latest postulates of the 1984 version of the *deep ecology movement*, give some signs about those tendencies to permit concrete actions:

> 4. The flourishing of human life and cultures is compatible with a substantial decrease of human population. The flourishing of nonhuman life requires such a decrease. 5. Present human interference with the nonhuman world is excessive, and the situation is rapidly worsening. 6. Policies must therefore be changed. These policies affect basic economic, technological, and ideological structures. The resulting state of affairs will be deeply different from the present. 7. The ideological change is mainly that of appreciating life quality (dwelling in situations of inherent value) rather than adhering to an increasingly higher standard of living. There will be a profound awareness of the difference between big and great. 8.Those who subscribe to the foregoing points have an obligation directly or indirectly to try to implement the necessary changes (Næss and Sessions 1985, p. 70).

Adherence to the postulates somehow implies sharing a diagnostic of the situation. Sharing tenets also testifies to an agreement about shared values, as the intrinsic value of every form of life.

Lacking emotions there are no changes. Empowered by a desire promoted by joy, we may change our actions and feel different. But is it just a question of promoting the emotions of the family of joy? As opposed to Spinoza, Næss attributes instrumental value and enormous usefulness to negative emotions, such as anger, rage, "fury," horror, etc., insofar as they may be used in a constructive style. And he quotes the example of conservation biologist Michael Soulé who, moved by profound frustration and despair in the face of the growing deterioration of the biosphere of the Earth, became one of the most conspicuous scientific environmentalists (Næss 2003, p. 78).

Næss retrieves the idea that it should be a collective task, associating or creating consensus that ease or multiply the intentions of the promoters. A grassroots movement from a wide cultural spectrum would contribute, according to Næss, to their effort or strength in following the highest vital aims for an ever-larger number of beings. It does not aim to become a closed paradigm, shared exclusively by a few specialists but a grassroots movement with wide aims, not exclusive for participants of the first world or the most developed countries.

The aim of *deep ecology* is not to establish yet another environmental ethics. Its interest is connected to a psychological approach rather than a moral one. It suggests that interests or desires should be modified and especially the current inadequate perception of reality, rather than submitting inclinations or interests to an ethics imposed from outside. Voluntary simplifications of lifestyles promoted by some radical environmentalists are probably within our capacities but are they also within our desires? And desires count.

Nothing will necessarily force us to live in a humbler way. One of the distinguishing characteristics of deep ecology and, particularly, ecosophy is that it offers a joyful way and authentic fullness as an alternative to mere consumerism and materialism. It is not exclusively a road of deprivation and losses. For many followers of the movement, natural wealth contributes to personal well-being insofar as each one will identify and display his or her own self in that diversity that enriches each one's being. In the face of wonder, amazement, and the aesthetic and cultural value of diversity, one can strengthen love or sensitivity toward all existents. Although the global approach is essential, regional differences must determine environmental policies of the coming years. The seven principles are presented as vague generalizations that should become more precise in some direction when it comes to a practical analysis. Briefly, along the world and amid cultural diversity, the inspiration of ecology seems to be reaching notable agreements, in Næss' view. The *deep ecology movement* only aspired to be an expression of one of those agreements.

11.5 Concluding Remarks

The principles of the *deep ecology movement* in any of its versions, are not a new catechism or an unquestionable dogmatism. The point is, rather, to put living options into practice at one's risk, that is, under one's own responsibility at one's risk, in the face of an eventual contrast with its future consequences (some sort of progressive rhetoric, in words of Valera 2016). They are provisional, subject to modifications or plain disappearance, should the particular conditions of the context require it.

Consequently, the ultimate end of the proposal of this wide-based social movement is to contribute to saving the biosphere from future devastation that not only goes against the interest of humans and non-humans, but they also bring an impoverishment of the real possibilities of a joyful life for everyone. In view of its design, it might be upheld by social activists of all hues in the political spectrum. The contrast shallow-deep is not analogous to left-right and crosses through many conventional distinctions.

References

Aristotle (1959) Ars Rhetorica. Oxford University Press, Oxford

Bugallo AI (2011) Ontología relacional y ecosofía en Arne Næss. Revista de Filosofía Nuevo Pensamiento 1:1–33

Bugallo AI (2015) Filosofía ambiental y ecosofías. Arne Næss, Spinoza y James. Prometeo, Buenos Aires

Bugallo AI (2018) Prólogo. In: Næss A. Ecología, comunidad y estilo de vida. Prometeo, Buenos Aires, pp 13–35

Buganza J (2011) El carácter retórico de la ética a partir de Aristóteles. Devenires XII(24):26–145

Næss A (1973) The shallow and the deep: a long-range ecology movements. Summ Inq 16:95–100

Næss A (1975) Freedom, emotion and self-subsistence. The structure of a central part of Spinoza's ethics. Universitetsforlaget, Oslo

Næss A (1978) Spinoza and ecology. In: Hessing S (ed) Speculum Spinozanum 1677–1977. Routledge, London, pp 418–425

Næss A (1989) Ecology, community and lifestyle. (trans. Rothenberg D). Cambridge University Press, Cambridge

Næss A (1992) Spinoza and the deep ecology movement. Eburon, Delft

Næss A (1995a) The shallow and the deep: a long-range ecology movement. A summary. In: Drengson A, Inoue Y (eds) The deep ecology movement: an introductory anthology. North Atlantic Books, Berkeley, pp 3–9

Næss A (1995b) Ecosophy and gestalt ontology. In: Sessions G (ed) Deep ecology for the twenty-first century. Shambhala, Boston, pp 240–245

Næss A (2003) Life's philosophy. Reason and feeling in a deeper world. (trans. Huntford R). Georgia University Press, Athens

Næss A (2005a) How my philosophy seemed to develop. In: Glasser H, Drengson A (eds) The selected works of Arne Næss, vol IX. Springer, Dordrecht, pp 301–316

Næss A (2005b) Reflections on gestalt ontology, part 1. Trumpeter 21(1):119–123

Næss A, Sessions G (1985) Platform principles of the deep ecology. In: Devall B, Sessions G (eds) Deep ecology: living as if nature mattered. Gibbs Smith, Salt Lake City, pp 69–73

Santibáñez C (2012) Teoría de la Argumentación como Epistemología Aplicada. Cinta de Moebio 43:24–39

Spinoza B (1941) Ethics: Ethica Ordine Geometrico Demonstrata. Dent, London

Valera L (2016) Retórica y ética ambiental: la contribución de Arne Næss. Medicina y Ética 4:587–607

Part II
Current Ecological Concerns and Cosmologies: Exploring Pantheism

Chapter 12
Raimon Panikkar's Sacred Secularity: An *Advaita* Interpretation to Understand the Sacredness of Nature

Jessica Sepúlveda

Abstract The emergence of an ecological consciousness has brought forward the perception of nature's sacredness. This appreciation is not indifferent to an intellectual sector, which has problematized it by its reference to the latest dichotomies of reality, such as mater and spirit, or time and eternity. In this chapter I address the theme of nature's sacredness starting from the neologism of *sacred secularity* coined by Raimon Panikkar. In his reflection, Panikkar enriches the meditation about the relationship of both orders of reality included in his neologism. He explains this relationship through the Hindu concept of *advaita*, which leads him to avoid monist and dualist interpretations. Panikkar recognizes the ultimate character of the temporal structures of reality, but without making them absolute. This leads us to propose new hierophanies where the very existence of human beings and Earth needs to be experienced as sacred realities.

Keywords Sacred secularity · Raimon Panikkar · Hindu *advaita* · Nature's sacredness

12.1 Introduction

The emergence of an ecological consciousness has brought forward the perception of nature's sacredness. This appreciation is not indifferent to an intellectual sector, which has problematized it by its reference to the dichotomies of reality, such as mater and spirit or time and eternity. Beyond the philosophical discussion, at an international scale a series of relevant milestones have been achieved since the end

J. Sepúlveda (✉)
Deparment of Applied Ethics, Faculty of Religious Sciences and Philosophy, Catholic University of Temuco, Temuco, Chile

Cape Horn International Center – CHIC, Puerto Williams, Chile
e-mail: jsepulveda@uct.cl

L. Valera (ed.), *Pantheism and Ecology*, Ecology and Ethics 6,
https://doi.org/10.1007/978-3-031-40040-7_12

of the twentieth century. International organizations such as the United Nations Environment Programme (UNEP), the International Union for Conservation of Nature (IUCN), or the United Nations Education, Culture and Science (UNESCO) have made progress in recognizing and incorporating spiritual values for the conservation of protected areas (Mallarach et al. 2012). In this vein, philosopher Henryk Skolimowski (2017, pp. 306–313) proposes that spirituality is one of the characteristics of the nascent ecological consciousness, which in contrast to technological consciousness, is related to a reverential attitude towards the universe.

To understand the emergence of nature's sacredness, we should remember that according toMircea Eliade (2017, p. 87), the phenomenon of nature desacralization is a recent characteristic of modern human beings and science. In this regard, Raimon Panikkar (1999b, p. 43) clarifies that the consciousness of primordial (not primitive) human beings -which still persists in certain cultures- synthesizes the attitude of the *"Man" of nature*, for whom the latter was coextensive with human beings. In this type of consciousness, nature was not merely "natural," it was also sacred, equating in this way ontophany with hierophany (Eliade 2017, p. 87). The shift of this consciousness occurred with the rise of *scientific* humanism, in which the the attitude of *"Humanity" over nature* was developed (Panikkar 1999b, p. 52). This attitude led to the desacralization of nature that was now understood as inert matter. It is paradoxical that under the eaves of scientific humanism -still current- the contemporary ecological consciousness was born. This is not a minor fact, since it requires us to explain the resurgence of the sacredness of nature through new coordinates, which are different from those found in premodern consciousness.

Raimon Panikkar's choice to carry out this reflection derives from his meditation on his neologism *sacred secularity*; specifically, the *relationship* of both orders of reality that this neologism entails but avoiding monist and dualist interpretations. To understand his neologism, we need to consider that Panikkar's incursion into this topic is linked to the profound crisis that marks our current era as manifested in three radical human experiences (Panikkar 1999b).

The first radical experience is awareness of Earth's limits. Its exploitation in benefit of humans (and only a few) is limited. This experience is at the base of the emerging ecological consciousness. The second radical experience refers to the awareness of human limits in all its radicalism; that is, despite the great technological development and the sophisticated theoretical elucubrations to solve problems, the human being "feels more than ever in the clutches of a destiny that he cannot control it in any way" (Panikkar 1999b, p. 58). The third experience refers to a crisis of the traditional idea of God and of nature. In the case of the first experience, there is awareness that "the God of history remains idle, the God of philosophers is indifferent, and the God of religion no longer seems very concerned with the human condition" (Panikkar 1999b, p. 60). Regarding nature, traditional conceptions about it have been questioned by contemporary science, thus giving place to interpretations of nature revival and abandonment of mechanicists parameters to explain life. Nowadays we know that nature possesses degrees of freedom and self-organization. Therefore, the human intellect is no longer capable of fully mastering it as previously thought.

This context of profound crisis affects the very roots of reality. Even though ecological awareness implies an attitude change towards nature, the change must be directed towards a radical *metanoia* which "is less a new policy of men *towards* nature than a conversion that recognizes their common destiny" (Panikkar 1999b, p. 67); that is, the constitutive implication between the human being and nature must be recognized. On the other hand, if the nascent ecological consciousness is accompanied by spirituality, it is necessary to add the divine to this constitutive relationship . This *metanoia*, which implies the awareness of a common destiny, led Pannikar to revise the traditional notions of God, nature, and the human being. He gives us a new understanding of them through his neologism, *sacred secularity*, where the four sources of knowledge that nourish his thought converge: Christianity, Hinduism, Buddhism, and science.

12.2 Sacred Secularity

In the inquiry we are carrying out on the notion of sacred nature, the contribution made by Panikkar with his sacred secularity neologism is valuable. This compound term establishes an intrinsic relationship between both orders of reality that have traditionally been conceived separately (dualism) or together (monism).[1] Prior to this explanation, it should be clarified that our author, did not make major distinctions between nature, cosmos, and secularity. On the contrary, his emphasis is on highlighting the temporal structure of the three; with it, he justifies its character of ultimate reality. To explain the sacred secularity neologism, we will first clarify each separate concept, and then we will explain its intrinsic relationship where Panikkar will argue from the Hindu *advaita* thought.

For Panikkar (1999a) the Latin origin of the word secularity, *saeculum*, takes us to the triple composition of space/time/matter. These three factors are independent and make up corporeity, finitude of things, that is, *dis*-tension and *ex*-tension of reality. Due to their relationality they cannot be separated, but it is undoubtedly the peculiar experience of time that Panikkar wants to highlight in the notion of secularity, thus recovering the vital and temporary nature of reality. With this idea, R. Panikkar wants to give the character of ultimacy to this dimension, although without making it absolute. It is important to note that Panikkar distinguishes the meaning of secularity from that of secularization and secularism. Secularization is a historical process of loss of power of religious institutions, and secularism is an ideology that denies the reality of a supra-natural world, absolutizing the empirical world (1999a, pp. 26–27).

The notion of sacredness has its etymological origin in the Indo-European root *sak* whose meaning is related to the fact of conferring reality. This way, Panikkar

[1]When using the concepts of monism and dualism, R. Panikkar follows the clarification of Christian Wolff, whom introduced them in the XVIII century to refer to metaphysical doctrines regarding the existence of one or two types of the universe's fundamentally irreducible substances.

(1999a, p. 42) indicates that the first characteristic of the sacred is being the most real thing. In the same vein, Eliade (2017, pp. 14–16) highlights that the sacred is the reality *par excellence*, and it manifests itself in human consciousness through hierophanies. The manifestation of the sacred, its appearance in the consciousness of *homo religiosus*, implies that its center of gravity does not lie in the human being; therefore, it resists all human manipulation (Panikkar 1999a, p. 42). On the other hand, recognizing that the sacred manifests itself in human consciousness, implies that it constitutes a modality of being in the world that differs from profane existence. An important idea in this conception of the sacred is that every hierophany presents a paradoxical relationship. Although the sacred refers to a realm of reality that transcends the human being, its revelation occurs in the concrete of existence, where it is manifested and limited (Allen 1985, p. 110). That is, its scope of concretion is immanence.

The question that rises from this paradoxical relationship is if the immanent, the concrete, can be understood either as ultimate reality (Panikkar's argument in his sacred reality neologism), or as subject to the foundation of transcendent reality. The latter can be inferred from Eliade (2017, p. 22) for whom "the manifestation of the sacred is ontologically grounded in the world." What is important here is to highlight the argumentative emphasis made by Panikkar (1999a, p. 34) when he recognizes the ultimate character of the temporal structures of existence, by affirming that "they belong to the warp and woof of the very fabric of reality." Hence, secular values are not mere means to a higher end but rather they are ends in themselves; therefore, they are real. When recognizing secularity as ultimate reality, Panikkar (1999a, p. 43) follows those religious traditions and sages that have "experienced the ultimate reality of mundane things without reducing them to what is empirically given." He recognizes sapiential formulas that account for it, such as *Saṃsāra*/ *nirvāna*, *ātman*/*brahman*, Incarnation, nature of Buddha, among others. Zen Buddhism, as explained by Byung-Chul Han can be qualified within these traditions, as it does not know the idea of escaping from the world and the "Buddhist expression 'nothing sacred' denies any extraordinary place, extraterrestrial" (Han 2015, p. 43).

The separate review of each term of sacred secularity, allows Panikkar to conclude that if secularity is part of the ultimate warp of reality, it means that it is real and since the sacred is real, he infers from this that secularity is sacred. In his own words "The *saeculum* itself, and not only what it can lead to or point to, is 'real', that is, sacred" (Panikkar 1999a, p. 43).

Beyond the former formulation, Panikkar delves into the intrinsic relationship between secularity and the sacred, avoiding monist and dualist interpretations of reality. For this, he will resort to the trinitarian thought that he finds in the Trinity of Christianity, in the intuition of the *pratīyasamutpāda* of Buddhism, and the Hindu *advaita*. In this chapter, we will focus on the Hindu concept of *advaita*. But before explaining it, it is convenient to problematize the dangers that the Panikkar (1999a) observes in the monistic and dualistic visions of reality which do not allow us to understand the paradoxical relationship implicit in the neologism of sacred secularity.

In the case of dualistic interpretations in the religious sphere, it can be observed how certain traditions view the experience of time as something provisional, as a means to achieve eternal life (Abrahamic religions) and others, as an illusion, whether positive (Buddhism) or negative (some Hindu spiritualities). On the political-religious level, this dualism is dangerous because it understands that political power is separated from religion and that therefore "Religion only serves the 'salvation of the soul' and is only valid in the 'sublime' sphere of the divine." In this way it leaves a free rein for the exploitation of the land, the poor, the weak and other cultures" (1999a, p. 47).

On the other hand, the author is also aware of the dangers of monism, since it leads to a single order of reality dominating human life, be it, in this case, the secular or the sacred world. The reduction of everything real to the merely secular stifles reality, depriving it of infinity and freedom, and, on the contrary, if everything is sacred, the real character of the secular world is denied and human life and nature are degraded, and at the same time, the divine loses its specificity.

As we can see, for Panikkar, both monism and dualism do not account for a constitutive relationship of both orders of reality. For this reason, to explain sacred secularity, he resorts to Hindu *advaita*, whose relational awareness distinguishes both orders, but it does not separate them.

12.3 Hindu *Advaita*: An A-Dual Interpretation of Reality

Advaita Vedānta is one of the most widespread Hindu schools of philosophy today in many spiritual circles and is fundamentally based on the interpretation that Śankara (788–820) made of the *Upanisad* and the Brahma-sūtra, whose doctrine interprets "the 'supreme experience' of non-duality, meaning the essence of non-separability between the Self (*ā-tman*) and 'God' (*Brahman*)" (Panikkar 2007, p. 299). Panikkar distinguishes between the *advaita* as fundamental principle of a-dualism (etymological sense) and the Advaita Vedānta. In the case of his thought, he will use *advaita* in its original etymological sense applicable to the a-dual experience of reality,[2] present in other religious traditions such as Buddhism, Taoism, and Christian mysticism. Our author also clarifies that *advaita* has been habitually mistranslated as "non-duality" due to the dialectical mentality of European Indologists, who interpreted *a-* as a negative particle. *Advaita* intuition *a-* does not represent a dialectic negation, it is more a primitive prefix that aims to an "absence of duality." In words of Panikkar (2005, p. 205): "Advaita intuition does not consist in affirming Unity, nor in denying duality, but precisely, with a vision that transcends the intellect, in recognizing the absence of duality in the background of a reality that lacks duality itself; namely, that it is not numerical since it does not have a two. In other words, neither unity nor duality correspond to the proper structure of reality."

[2] Panikkar was introduced to *advaita* by the French Benedictine monk Benri Le Saux (1910-1973), who was a disciple of Ramana Maharshi (1879-1950), the most prominent representative in our century of Advaita Vedānta.

The absence of duality implies the paradox of thinking about unity from a one that does not mutilate the multiplicity and that, in turn, does not make reality something monolithic. *Advaita* wisdom has synthesized this paradox in the definition of reality, the Absolute, Brahman, as "one without a second" expounded in its Vedic texts. One of these, the *Kārikā* of *Gaudapāda* notes that the sound *aum* that refers to the supreme and non-supreme Brahman "does not have a cause and is without a second. Nothing exists outside of it or as its effect. It is beyond any change" (I, 26). From this quote we can interpret a monistic approach to reality; however, Consuelo Martín (1998), a Śankara scholar clarifies that *advaita* metaphysics is not monist nor dualist given that "it goes beyond the distinction of unit-multiplicity, as it goes beyond every duality. *Brahman* is not one, by exclusion of the multiple, nor a reduction of every being into a single Being" (1998, pp. 46–47).

Without going into Śankara's metaphysical interpretation of "one without a second," we will focus on that of Panikkar, to whom "one without a second" is a state of consciousness that enables the primary vision of relationality as such, emphasizing in this conception the "constitutive relationship of everything with everything" (1996, p. 234). The same author, in the previous quote, notes that we are not talking about a numerical reality and that unity and duality do not correspond to the structure of reality itself. To better understand this aporia, Panikkar (2005) performs an equivalency exercise between the *advaita* and the harmony notion developed by the Greeks: "Harmony would be the homeomorphic equivalent of *advaita*, as we will see further. It is the experience of *nirdvaṃda* or overcoming of all *dvaṃda*, of all opposite pairs, of all duality, without falling into monism on the other hand. The opposites, the *dvaṃda* appear as contradictory only to reason, cold and heat, pain and joy" (2005, p. 188).

The vision of harmony accounts for the bipolar structure of reality, and as indicated by Heraclitus, this structure manifests itself in a rhythmic alternation of a game of opposites. However, under the guise of opposites, there is an invisible unity that transcends duality, and that instead of seeing opposites as exclusive, sees them as interdependent. In this way, the vision of harmony allows the simultaneous perception of the poles – overcoming the opposites – which brings with it the relational consciousness. That is to say, in this consciousness both the principle of identity and the principle of difference operate, since beings are not a reality "in themselves" separated ontologically, nor are they absorbed by a unit that denies their being. It is the identity-in-difference, as there is an infinite difference between each being, and in turn, paradoxically, an identification between them since they are a relationship. In this way, the *advaita* relational consciousness unites and distinguishes at the same time, and, therefore, it is what allows us to see the divine and the cosmic as poles and not as things "in themselves." From this vision, the *advaita* interpretation of sacred secularity that is explained below can be understood.

12.4 *Advaita* Interpretation of Sacred Secularity

In the *advaita* interpretation of sacred secularity, we can distinguish three ideas that are relevant in the understanding of the notion of sacred secularity.

First, *advaita* sacred secularity means transcending a model of the universe shared on two levels: the secular and the sacred, where "the 'supernatural' is not a superstructure of the human; the divine is not foreign to the human" (Panikkar 1999a, p. 51). From this it can be inferred that political injustice, the exploitation of human beings and of Earth, are no longer a purely profane issue that requires only technical solutions, but on the contrary "sacred secularity makes human problems ultimate" (Panikkar 1999a, p. 48), and the same rules for the Earth and its sentient beings. In this way, there are no two separate kingdoms, since "the temporal is also religious and the sacred is also secular" (Panikkar 1999a, p. 47); but, without reducing it to a monolithic unit.

Second, *advaita* highlights that sacred secularity carries with it the interpretation that the divine[3] is not separate from the world. However, it is not identified with it either: "The non-dualist vision *(advaita),* in which divinity is not individually separated from the rest of reality, nor is it totally identical to it (...); transcendent but immanent in the world; in-finite but bounded in things" (Panikkar 2001, p. 86).In this way, the divine dimension is not a superstructure superimposed on beings, or an absolute Other, but is the constitutive principle of all things. Hence, the reference of the divine to the infinite inexhaustibility of every being that gives it its always open and never finished character. From the *advaita* comprehension of the divine with the world, it can be understood that God fulfills cosmological functions, since the world does not expand mechanically. Rather, on the contrary, the world generates forms of self-organization, and freedom unfold in it. This same interpretation is what allows Panikkar to affirm that, just as there is neither a God nor a disembodied Human, also there is no world without the conscious and divine dimensions. To illustrate this vision of seeing God in the sameness of things (in its most intimate core), the following quote by aimon Panikkar can be understood without inferring a pantheistic interpretation from it:

> I see, and I'm not ashamed to say it, God in Nature, in animals, in people and obviously in beauty (...). That is why I am so amorous, because I see the divine dimension in all things. We are bordering on transcendence, but if God is not also immanent, the rest is an idol. Immanence and transcendence go together (Panikkar in Abumalham 2001, p. 12).

Third, sacred secularity brings with it a reconsideration of the sameness of each being, since the sacred is not an ontological reality separated from things, it is rather an aspect of all things by the very fact of being real. With this, it is affirmed that sacred secularity does not eliminate transcendence but rather discovers its immanent *locus*; that is, things are "more" than they appear to be, they have a dimension

[3] Panikkar uses the words God, the divine and divinity. Without entering these distinctions, the important thing to highlight is the author's sense of the divine that is explained in this chapter of the book.

of transcendence that is immanent to the thing itself (Panikkar 1999a, pp. 54–55). In this way, sacred secularity does not put a sacred "other" before the concrete individuality of a thing, nor does it suffocate it by isolating it in an "in itself." What a thing really is, is not what differentiates it from others (ontological separation of each being), but rather what identifies the thing with what it really is, as "it is itself." On the contrary, for Panikkar, in the "in itself" operates the principle of non-contradiction that has mainly governed Western civilization and that has led us to know by differentiation, namely, the more A is A, the more I can distinguish it and separate it from not-A. This has meant the mutilation of A's relationality with other entities. At the opposite end of this conception, sameness understood as "is itself" operates the principle of identity that has predominated in Indian culture where "A is more A the more it identifies with itself. We know here by identification, participation, union" (Panikkar 1999b, p. 89). Here he does not refer to an individualistic and closed principle of identity, but to an identity that takes shape -and becomes real- to the extent that it is in relation to all of reality and is suspended on "an unfathomable foundation that makes growth, life and liberty possible" (Panikkar 1999b, p. 82).

This sameness "in itself" settles us in a relational ontology, where it is inferred that each being is real, and therefore sacred, to the extent that it is understood in this interrelation. In the words of Panikkar (1999a, 50): "Everything is interconnected and interrelated. And this very connection implies the sacred dimension." Along the same lines, Vandana Shiva, one of the main exponents of ecofeminism, advocates the sacralization of Earth. For her, the sacred denotes "a relationship between the part and the whole," for which the conservation of biodiversity understood as a relational category that recognizes and protects the integrity of ecosystems, it is sacred and inviolable (Mies and Shiva 2014, p. 282). From this conception of the sacred, Shiva distinguishes between sacred seeds and profane seeds. The latter violate ecological cycles and fragment agricultural ecosystems, unlike sacred seeds where they are considered a microcosm of the macrocosm:

> The sacred seeds are considered a microcosm of the macrocosm and the *navdanya* symbolizes the *Navagraha*. The influences of the planets and the climate are considered essential for the productivity of plants. In contrast, high-yielding varieties break all ties to seasonal and cosmic weather cycles (Mies and Shiva 2014, p. 282).

12.5 Conclusions

As we have made clear, the compound term *sacred secularity* offers us new understandings about the divine, the world, and the human being. Panikkar's interpretation of the *advaita* enriches the understanding of them seen from a relational consciousness, which unites and distinguishes at the same time, without falling into monism or dualism. On the other hand, recognizing the character of ultimacy of the temporary structures of reality, without making them absolute, leads us to propose new hierophanies where the very existence of the human being and of Earth need to be lived as sacred realities; even more so in the present context, when the possibilities of planetary ecocide are real. It is in this scenario that the words of Raimon Panikkar (1999a, p. 65) resonate strongly when he pointed out that "the sacred-secular conception, which does not believe in 'second' chances or in 'another' world, can activate forces of salvation that are dormant in the human race."

Acknowledgements This research was supported by the project Cape Horn International Center (ANID/BASAL FB210018).

References

Abumalham M (2001) Conversación con Raimon Panikkar. Revista de ciencias de las religiones. Anejo VI:7–26
Allen D (1985) Mircea Eliade y el fenómeno religioso. Cristiandad, Madrid
Eliade M (2017) Lo sagrado y lo profano. Paidós, Barcelona
Han B-C (2015) Filosofía del budismo zen. Herder, Barcelona
Mallarach JM, Coma E, de Armas A (2012) El patrimonio inmaterial: valores culturales y espirituales Manual para su incorporación en las áreas protegidas. Fundación Fernando González Bernáldez, Madrid
Martín C (1998) Consciencia y Realidad. Estudio sobre la metafísica advaita con la Māndūkya Upanisad, las Kārikā de Gaudapāda y comentarios de Śankara. Trotta, Madrid
Mies M, Shiva V (2014) Ecofeminismo. Teoría, crítica y perspectivas. Icaria Antrazyt, Barcelona
Panikkar R (1996) El silencio del Buddha. Una introducción al ateísmo religioso. Siruela, Madrid
Panikkar R (1999a) El mundanal Silencio. Una interpretación del tiempo presente. Martínez Roca, Barcelona
Panikkar R (1999b) La intuición cosmoteándrica. Las tres dimensiones de la realidad. Trotta, Madrid
Panikkar R (2001) Iconos del misterio. La experiencia de Dios. Península, Barcelona
Panikkar R (2005) Espiritualidad hindú. Sanātana dharma. Kairos, Barcelona
Panikkar R (2007) Mito, Fe y Hermeneútica. Herder, Barcelona
Skolimowski H (2017) Filosofía viva. La ecofilosofía como un árbol de la vida. Atalanta, Girona

Chapter 13
Spinozism and Native Americans on Pantheism and Panentheism

Joel Álvarez

Abstract Baruch Spinoza famously said, "Whatsoever is, is in God, and without God, nothing can be, or be conceived". This form of Pantheism is quite like eastern Pantheism, where in Hinduism they assert that "everything is Brahma", or in Taoism, where Lao Tzu says, "Heaven and I were created together, and all things and I are one". Although the western and eastern world shared their respective ideas of Pantheism, Native Americans also contributed to such discussion. However, comparative philosophy between western and Native American philosophy has been far neglected, and the same neglect has been shown regarding Native American contribution to Pantheism. Many Indigenous religions believe that there is a Great Spirit that is manifested as a life force or energy that communicates and speaks through nature. In other words, some Native American tribes believe that nature, or in extension the universe, is part of a Great Spirit that communicates to individuals in a naturalistic way. Such a Great Spirit created the world and is in everything that the Great Spirit created. Moreover, there are other Indigenous beliefs that not only hold a Pantheistic view but hold a Panentheistic perspective where God is a being that is present in creation but transcends from it. Therefore, although Indigenous beliefs about Pantheism or Panentheism have been much neglected, there is much to say about it. With such input from Native American religious culture, this paper seeks to show and compare Spinoza's Pantheism and that of Native Americans. In addition, I wish to investigate in what ways Native American philosophy contributes to the discussion of cosmology, nature, and Pantheism. In particular, how do some Native American tribes understand God in a Pantheistic sense where nature and cosmology are one.

Keywords Pantheism · Panentheism · Great Spirit · Nature · Cosmology

J. Álvarez (✉)
Department of Philosophy, University of South Florida, Tampa, FL, USA
e-mail: joelalvarez@usf.edu

L. Valera (ed.), *Pantheism and Ecology*, Ecology and Ethics 6,
https://doi.org/10.1007/978-3-031-40040-7_13

13.1 Introduction

There have been various philosophical contributions between the Western and the Eastern world regarding Pantheism. However, there has been no mention or comparison of some sort between Native Americans and the forms of Pantheism in Western and Eastern philosophy. That is quite surprising since Native American philosophy has much to say about it. In particular, Native Americans believe that a Great Spirit manifests as a life force or energy in nature. Moreover, many Native Americans also hold a Panentheistic perspective where God is a being that transcends creation and communicates with it. With such input, this paper seeks to show and compare Spinoza's Pantheism and that of Native Americans. I will show how Indigenous Pantheism, such as the Great Spirit, can have a form of Spinozism regarding infinite and finite modes. However, I argue that the Native American departs from Spinoza's Pantheism and instead arrives at a form of Panentheism because the Great Spirit is involved with nature and individuals.

13.2 Spinoza's Pantheism

Before we begin discussing Native American Pantheism and Panentheism, let us first explain Spinoza's form of Pantheism. As we see in Spinoza's Ethics: Concerning God, he goes against the traditional Abrahamic (Judaism, Christianity, and Islam) understanding of God's nature and states that "nothing can be or be conceived without God, but that all things are in God" (E1P17 dem). In particular, this stemming away from traditional Abrahamic religion is that in God, there are modes. And these modes are inside of God. It is Spinoza's assertion of modes and their residing in God that classifies Spinoza as a Pantheist. As such, I pursue here to explain what Spinoza means when he says, "things are in God," and how exactly Spinoza's modes correlate with God.

13.2.1 God and Substance

Spinoza, like theists, believes that everything comes from or is caused by God (E1P8chS2); however, theists assert that everything comes from God in the sense that God created all that exists that is external to Him. Moreover, these existing things for the theist are substances. For Spinoza, this is not the case. Instead of God creating everything external to Him, Spinoza and Morgan (2002) argues that all things exist in God and that "apart from God there can be no substance" (E1P14). This take on Spinoza derives from one of his definitions where he says, "by substance I mean that which is in itself and is conceived through itself" (E1D3). In other words, for something to be a substance, it needs to have the power to conceive itself. The point being made here is that everything that God causes cannot be a substance since these particular things cannot conceive themselves. Therefore, unlike the theist, Spinoza classifies God as the only substance that exists since God is the only thing that can conceive itself.

13.2.2 Infinite and Finite Modes

Now, although for Spinoza, God is the only substance, there needs to be an explanation of some sort that shows how existing things can be part of God since this essentially is what classifies Spinoza as a Pantheist.

We can get such an explanation from Spinoza when he speaks of modes. However, what exactly, is a mode? For Spinoza, these particular modes are either infinite or finite modes, and most importantly, these modes are existing things. In fact, modes, along with substance (God), are the only two things that exist. That is, "nothing exists except substance and modes" (E1P15 proof). But although modes, like substance, are existing things, modes are prior to substance (E1P1) and depend on substance because they "cannot be or be conceived without substance" (E1P15 proof). In other words, since modes are in the Divine nature, they must and can only be conceived through God.

Now, as already mentioned, there are two types of modes: infinite and finite. Regarding infinite modes, they can be understood as the laws of nature. For instance, the laws that control gravity, motion, force, or rest are laws of nature. Spinoza makes this clear in his *Theological-Political Treatise* when he states, "those features that are most universal and common to the whole of Nature, to wit, motion-and-rest and the rules and laws governing them which Nature always observes and through which she constantly acts" (TTP7). Now, for Spinoza, laws of nature do not only govern these things, but for Spinoza, it consists of everything that happens (E1P15 schol; E3 preface). And most importantly, these infinite modes, Spinoza claims, are the determined acts of God that are expressed infinitely (TTP6).

However, regarding finite modes, they are things that depend on God and the laws of nature. But in addition to this, these modes are individual finite things (E2Exp. 3) that are classified as bodies (E1D2). They are "things" in the sense that they have properties such as length, breadth, depth, and shape (E1P15 schol). This would entail that bodies or individual things are in extension since, for Spinoza, "bodies are extended things" (E2D1). For instance, humans, animals, rocks, or celestial bodies are extended things since they have "length, breadth, and shape." Simply put, finite modes must reside in extension because they have properties.

13.2.3 God, Modes, and Spinoza's Pantheism

It is from these two modes that one can interpret Spinoza as a Pantheist. Essentially, the crux of Spinoza's Pantheism is that finite modes are affections of God, and concerning infinite modes, they are "what necessarily follows from the absolute nature of some attribute of God" (E1P21Dem) (these particular attributes will be discussed in Sect. 13.3). In other words, these finite modes are "particular things that are affections of the attributes of God" (E1P25 Cor). Whereas infinite modes are eternal and can be considered features of God.

Regarding finite modes, Spinoza mentions in the fifth definition in the first part of his Ethics that modes are the affections of substance (God) (E1D5). Furthermore, in Spinoza's letter to Lodewijk Meyer, he called affections modes (EP12). What this entails is that every finite mode and their way of interacting in extension are all affection of the attributes of God. For instance, if a human were to eat vegetables or if a rock were to cause an avalanche, these particular events would be affections of God's attributes. Overall, all events that occur in extension are the affections of God. Thus, these modes that are affection are not external to God but instead are occurrences that are internal to the substance. Now besides finite modes, infinite modes are also part of Spinoza's Pantheism. As previously mentioned, infinite modes are laws of nature that govern finite modes. They are metaphorically the organs of Spinoza's God, allowing finite modes to be expressed in God. One can think of the human body as an example. For instance, suppose x is God the substance, and the white or blood cells of x would be the finite modes of x. What moves the cells are the organs (infinite modes) of x. In other words, the organs would be the infinite modes that govern and moves the cells (finite modes). Although these two modes function differently, they are both in x. Similarly, although not entirely identical, infinite modes for Spinoza are the driving force of things in extension.

As shown, Spinoza's Pantheism consists of God as the substance that has inhered infinite and finite modes. In particular, the former governs and allows things in extension to do what is determined to do, and in the case of the latter, they are things in extension that are affections of God. These modes are the pillars and foundation of Spinoza's Pantheism, and Native Americans have quite a similar perspective where the Divine has attributes that showcases its power.

13.3 Native American Pantheism

Unlike western philosophy, where things are written and preserved in books, Native Americans share their ideas orally. They are heavily involved in oral traditions, where stories and legends are passed down from generation to generation. Various philosophical ideas exist among these stories, and many speak of a Great Spirit with similar views of Spinoza's God. In particular, the similarity is from some of the Native American legends that describe the great spirit as a force that has within itself attributes. And these attributes can be the laws of nature or the things that the laws of nature govern. These same attributes interestingly correlate to Spinoza's infinite and finite modes, where each showcases the Divine.

13.3.1 The Great Spirit

In many Native American tribes, they believe that there is a Great Spirit that underlies all of creation and this Spirit has various names. For instance, the Great Spirit is called Orenda, Manitou, or Wakan Tanka. Moreover, the naming of such Spirit can also be called "Gigantic Spirit," "Superior Lake," "Huge Lake," or "Great Mystery." Paula R. Hartz in *Native American Religions* gives a similar description stating: "The Algonquins speak of Manitou, the Iroquois of Orenda, and the Lakota of Wakan Tanka, words usually translated as "Great Spirit" or "Great Mystery." These words all refer to the indefinable power that underlies all creation" (Hartz 2009, p. 20). Overall, the Great Spirit has various names, but regardless of their different names, they all are attributed as a force or energy that "animates all things" (Wilshire 2000, p. 64).

In addition, it is described that the Great Spirit is the giver of breath and spirit in which nothing can exist without it. As described by Hartz (2009, p. 23): "Everything in the world that can be seen or touched is alive with spirit or breath." This is in connection with what Jamil Nassar in *Spirit Talks* conveys, "We refer to the originator of life as the Great Spirit! Nothing can have life within itself without the active life of spirit in it!" (Nassar 2016, Ch. 3). Each thing has spirit or breath that the Great Spirit gives, and most importantly, without these things, nothing can exist. In other words, nothing inside of the Great Spirit can exist unless breath and spirit are given. Thus, like Spinoza, nothing can be or be conceived without the Great Spirit.

13.3.2 The Great Spirit and Infinite and Finite Modes

This Great Spirit, like Spinoza's God, has within it all that exists where each thing is a sort of attribute of the divine. For instance, some Native American tribes believe that the Great Spirit has Earth and the Sun as its attribute. Such a description is mentioned in The Native American Book of Wisdom, stating:

> The Great Mystery is both male and female, so many Native Americans think of the Sky, or sometimes of the Sun, as Father, and the Earth as Mother. These are the male and female qualities of the Great Mystery. Both are equally important, although each has a different role (White Deer of Autumn 2009).

As noticed, the Earth and the Sun are not the Divine, but they are qualities or attributes of it. For this reason, one can incorporate Spinoza modes here, where the Earth and the Sun are the modes of the Great Spirit.

But besides the Earth and Sun being part of its attribute, the Great Spirit also has other attributes. In The Old Lady Trill, the Victory Yell, Patrice E.M. Hollrah mentions other attributes when she says:

> The reverent and poetic natures of these forest children feel the benign influence of the Great Spirit; they hear his voice in the wind; see his frown in the storm cloud; his smile in the sunbeam. Thus in reverential awe the Red man lived. His was the life that is the common lot of humankind (Hollrah 2016, p. 33).

Not only does the Great Spirit have Earth and the Sun as attributes, but the wind, storm cloud, and sunbeam are part of its attributes. The interesting aspect here is that both the sun and the sunbeam are attributes of the divine, but the two are entirely different. This suggests that there are multiple things within things that are qualities of the Great Spirit. Such classification can indicate that these can work as finite modes. For example, for Spinoza, finite modes consist not only of humans, rocks, or water but also of celestial bodies. As such, the sun as a heavenly body can be a finite mode that brings forth a sunbeam that is another finite mode.

Now concerning infinite modes, these modes can be, for Native Americans, the attributes of the divine that keeps things in motion. For instance, Hartz describes the Great Spirit in such a way, saying, "almost all Native American peoples believe in a great sacred force from which all things come and which keeps the universe in motion" (2009, p. 20). The force that keeps things in motion can have the same meaning as that of infinite modes, where it governs everything inside the Divine. Now, although the Great Spirit as a force keeps things in motion, it's essential to discuss the details that keep things in motion. Some possible explanations exist, such as energy, spirit forces, or cultural hero.

Regarding the first, some Native Americans believe that the law of nature is the energy of the Great Spirit, who governs everything inside it. For instance,

> Wakan-Tanka is an American Indian word for the Great Spirit in Manifestation. It is the universe and all that is. It has also been called The Great Everything. In order to manifest, the Great Spirit split in two, creating male and female energies. This is the principle of polarity or duality. It is the law of attraction that governs everything. This is why there is an inherent attraction to all things; this law governs every atom (Sacco-Belli 2012, p. 18).

Vincenzo Petrullo says something similar, but it's not called energy but Spirit-Forces. These forces manifest the power of the Great Spirit where they "reveal themselves to human beings, teaching them the ways of the Great Spirit" (Petrullo 1975, p. 29). It's the Spirit-Forces that showcase the Great Spirit's power. These particular Spirit-Forces are influential in Native American thought since they contributed to the creation of the universe. For example, these forces, also called cultural heroes, are the world's creators. Now it's important to mention that some Native Americans believe that the Great Spirit is not the creator of the world. Like Spinoza, the Divine does not create the world, but everything stems and comes from the divine. Instead, what contributed to the existence of things within the Great Spirit are the Spirit-Force, energy, or culture heroes, who have special powers that come from the Great Spirit (Hartz 2009, p. 23). Like infinite modes, they govern the whole universe, whether by creating the world, the way nature functions, or the movements of atoms. They function as infinite modes where it governs things that are finite modes, and most importantly, they are the very expression and attributes God.

13.3.3 Circles Within Circles and Spinoza's Modes

Now, as shown in Native American thought, they have within their philosophy infinite and finite modes. However, there is another aspect of their philosophy that would suggest modes existing in the Great Spirit. In particular, the Great Spirit is a "Great Circle" or "Hoop" that has within it smaller circles. For instance, The Native American Book of Wisdom illustrated a dialogue between Native American family members, saying:

> One day on the bus, Showanna Jones asked Jamie what his god, the Great Mystery, looked like. "I bet you can't even draw a picture of him", she said. "How can you draw a picture of all things living together for all time?" Uncle Nip asked later when he heard the story. "No, Jamie, no Indian ever tried to picture what the Great Mystery looks like." "We do, however, have a symbol that helps us to represent the Mystery –the Circle. It has no beginning and no end." The circle seemed like a fitting symbol to Jamie. He would always remember Uncle Nip explaining, "Is not the earth round? And does she not move, spinning slowly, in a circle around the sun, which is also round? And are not the stars round, and do not the great galaxies of stars spin in great spirals? The trunks of trees, the eagle's eggs, even our bodies are round. Do not the seasons move in circles, and life too? Circles within circles" (White Deer of Autumn 2009, p. 31).

As stated, the smaller circles are the earth, the sun, stars, galaxies, trees, eggs, and humans. This entails that everything inside the Great Spirit is a circle. The whole world, the cosmos, and the universe are circles (Wilshire 2000, p. 24). But these things are not the only circles; the Great Spirit is also a circle. Essentially, the circle is in everything and in each of us (White Deer of Autumn 2009, p. 32), and all these circles are within the Great Spirit. This, therefore, implies that all these smaller circles can be understood as modes where the laws of nature (infinite modes) are what control and govern the smaller circles (finite modes). And these modes, the laws of nature, and the smaller circles are the attributes or affections of the Great Spirit.

As shown, Native American Pantheism is quite similar to Spinoza, where the divine is a circle and a Great Spirit with various things within it. In particular, the Great Spirit has within it all that exist and the laws of nature that govern all creation. These laws of nature, such as the Energy, Spirit-Force, and culture heroes, can be understood as infinite modes which control all of existence that are finite modes. However, Although Spinoza and Native Americans have similarities regarding Pantheism, they differ in how the divine is active in nature. And in particular, it's through this difference where Native Americans would no longer be classified as Pantheists but rather Panentheistic.

13.4 Spinoza and Native Americans on God and Nature

Spinoza and Native Americans both implement in their philosophy that God is involved in nature. However, they have a different understanding of how exactly the divine is part of nature and the universe. For Spinoza, God is active in nature because our thoughts are included in the attribute of God called thinking, and modes in general as extended substance are in the attribute of God called extension. Whereas for Native Americans, God is involved in nature by spirits being messengers of the Divine and individuals praying and receiving blessings from the Great Spirit.

13.4.1 Thinking and Extension as Attributes of God

For Spinoza, there are two attributes that belong to God: extension and thinking (E1P1–2). These two particular attributes of God are involved in some ways with individuals that are finite modes. And it is from this involvement that some interpreters of Spinoza would classify him as Panentheistic. However, such a take is incorrect because, for Spinoza, God is not involved with humans in a personal matter, nor is God in everything. Instead, God is involved in human affairs in the sense that God's attribute of thinking and extension is connected with human thoughts and individual extended things.

Regarding God's attribute thought, Spinoza and Morgan (2002) argues that the thoughts of humans are modes that express the nature of God. In other words, our thoughts or ideas are modes that showcase God's nature. Spinoza makes such a claim when he states, "Individual thoughts, or this and that thought, are modes expressing the nature of God in a definite and determinate way" (E2P1 dem). Now, given that individual thoughts are modes, there must be an attribute of God to which all our thoughts belong. For Spinoza, such belongingness of individual thoughts belongs in God's attribute thought. This is the case for Spinoza since everything originates in God, for everything cannot be conceived without such a being (E1P17 dem). For Spinoza, such thoughts of individuals must belong in God's attribute "thought" since God ultimately is a thinking thing. Spinoza asserts this when he says:

> Therefore, there belongs to God (P1Def. 5, I) an attribute the conception of which is involved in all individual thoughts, and through which they are conceived. Thought, therefore, is one of God's infinite attributes, expressing the eternal and infinite essence of God (EID6); that is, God is a thinking thing (E2P1 dem).

As stated, by Spinoza, since all things are in God as modes, then our thinking as a mode must also belong in God. In connection to this, Spinoza also asserts in the second part of his ethics that thoughts are in God, "idea of both thought and its affection –and consequently of the human mind as well– must necessarily be in God" (E2P20 dem). Overall, this signifies that our thoughts and ideas are modes of God that are subject to God's attribute thought.

Now, although our thoughts belong in God's attribute "thought," our body as an extended thing also belongs in one of God's attributes. Mainly, the body or any extended thing Spinoza and Morgan (2002) argues belongs in God's attribute extension. And these particular extended things, like thoughts of individuals, are modes of the Divine. However, unlike modes of thought, extended things are modes of God's attribute extension.

13.4.2 *God's Intellect and Nature*

Given that our thoughts and extended things are involved in God as modes, the question at hand is how exactly are his attributes involved in nature? In other words, how does Spinoza's God extend itself to nature and individuals? One suggestion in trying to answer such a question may come from Spinoza's second part of his ethics, when he speaks of ideas, saying, "In God there is necessarily the idea both of his essence and of everything that necessarily follows from his essence" (E2P3). Additionally, regarding extension, he says that extended things necessarily follow from God's action (E2P3Sch). As Spinoza said in the former and the latter, the infinite number of ideas and extended things stem from and are caused by God's attribute thought and extension. And it is from here that one can connect God and nature.

Notably, God and nature for Spinoza are connected since ideas, and extended things are one with God. Not only are they one with God, but both thoughts and extended things are connected and the same. This may sound unclear at first, but this can be clarified if we understand Spinoza as a parallelist. Essentially, ideas are connected to extended things and cannot be separated from each other. Spinoza asserts it this way: "The order and connection of ideas is the same as the order and connection of things" (E2P7). One can use formal and objective essences to affirm their connection. Objective essences are ideas of things, whereas formal essences are things in extension. According to Spinoza, ideas and extended things are structurally connected, where the idea of a thing is also the very thing in extension. Spinoza uses Peter as an illustration saying that the idea of Peter is also what Peter is in extension. In Spinoza's words:

> Now the true idea of Peter is the objective essence of Peter and is in itself something real, something entirely different from Peter. So since the idea of Peter is something real, having its own individual essence, it will also be something intelligible, that is, the object of another idea which has in itself objectively everything that the idea of Peter has formally. And in turn the idea of the idea of Peter again has its own essence, which can also be the object of another idea, and so on without end (TIE 34).

In other words, the idea of the mind thus has to represent the body that is in extension perfectly. And interestingly, the ideas of things and their extended representation are both in unison with God. Therefore, God's involvement in nature is by formal and objective reality being inside God and both realities being subject under the attribute thought and extension.

13.4.3 Native American and Panentheism

Unlike Spinoza, Native Americans' take on God and nature is entirely different since some Native American tribes understand the Great Spirit to be personally involved in human affairs. This would thus classify these particular Native tribes as Panentheistic since God transcends the universe but is also involved in human activity. There are many ways that the Great Spirit is involved in nature and too many to describe, but there are precisely three that I wish to discuss here. In particular, I want to discuss the messengers of the Great Spirit, praying and communicating to the divine, receiving blessings from the great spirit, and the Great Spirit's omniscience and omnipresence.

13.4.4 Communication with the Great Spirit

Regarding speaking with the Divine, various Native American tribes have a form of communication with the Great Spirit. One form of communication is through prayer. In some tribes, prayers are spontaneous; in many others, they are ritualized because they need to be memorized. But regardless, these forms of prayers are so that they can communicate with the Great Spirit. But in addition, Native American tribes would also commence rituals dedicated to the Great Spirit, showcasing their thankfulness for the Great Spirits' provisions of nature (Hartz 2009, p. 72). But besides prayers, other tribes would go to sacred places of nature to communicate with the Great Spirit (Hartz 2009, p. 27). Now, although there are various Native American tribes that believe one can communicate with the Great Spirit by going to sacred places or doing rituals, there are some that believe there are animals that are the messengers of the Great Spirit, such as the buffalo (Hartz 2009, p. 25, 36) and the eagle (Hartz 2009, p. 136). For instance, it is believed that the Eagle carries the people's prayers and sends them to the Great Spirit (Hartz 2009, p. 60). In addition, the Great Spirit would send its messengers to communicate to individuals how they ought to behave, what rituals they should practice, and what is expected from them (Petrullo 1975, p. 29). Overall, this is quite different from Spinoza since the Great Spirit has a sort of persona where it can communicate with individuals in various ways. However, the Great Spirit does not speak but uses the very things in nature to share what it wants to convey.

13.4.5 Ecology and Blessings from the Great Spirit

Now, prayers are not the only aspect of Native American culture that showcases how the Great Spirit is connected to nature and individuals. The Great Spirit is also involved with nature and individuals by providing blessings to humans and providing blessings to nature by placing humans on Earth to take care of it. An example can be made in *The Sacred Wisdom of the Native American,* where Larry J. Zimmerman describes an anguish of an individual from the Cayuse tribe stating:

> "I wonder if the ground has anything to say?" I hear what the ground says. The ground says, "It is the Great Spirit that placed me here. The Great Spirit tells me to take care of the Indians, to feed them properly." The water says the same thing. The grass says the same thing. "Feed the Indians well," the ground says, "It was from me man was made. The Great Spirit, in placing men on Earth, desired them to take good care of the ground and to do each other no harm" (Zimmerman 2011, pp. 105–106).

As shown here by Zimmerman, particular things in nature are placed on Earth to bless humans, and in return, humans are placed on Earth to take care of Earth. Everything overall on Earth is considered a blessing that the Great Spirit gave (Hartz 2009, p. 102). And interestingly, for many Native Americans living a long life was also considered a blessing from the Great Spirit (Hartz 2009, p. 98). Not only was long life a blessing, but also the provision of food was a blessing from the divine (Hartz 2009, p. 26).

13.4.6 The Great Spirit's Attributes and Its Dichotomy from Spinoza

The Great Spirit overall is involved in nature, and how it is involved would classify some Native Americans as Panentheistic. This is the case since the Great Spirit communicates with individuals. Whereas for Spinoza, God is only involved in nature in the sense that the ideas of nature and its extension are in God as modes that are subjected under God's attribute thought or extension. This would thus label Spinoza more as a Pantheist rather than Panentheistic since God is not involved in nature on a personal level.

But amongst these differences, Native Americans also differ in describing God's attributes and how God is involved in us. For instance, the Lakota tribe believes that the Great Spirit is omnipresent and omniscient. Zimmerman gives such description saying, "For the Lakota, every human act is imbued with spiritual significance and pays homage to the omnipresent and omniscient Great Mystery, or Great Spirit, a transcendent god-like power pervading people, animals, places and phenomena" (Zimmerman 2011, p. 84). As mentioned here, the Great Spirit is present in all of creation and pervades in us all. However, it is unclear how exactly the Great Spirit is all-knowing. One thing we can consider to answer such unclarity is that everything has in it the power and presence of the Great Spirit. Thus, perhaps one can understand God's omniscience in the sense that God is present in all things, and through his presence within things, the Great Spirit also simultaneously knows all.

Now, as seen in Sect. 13.3, the Great Spirit is similar to Spinoza's God regarding infinite and finite modes. However, for Spinoza, these modes are in God. For Native

Americans, these modes are in the Great Spirit, but in the same way we are in the Great Spirit, the Great Spirit is within us all. As stated by White Deer of Autumn, "where is the center of the circle of an endless universe? It is everywhere and within each of us, making us all relatives of equal value and importance" (White Deer of Autumn, p. 33). Spinoza does not give such an assertion. Instead, he says all things are in the Divine. Therefore, many Native Americans differ from Spinoza in that they have a Panentheistic understanding of the Great Spirit, where the Great Spirit can communicate with us and are also in us. Whereas Spinoza is classified as a Pantheist since God is not in all, but everything is in God.

13.5 Conclusion

It's pretty fascinating how much Native American Philosophy can contribute to philosophical discussion. And here, precisely, we see their contribution to Pantheism and Panentheism. One can pinpoint here how similar is the Great Spirit with that of Spinoza's God. In particular, everything in the Great Spirit can be applied to infinite and finite modes, where infinite modes are the laws of nature that govern all things in the Great Spirit that are finite modes. However, Spinoza and Native Americans differ because Spinoza's God is not involved in nature on a personal level. Instead, God is in unison with individuals and nature by both the ideas of things and things in extension being under God's attribute thought or extension. For Native Americans, on the other hand, the Great Spirit is involved in nature, where he blesses both humans and the Earth. And in addition, the Great Spirit is connected to nature by things not only residing in God like Spinoza and Morgan (2002) suggests but that God is in everything. Moreover, for Native Americans, the Great Spirit can have some persona where it can communicate to individuals not in the form of language but by the Great Spirit speaking through nature. But overall, between the two, Spinoza's God and the Great Spirit for Native Americans assert that God's attributes are showcased in nature. However, for the former, God is taken in a more Pantheistic sense, whereas, for the latter, the Great Spirit has characteristics of Panentheism.

References

Hartz P (2009) Native American religions. Chelsea House, New York

Hollrah P (2016) Old lady trill, the victory yell: the power of women in native American literature. Routledge, London

Nassar J (2016) Spirit talks: Redman's truths. Page Publishing Incorporated, New York

Petrullo V (1975) The diabolic root: a study of peyotism, the new Indian religion, among the Delawares. Octagon Books, London

Sacco-Belli J (2012) The Great Spirit says a rainbow Warrior's journey. Balboa Press, Bloomington

Spinoza B, Morgan ML (2002) Spinoza: complete works. Hackett

White Deer of Autumn (Horn G) (2009) The native American book of wisdom. Beyond Words Publishing, Hillsboro
Wilshire BW (2000) The primal roots of American philosophy pragmatism, phenomenology, and native American thought. Pennsylvania State University Press, University Park
Zimmerman LJ (2011) The sacred wisdom of the native Americans. Chartwell Books, New York

Chapter 14
Ground of Being: The Panentheism of Paul Tillich, Earth Care, and Intercultural Dialogue

Roy H. May Jr

Abstract "Ground of being" as Paul Tillich understood the term is presented as an expression of panentheism and its rich possibilities for earth care and intercultural dialogue are explored. Whereas pantheism proposes that "all is God" panentheism argues that "all is in God." The salient distinction is important because panentheism allows for indeterminacy and value judgments that are not possible when all is divine. Tillich finds that ground of being connotes ultimate source and potentiality and therefore is an apt metaphor for God and divine-filled existence. It also binds existence into an integral whole. This vision of existence makes environmental crisis an existential one because it puts into question existence itself. Any assault on nature, then, is an assault on the divine and existence itself. These ideas resonate with similar metaphors prominent in Indigenous cultures. Such affinities of thought can be avenues for earth care and intercultural dialogue and thus facilitate cross-cultural collaboration and understanding.

Keywords Existence · Indigenous cultures · Ontology · Symbolic language · Ultimate reality

14.1 Introduction

Classical theism, which posits a substantialist concept of God or the Divine as totally transcendent, is challenged because it separates the Divine from participation in material existence and for proposing "God" as supernatural being. Transcendence suggests the Divine as aloof, unconcerned with the material world, and a supernatural being not credible for contemporary knowledge. Likewise, the idea of God as totally immanent –pantheism– is criticized because it universally divinizes material reality by collapsing into one the creator and created: "All is God" (Nikkel 1995, p. 2). Transcendence disappears, distinctions are erased, and value judgments are

R. H. May Jr (✉)
Departamento Ecuménico de Investigaciones, San José, Costa Rica

L. Valera (ed.), *Pantheism and Ecology*, Ecology and Ethics 6,
https://doi.org/10.1007/978-3-031-40040-7_14

not possible because everything is inherently divine. Classical theism is dualist. Pantheism is monist. This discussion is pertinent not only for theology (and philosophy), but also because it has implications for earth care and intercultural dialogue and collaboration. To respond adequately, surely theology is tasked with understanding the Divine as both transcendent and immanent.

Paul Tillich (1886–1965), one of the most influential Christian theologians of the twentieth century, undertook such a project. Born in Germany, the Lutheran theologian was profoundly traumatized by the horrors of the trenches during World War I when he was an army chaplain. He also was a romanticist who found great beauty in the natural world and was deeply concerned about humanity's attempt to master nature through technology (Drummy 2000). And in great art he perceived the divine (Tillich 1989). Late in his career he became deeply interested in intercultural dialogue (Foerster 1990).

Existence in the face of non-existence, or put another way, the meaning of life, deeply troubled him. He was unable to escape the question of what life was about. So as theologian, philosopher, and art critic he sought to understand the meaning of existence. This profoundly existential question was the organizing theme of his thought. He concluded that only by embracing what concerns us ultimately can the meaning of existence be found. He believed that to live authentically or meaningfully is to be rooted in ground of being or Ultimate Reality (Tillich 1952). For Tillich, to show this was the theological task (1967, p. 12, 21). Methodologically, his "method of correlation" joins existentialism (the question of meaning) and ontology (the structure of existence) to propose ultimate reality or ground of being as empowering meaningful life, but only if ground of being or ultimate reality is joined with existential situations to give answers to life's deep questions of being versus non-being (Tillich 1967, p. 8).

Doing this, he understood, required radically reconceptualizing theological propositions and religious language. Both would need to resonate with contemporary thought and experience. Above all, literal interpretations and religious language would have to be reframed as symbolic representations. Indeed, Tillich warns that the word "God" is a symbol and does not refer to an anthropomorphic or literal thing or entity. "[A]ny concrete assertion about God must be symbolic," he asserts (Tillich 1967, p. 239) because "symbolic language alone is able to express the ultimate" (Tillich 1957, p. 41). Traditional religious language, Tillich believed, is symbolic representations of deep, existential truths. Only by grasping the meaning of the symbol, can such language be relevant because "[t]he truth of a religious symbol has nothing to do with the truth of the empirical assertions involved in it, be they physical, psychological, or historical" (Tillich 1967, p. 240). His theological work endeavors to do this (see Golijanin 2019; Nikkel 1995, pp. 13–19).

As a founder of the Frankfort School and active in Christian Socialism, he was among the first German academics to be expelled from the academy when Hitler rose to power in 1933. Tillich immigrated to the United States where he had distinguished careers at Union Theological Seminary (New York), Harvard University, and the University of Chicago.

In this chapter I present "ground of being" as Tillich understood the term as an expression of panentheism, and I explore its rich possibilities for earth care and intercultural dialogue. The first two sections lay out Tillich's philosophical/theological (and very abstract) meaning of ground of being in which I show that the term implies panentheism and functions to unify existence. I then reflect on its rich potential for theological/philosophical undergirding of earth care and argue that ground of being resonates with Indigenous thought about existence, thus indicating the value of the concept for intercultural dialogue and collaboration. Finally, I propose that Tillich's understanding of ground of being, with its inherent panentheism, offers a healing framework for a world in which the integral unity of God-Self-World is sadly marked by deep estrangement.

14.2 Ground of Being: Existence as Panentheistic

Tillich (1967, pp. 11–12ff, 235–238) understands "God" as Ultimate Concern or Ultimate Reality. "Ultimate" connotes depth, for "ultimate truth is deep or profound" (Robinson 1963, p. 45). For Tillich, Ultimate Reality or Ultimate Concern –he uses both terms interchangeably – is the depth of existence and "the name of this infinite and inexhaustible depth and ground of all being is *God*" (Tillich 1948, p. 57; italics in the original). In this sense, "God" as ultimate depth is the

> ground of our being, that which determines our being or not-being, the ultimate and unconditional power of being …[that] expresses itself in and through the structure of being. Therefore, we can encounter it, be grasped by it, know it, and act toward it (Tillich 1967, p. 21).

Still, exactly what Tillich means by "ground of being" is not easy to grasp. He draws on the Christian mysticism of Jacob Boehme of the seventeenth century, especially his reflections on being and non-being, then Friedrich W. J. Schelling's philosophy of nature as an expression of the divine in the nineteenth century, and in turn is influenced by the ontology, existentialism, and deep psychology of the twentieth century (Emmet 1964, pp. 280–282; Drummy 2000, pp. 36–41). As one observer notes, Tillich "is disconcertingly hospitable to all sorts of suggestions as to what he might mean" (Emmet 1964, p. 289). However, this hospitality to interpretations gives the metaphor rich potential for understanding the meaning of existence in different contexts and times.

Certainly "grounding" is a powerful metaphor. The word "ground" originated in Proto-Germanic languages and Old English by the twelfth century to mean bottom, foundation, abyss, or deep place, as well as source, origin, and cause, always designating fastened or supported (Ground 2022). The metaphor served Tillich well because it connotes the "infinite power of being," or "the power of being in everything and above everything" (Tillich 1967, p. 236) that originates, foundations, and holds everything together.

As ground of being, Tillich proposes that God "is the structure of being. He [sic] is not subject to this structure; the structure is grounded in him. He is the structure, and it impossible to speak about him except in terms of this structure" (Tillich 1967, p. 238). Therefore, all existence participates in that grounding power that makes existence possible (Tillich 1967, p. 245). Although the language is highly symbolic, Tillich (1967, p. 245) says, as ground of being "divine life participates in every life as its ground and aim. God participates in everything that is; he has community with it; he shares its destiny." Therefore,

> the world process means something for God. He is not a separated self-sufficient entity who, driven by a whim, creates what he wants and saves whom he wants. Rather, the eternal act of creation is driven by a love which finds fulfilment only through the other one who has the freedom to reject and to accept love (Tillich 1967, p. 422).

This, Michael Drummy (2000, p. 99) explains, "expresses the simultaneous reality that the universe is created for fulfilment in God and that God's most profound concern is that all creation shares fully in the Divine Life." God needs the world. For Tillich, Drummy (2000, p. 98) says, "God is not who God is without the world, and the world is not what it is without God." There is no ground of being without being; otherwise it would be ground of non-being which is nonsensical. In this sense, for Tillich "God" is the structure of being or existence, as being itself –not the structure of *a* being. Therefore, "divine life" participates fully in all life as it is at the same time its goal because the *telos* of "creation" is that every form of being fulfil its own nature since the end of creation is the creature: material reality.

Although Tillich remains within the substantialist ("being") confines of classical theism (Ogden 1963, p. 55), he breaks with classical theism by arguing that this is a constant, never-ending dynamic. Indeed, creation "is not the story of an event which took place 'once upon a time.' … God *has* created the world, he *is* creative in the present moment, and he *will* creatively fulfil his *telos*. Therefore, we must speak of originating creation, sustaining creation, and directing creation" (Tillich 1967, pp. 252–253; italics in the original). Thus, God not only creates, but God also participates in the creative process itself. Ultimate Reality or ground of being, is not static but dynamic. Tillich thus affirms the material and recognizes that creation/nature is dynamic, changing, and evolutionary, and proposes a dynamic panentheism that imbues nature with the quality of the Ultimate.

14.2.1 Existence as Panentheistic

Ground of being as theological ontology implies that the divine is implicitly present in material existence. Although Tillich does not use the term until the end of his systematic theology, this is panentheism, the affirmation "that although God and the world are ontologically distinct and God transcends the world, the world is 'in' God ontologically" (Cooper 2006, p. 18). "All is in God" according to panentheism (Nikkel 1995, pp. 1–2). That is, the world is not divine but the divine is in the world.

This allows the world –material existence– freedom or indeterminacy even though empowering being is integrally present in the world. The non-human has agency; it is subject. Existence in any form is not pre-determined or pre-destined in any absolute fashion.

The origin of the term "panentheism" only dates from the nineteenth century, but the idea has a long philosophical and theological history, albeit with widely varied usage and meaning (Cooper 2006). However, common to all usage and definitions is the avoidance of identifying the divine and the material as one and the same as in pantheism. The power of panentheism is its conviction that existence can never be reduced to mere substance because it is imbued with Ultimate Reality; it participates in ground of being. It can never be just inert matter, devoid of meaning. It is "vibrant," as the new materialism would say (Bennet 2020) or that it is imbued with the divine as theology would affirm. For Tillich this means that God simultaneously creates material existence for fulfillment in God and that material existence fully participates in divine life (Drummy 2000, p. 99). Existence is panentheistic.

14.3 Ground of Being: Existence as an Integral Whole

Ground of being implies unity. Since "ground" is foundational to "being," logically it also holds together all that exists; it "fastens" reality. It is necessarily unifying, even though being is manifested multidimensionally, because ground –God– also "has the power of determining the structure of everything that has being" (Tillich 1967, p. 239). Ground of being glues existence into a coherent and integral whole. Ultimacy (ground) and concreteness (being) come together. Tillich believed that "trinitarian monotheism" –the Trinity of orthodox Christianity—expresses this unity of the ultimate and the concrete. Throughout his work Tillich insists on the unity of existence.

In this sense, tellingly Tillich affirms that there is no absolute discontinuity between animals and humans and that it is impossible to say at what point animal nature is substituted by human nature (Tillich 1967, p. 41). "Man [sic] reaches into nature," Tillich says, "as nature reaches into man. They participate in each other and cannot be separated from each other" (Tillich 1967, p. 43). He recognizes that there are differences among life forms and that life is multidimensional. He refers to other-than-human life as "subhuman." However, he is clear that this term is not to be understood as value judgement:

Subhuman does not imply less perfection than in the case of the human. On the contrary, human as the essentially threatened creature cannot compare with the natural perfection of the subhuman creatures. Subhuman points to a different ontological level, not to a different degree of perfection (Tillich 1967, p. 260).

His point is that there is no marked division between humans and non-humans. Nor is there a sharp dichotomy between the "organic" and the "inorganic." Tillich (1967, p. 19) even proposes a "theology of the inorganic" because "it is the first condition for the actualization of every dimension." All participate in ground of being. Even the "Kingdom of God" is universal. "It is a kingdom not only of men," Tillich affirms (1967 III, p. 359), "it involves the fulfillment of life under all dimensions." Ground of being implies ontological interrelatedness that binds the whole of reality into a single unity.

14.4 Ground of Being: Earth Care and Intercultural Dialogue

Ground of being and panentheism have rich potential for earth care and intercultural dialogue. For Tillich (1948, p. 82; 1945, p. 302), humankind and nature share the same ground: the ground of being/the ground of nature is the same ground. Many Indigenous cultures understand the earth as the foundation of existence. Ground of being/ground of nature can connect with this ancient, but still living, cultural rootedness in the earth. Earth care and intercultural dialogue are facilitated as the commonality of concerns emerge.

14.4.1 Earth Care

Tillich was worried about the state of nature and the environment. For him, the deteriorating state of nature provoked angst and brought on existential crisis because "the survival of civilized mankind, or of mankind as a whole, or even of life altogether on the surface of this planet" was at stake, thus producing "tormenting anxiety" in the face of "the imminent danger of a universal and total catastrophe" (Tillich 1963, pp. 66–67). Underlying this concern is Tillich's ground of being/panentheistic framework.

"Ground" commonly means "soil" or "the surface of the earth" and place of "abode" (Webster 1956), thus integrally associating the term with nature. Ground of being, then, also connotes the Earth and the natural world. In this "same Ground [...] the universe and all its galaxies is [sic] rooted," Tillich (1963, p. 72) explains. The ground of being and the ground of nature are the same. The Earth and the natural world participate in divine being and so are imbued with ultimate importance. Thus, environmental destruction is nothing less than an assault on the Divine, on Ultimate Reality or ground of being as the very basis of all creation. It tears apart "the interdependent fabric that unifies all life" (Drummy 2000, p. 62).

Since, by definition, divine love includes all creatures (Tillich 1967, p. 282), ontologically God needs the world, humans, and other-than-human creatures. Likewise, according to Tillich, humans need other creatures because "proper human self-love includes everything with which man is existentially related" (Tillich 1967, p. 282). In Tillich's theology, the Earth is object of divine love and, correspondingly, the Earth is object of human love. Caring for the Earth is axiomatic in Tillich's ground of being/panentheistic framework.

Tillich (1963, p. 69; 1962, 1948, p. 79) argues that the origin of the environmental crisis or "predicament" is humanity itself, "brought about chiefly by the scientific and technical development of our [twentieth] century." Such intellectual hubris has led humanity to deny its interrelatedness with nature. This denial is the root of the environmental crisis (Drummy 2000, pp. 81–82). A fundamental concept that permeates Tillich's writings is the existentialist concept of estrangement or separation. "Sin" is "estrangement" from that which we properly belong: ground of being in its dimensions of God-Self-World (Tillich 1967, pp. 44–47) "and world means nature as well as man" (Tillich 1948, p. 77). By misusing its God-like powers, humankind has become estranged from nature –the "ground" of ground of being–, and so "[i]t is possible that the earth may bear us no longer," Tillich fears (1963, p. 68). Only by repairing the breach, by re-uniting with nature/ground of being, will the crisis be healed. There are innumerable opportunities to do this and "we must fight with all our strength" to take advantage of them (Tillich 1963, p. 68). Caring for the Earth repairs the broken relationships among God-Self-World and brings about existential healing and the well-being of the Earth because the salvation of humanity and the salvation of nature are bound together (Tillich 1948, p. 84; 1945, p. 304).

14.4.2 Intercultural Dialogue and Collaboration

Although symbolic language, ground of being and panentheism are concepts proper to Western philosophy and theology that can be avenues for intercultural dialogue and cooperation. Albeit differently expressed, non-Western, Indigenous cultures use analogous concepts for understanding existence. This makes dialogue and cooperation possible.

Like Tillich, Indigenous natural philosophy is expressed through symbolic or metaphorical language (Cajete 2000, p. 45). Myths, images, and especially metaphors are basic forms of communication (Kidwell et al. 2001, p. 53). Recognizing that religious or theological language is not literal (as Cajete, Tinker, and Tillich argue) can open dialogue leading to the discovery of meanings that conversation partners may hold in common. Metaphorical language moves one away from form to content.

The metaphor ground of being and the concept of panentheism can be particularly helpful for intercultural dialogue. For many Indigenous cultures, the Earth, imbued with sacredness, is fundamental to their religious worldview (Cajete 2000, pp. 177–213). Zuñi (New Mexico) potter, Noreen Simplicio, explains, "Clay is sacred to us. If you give it respect, it will be good to you. … All it needs is water and water is life" (Museum display, Museum of Northern Arizona, Flagstaff, Arizona). Native American scholar Vine Deloria, Jr. (2003, p. 61) explains that "land" is a constant point of reference in Native American thought. Features of the landscape have deep symbolic importance and are perceived as having agency. "The earth is alive and everything related to it is also alive," Deloria (2003, p. 146) affirms. During a mid-nineteenth century treaty negotiation, a chief asked, "I wonder if the ground has anything to say?" (Deloria Jr. 2003, p. 80). The Tewa intellectual from the Santa Clara Pueblo (New Mexico) Gregory Cajete (2000, p. 273), even uses the term "ground of existence" for a Native American theology of nature. Earth, soil, landscape –creation– or ground of existence, imply that all function as an integral whole (Deloria 2003, p. 80). In this sense, ground (of ground of being or ground of existence) as soil, the surface of the earth, or place of abode resonates with Tillich's concept of ground of being/ground of nature which he also connects to soil as the ground of life (Tillich 1967, p. 19). The sacredness of the Earth can be identified as an expression of panentheism.

Lutheran theologian George Tinker, a member of the Osage Nation (USA), is illustrative as he theologizes from within the religious traditions of Native Americans. He does not use the term ground of being, but in many ways his thought resonates with it. Furthermore, he insists on the importance of religious language as symbolic language and proposes a panentheistic ontology of existence.

He explains that in Indigenous religious thought, "God" is not "a person" but "a spiritual force [that] permeates the whole world and is manifest in countless ways in the world around us at any given moment and especially in any given place" (Tinker 1996, p. 164). This spiritual force is deeply metaphorical because "God appears in many forms apparent in the environment. Descriptions of nature embed an understanding of spiritual power in description of natural events (Kidwell et al. 2001, p. 53). For this reason, the beginning point for knowing "God" is creation, that is, the world of nature, because it is in that spatial holism that life is lived out and the divine is experienced (Tinker 2008, p. 37, 38). Ultimately "an unknowable mystery," the divine is manifested in multiple ways and places (Tinker 2008, p. 64) as harmony, balance, and, especially, as the "interrelatedness" of creation. Thus "God" is "revealed in creation, in space or place, and not in time" (Tinker 2004, p. 104) as "the physical forces in the environment –the sun and moon, the blue of the sky, things that move, like wind, and things that endure, like rock" (Kidwell et al. 2001, p. 64). However, creation is not a one-time event because "the world [is] in constant creative process that requires our continual participation" (Tinker 1996, pp. 157–158; Tinker 2008, p. 65). Metaphorically, "God" is the Sacred Other, the Source of Life (Kidwell et al. 2001, p. 35) invisible, creative power that brings every living thing into existence (Kidwell et al. 2001, p. 57) and is present in the material world (Kidwell et al. 2001, p. 61). "God", then, is active presence in an on-going creative and life-sustaining process experienced as reciprocity and balance that ties together the whole of existence (Kidwell et al. 2001, pp. 46–47). "God" empowers the Earth to come alive.

Clearly, ground of being and panentheism resonate with the religious worldview of Indigenous cultures. They too, using symbolic language to communicate deep truth, understand God as originating or foundational creativity and conceive existence as imbued with the sacred or the divine, yet do not collapse existence into an ontological whole in which all is Sacred Other and Sacred Other is all. The divine presence makes existence an integral and interrelated whole that keeps together different dimensions of life, not as contradictions but as complements. Both humanity and other-than-humans share intimately this holistic existence. Obviously integral to their worldview is respect for the natural environment and the desire to honor and protect it. This understanding of the world has long made Indigenous people keepers of the Earth.

The evident commonality of the Indigenous worldview with the Western concepts of ground of being and panentheism could open avenues for intercultural dialogue that facilitate cross-cultural collaboration and ultimately enrich the lives of conversation partners. Most certainly it can draw together diverse cultures for the defense of the environment and together heal the existential crisis provoked by the ever-deteriorating health of the Earth.

14.5 Concluding Thoughts

Today's world is marked by deep divisions, value is reduced to that of commodities, tragically revealing profound alienation or "estrangement from that to which one belongs –God, one's self, one's world" (Tillich 1967, p. 46), the integral unit that bonds humankind not only to all biota but even to geographic landscapes. This estrangement or dis-integration manifests itself in anthropological and existential dimensions, as well as in political economy: mode of production and models of development and well-being. This alienation/estrangement configures the shape of the God-self-world unity, forming mind sets and excluding alternative ways of conviviality. Tillich's concept of ground of being as the integral unity of existence, with its inherent panentheism, offers a framework for healing in our day and age. It recovers the integral unity of existence and the deep values that only can be described as divine. Existential healing will recover the connection between the ground of one's own being and the ground of nature.

References

Bennet J (2020) Vibrant matter. A political ecology of things. Duke University Press, Durham

Cajete G (2000) Native science. Natural laws of interdependence. Clear Light Publishers, Santa Fe

Cooper JW (2006) Panentheism. The other god of the philosophers. From plato to the present. Baker Academic, Grand Rapids

Deloria Jr. V (2003) God is red. A native view of religion, 30th Anniversary Edition. Fulcrum Publishing, Golden

Drummy MF (2000) Being and earth. Paul Tillich's theology of nature. University Press of America, Lanham

Emmet D (1964) The ground of being. J Theol Stud 15:280–294

Foerster J (1990) Paul Tillich and inter-religious dialogue. Mod Theol 7(1):1–27

Golijanin V (2019) Paul Tillich's theory of religious symbolism: meaning, significance, potential. In: Šijaković B (ed) Ad Orientem: essays from Serbian theology today. Faculty of Orthodox Theology, Belgrade
Ground (2022) Online etymology dictionary. https://www.etymonline.com/
Kidwell CS, Noley H, Tinker GE (2001) A native American theology. Orbis, Maryknoll
Nikkel DH (1995) Panentheism in Hartshorne and Tillich. Peter Lang, New York
Ogden SM (1963) The reality of god and other essays. Harper & Row, New York
Robinson AT (1963) Honest to god. The Westminister Press, Philadelphia
Tillich P (1945) The redemption of nature. Christendom 10:299–305
Tillich P (1948) The shaking of the foundations. Charles Scribner's Sons, New York
Tillich P (1952) The courage to be. Yale University Press, New Haven
Tillich P (1957) Dynamics of faith. Harper & Row, New York
Tillich P (1962) Man and Earth Christianity and Crisis 22:108–112
Tillich P (1963) The eternal now. Charles Scribner's Sons, New York
Tillich P (1967) Systematic theology, three volumes in one. University of Chicago, Chicago
Tillich P (1989) On art and architecture. Crossroad, New York
Tinker GE (1996) An American Indian theological response to ecojustice. In: Weaver J (ed) Defending mother earth. Native American perspectives on environmental justice. Orbis, Maryknoll, pp 153–176
Tinker GE (2004) Spirit and resistance, political theology and American Indian liberation. Fortress Press, Minneapolis
Tinker GE (2008) American Indian liberation, A theology of sovereignty. Orbis, Maryknoll
Webster (1956) Webster's new collegiate dictionary. G&C Merriam Company, Springfield

Chapter 15
God, Home, and Thinking in the Place: What Kind of Pantheism Did Thoreau Endorse?

Jeyver Rodriguez

Abstract Scholars have drawn on Thoreau's thoughts and writings as a basis for environmental virtue ethics. His ecological vision of Home and 'self-in-the-place' have also inspired ecological restoration projects. However, a connection between his idea of God, Home, and what I call "thinking in the place" requires further exploration. In this chapter, I reconstruct several strands of his thought, seeking to answer the question: What kind of pantheism did Thoreau endorse? I will discuss his ideas about the restorative power of nature, his commitment to 'living deliberately,' his appreciation of simplicity, his advocacy of walking as a practice for peaceful living, and his disdain for materialistic living as an obstacle to the enjoyment of true happiness. The aim of the chapter is to inquire into the religious and ecological significance of pantheism in Thoreau's writings by analyzing the articulations between his visions of God, the Home, and Self. Thoreau's religious vision was deeply contextualized in certain episodes of enlightenment or ecological conversion that he commonly referred to in his writings as his ecstasies. These experiences accompanied him throughout his life and may help shed light on the kind of pantheism he espoused.

Keywords Thoreau · Pantheism · Panentheism · God · Place

15.1 *Walden* and *Walking*: "In Wildness is the preservation of the world"

One of the essential strands of Thoreau's thought is found at the end of *Walden*:

> We need the tonic of wildness,-to wade sometimes in marshes where the bittern and the meadow- hen lurk, and hear the booming of the snipe; [...] At the same time that we are

J. Rodriguez (✉)
Department of Applied Ethics, Faculty of Religious Sciences and Philosophy, Catholic University of Temuco, Temuco, Chile
e-mail: jeyver.rodriguez@uct.cl

L. Valera (ed.), *Pantheism and Ecology*, Ecology and Ethics 6,
https://doi.org/10.1007/978-3-031-40040-7_15

> earnest to explore and learn all things, we require that all things be mysterious and unexplorable, that land and sea be infinitely wild, unsurveyed and unfathomed by us because unfathomable (Thoreau 1971, pp. 317–318).

Thoreau seems to tell us that we constantly need a good dose of wild nature to regain our health and remain on guard against our own inner demons: arrogance, drowsiness, daze generated by wasteful lifestyles, laziness, and grief. As Smith (1985, p. 221) states: "Thoreau's trope of sleepiness is a major structuring motif in *Walden*, always associated with the wasteful, routine, unconscious, profane, and blindly conformist habits that make for death-in-life." In *Walking,* we find this same seminal vision expressed in a classic quotation: "The West of which I speak is but, another name for Wild; and what I have been preparing to say, is, that in Wildness is the preservation of the World" (1977, p. 227). Moreover, a few lines later he adds, "Life consists with wildness. The most alive is the wildest. Not yet subdued to man, its presence refreshes him." Hunt emphasizes that to properly understand Thoreau's philosophy it is necessary to pay careful attention to the distinction between Wilderness and *Wildness*: "Despite its fame, we sometimes hear this line misquoted as 'in Wilderness is the preservation of the World'. That would be a very different matter. A wilderness is an area, a sector of the world. Wildness is a characteristic, a feature of the world" (Hunt 2015, p. 211).

Thoreau reminds us of an essential fact that is frequently forgotten. Wildness puts us in touch with different sources of vitality, cheerfulness, and health. Nevertheless, it is important to stress that Thoreau did not think of wildness as something too far away, and totally foreign to us. "Thoreau clearly identifies 'wildness' not as a distant place but as a quality, something ineffable and strange and raw at the heart of the most common experience" (Walls 2000, p. 15). The *wildness* to which Thoreau refers is a reality that certainly has a more experiential and processual connotation than is often appreciated when we say that we need to expand natural parks and wildlife reserves: "wildness might be –in fact, is– something that is present in the world far outside the boundaries of those sacrosanct wilderness areas. There might be a wild aspect of the world that pervades most of the world, perhaps all of it, including the surroundings in which most people live their lives" (Hunt 2015, p. 111).

It is appropriate to emphasize that there is an aspect of wildness that only emerges in a dynamic of proximity and distance: this wildness is not fully available in wetlands, forests, and lakes, and neither is it a closed and inaccessible reality to human beings. When Thoreau thinks about wildness, he uses not only descriptive words, but also normative statements. It would be wrong to define wildness in Thoreau only as a determined "place" since wildness is also a *connection*, an *emotion*, and an *attitude* (Cafaro 2004; Schneider 2000; Thorson 2019; Walls 2000). Wildness is not a large area within a clearly defined geographical space but a processual and dynamic reality that can help to extend the encounter and the contact between self, nature, culture, and God. In a way, the places we inhabit and our connection with home are essential to our identity as embodied beings and our entanglement with place (Malpas 2004, 2014).

Thoreau arrived, through his immersive and spiritual experience in the forests of Walden Pond, at a vision of nature rich in textures and nuances; self-generating, mysterious in its infinitude and dynamic, a force that overwhelms one's attempts to make it something standardized and available for human purposes: "Nature is well adapted to our weakness as to our strength. The incessant anxiety and strain of some is a well-nigh incurable form of disease" (Thoreau 1893, p. 205). As Walls (2000, pp. 15–16) emphasizes: "Yet in the nineteenth century, it was entirely possible and even necessary to 'believe' in 'Nature', an affirmation inseparable from Nature's source in a Christian God and from science as the appointed interpreter of God's meaning." In Thoreau's view, the encounter with wildness is what gives rise not only to new modes of perception and awareness, but to a richer view of the self as embodied and rooted in place. As I will argue throughout this chapter, Thoreau was not advocating a naive pantheism; rather his view of the inextricable link between God, nature, and the self can be characterized as a pantheism linked to the experience of inhabiting the Cosmos with simplicity, independence, and magnanimity.

Thoreau's attempts to write about wildness have a further connotation: wildness is not only that mysterious and elusive reality that erodes our attempts to capture experience within the molds of scientific and literary explanation. Wild nature is always interwoven with sociocultural life, technology, and several strands of what we might call "anthropogenic impacts" (Spratt 2018). It is not, then, an "untouched" and entirely "free" or "pristine" nature, but rather a wild nature in which humans and non-humans meet and become entangled. Put differently, experiencing wildness is a way of being and positioning oneself in the face of events and facts that are part of that extremely fluid reality, and the multiple threads that we call "the wild." Far from being ecstatic and available, the wild, as Thoreau conceives it, would rather be a place of learning, attachment, and detachment, simultaneously making contact and distance possible with the eternal "Other."

Wildness is revealed as a complex property with multiple threads, that connects and communicates among the self, nature, and the material world. Thoreau (2017, IX, p. 45) suggests that, with enough attention, we could discover the sacredness of the wild both in our backyard and in the meteorite of Mecca. As he states (Thoreau 1971, p. 89), "Every morning was a cheerful invitation to make my life of equal simplicity, and I may say innocence, with Nature herself." The wild, as Thoreau envisioned it, was at once something distant, pure, and sacred, and something familiar, linked to our experience of the landscapes and places we inhabit and, therefore, an instance capable of being transformed by culture, technology, and history. "This polarity, faraway vs. home, travelling vs. dwelling, global vs. local, seems to assert an either/or dilemma, but Thoreau's solution is to assert both sides of the paradox at once: the deeper you dwell, the farther can be your reach; the wider you roam (even if only in imagination), the stronger must be your roots" (Walls 2011, p. 3). For Thoreau, the simple buzz of a mosquito could have cosmic significance, being a reminder of human vulnerability and fragility, but also of the fertility and potentiality of the world:

> I was as much affected by the faint hum of a mosquito making its invisible and unimaginable tour through my apartment at earliest dawn, when I was sitting with door and windows open, as I could be by any trumpet that ever sang of fame. [...]. There was something cosmical about it; a standing advertisement, till forbidden, of the everlasting vigor and fertility of the world (Thoreau 1971, p. 89).

Thoreau emphasizes that we need "new organs and senses" to pay attention to that wondrous, "invisible reality" that is daily out of reach due to the distresses and distractions of our daily life. Nevertheless, he is not referring to the fact that we should live in fear and trembling of the slightest movement of a butterfly's wings on the other side of the world, but that we should develop *attentiveness* to natural reality and to the social and natural processes and changes that unfold before us. Thoreau proposes that humans can come into contact with the wild to regain our health; he emphasizes that to plunge into the roots of this restorative power of nature, it is not necessary to be divorced from reality but to try to embody a way of life in *voluntary simplicity* (Cafaro 2012).

> Thoreau presents us not just with an idea, but with a practice. By this, we do not mean a prescription for living, which is what Thoreau is popularly regarded as furnishing. A prescription for living is, after all, just another idea. What we find in Thoreau is not a prescription but an enactment (Dustin and Ziegler 2005, p. 18).

The call to discover in the connection with the wild a way for "the preservation of the world" is at the same time a call to lead an unquenchable life focused on the "essential facts" of existence. In the final chapter of *Walden*, Thoreau (1971, p. 309) emphasizes that: "The earth is not a mere fragment of dead history, stratum upon stratum like the leaves of a book, to be studied by geologists and antiquaries chiefly, but living poetry like the leaves of a tree [...] –not a fossil earth, but a living earth." If the earth is not just "history and dead matter" but a living processual reality, in that case, the human being should make the most vigorous possible effort to learn to interpret the earthly book in which he is writing his own life: "Ultimately his theology provided a rationale and a method for seeking how to be wild or 'naturalized' in a civilized world" (Willsky-Ciollo 2018, p. 553).

The call to discover in the connection with wildness a path to the "preservation of the world" is simultaneously a call to lead an unquenchable life focused on the "essential facts" of existence. Through his untiring search to discover new ways of dwelling in the forest consciously, Thoreau embarks on an ecotheological quest whose final purpose is to achieve communion with the Nature of God. In other words, he attempts to embody an unquenchable, deliberate life, guided "by the highest purposes and laws": human beings need "not only to be spiritualized, but naturalized, on the soil of earth" he emphasizes in the final chapter of *A Week* (Thoreau 1900, p. 384).

15.2 The Intimate Interconnection Between God, Nature, and Self

In what kind of nature and what God did Thoreau believe? To what extent are humans "part and parcel of nature?" Finally, what is the pantheistic view that Thoreau espoused? Pantheism is the view that there is no clear dividing line between nature and God (Levine 2009, p. 138). If, however, we suppose that nature and the self are intertwined with and manifestations of the Absolute, then two related questions arise: How do we deal with that nature which is itself a manifestation of the Absolute? and Why would nature be as vulnerable or impotent as it appears to be in the face of human impacts? As Dewey notes, in the pantheism embraced by Spinoza, nature and the Self are both manifestations of the Absolute. In his words,

> The reconciliation of the elements here involved [God, self, and the world] leads to the third stage, where God becomes the Absolute, and Nature and Self are but his manifestations. This is Pantheism, and the viewpoint of Spinoza. Thought and being become one; the order of thought is the order of existence. Now a final unity seems obtained, and real knowledge possible (Dewey 1882, p. 149).

Nevertheless, is this the kind of pantheism that Thoreau advocated? This chapter scrutinizes this triad of concepts, God, nature, and self, and emphasizes how they can shed light on the type of pantheism that Thoreau advocated. A procedure frequently used in the literature to clarify the intricate connection between God, nature, and man is to ask how profound and lasting was Emerson's influence on Thoreau. Although it was decisive in the beginnings of Thoreau's thought, in later years he turned to an original search; and in his mature stage, little remains of the Emersonian vision of God and Nature. As Thoreau says in his journal (Thoreau 2007):

> By taking the ether the other day I was convinced how far asunder a man could be separated from his senses. [...] The value of the experiment is that it does give you experience of an interval as between one life and another—a greater space than you ever travelled. You are a sane mind without organs— groping for organs—which if it did not soon recover its old senses would get new ones. You expand like a seed in the ground.

In this passage, Thoreau evaluates his experiences of detachment and his attempts to overcome the limits of the senses through immersion in a world of tastes, smells, textures, and visions that can only be accessed through the experience of the expanded self. The reference to the "sane mind without organs" suggests a temporary detachment and disconnection from the demands and pressures imposed by the senses (Kopp 1963). The body without organs alludes to a consciousness that experiences itself neither as totally "autonomous" from previous embodied experience nor as devoid of physical and psychic experiences. It is in the interplay of the so-called "mind without organs" where "new organs" and a new type of *vitality* and *sensitivity* emerge, making it possible to live in that inter-space and in that between-time between one life and another. The "mind without organs" plausibly alludes to a transtemporal experience rooted in what Thoreau calls states of "ecstasy" or borderline experiences. Moreover, it is through this religious quest and in the attempt to transcend the limits of the self that he will come to conceive that his greatest desire is to remain alert to discover "God in nature" (Thoreau 1984). Thoreau's ecotheological quest, broadly understood, can be read as an original attempt to discover the interconnections between God, wild nature, and the self as expanded and embedded in place. Thus, it would be wrong to think that Thoreau's search fully coincides with Emerson's.

The other alternative to explore would be to see Thoreau as an advocate of Emerson's ideas, who would continue to write under the Transcendentalist umbrella until the end of his career: "The Transcendental notion of Correspondence, which Emerson called a 'slippery Proteus,' held that nature's signs were not indicative but suggestive; not emblematic but truly symbolic. And Thoreau's perception was that of a symbolist" (Moldenhauer 1959). Thoreau, according to Moldenhauer, would be a symbolist who sees the world through both the Platonic and the Aristotelian lenses:

> Through the cultivation of his 'outward' senses Thoreau hoped to perfect the inner ones which would enable him to grasp the entire range of correspondences between nature and spirit. Armed with an Aristotelian attention to detail but instructed by the Platonic epistemology, he went forth to wrestle with Proteus (Moldenhauer 1959).

In my opinion, however, the most plausible interpretation of these fragments, which sheds a clearer light on the pantheism endorsed by Thoreau, is the view that Thoreau's thought is incompatible with Emerson's. In his work, Moldenhauer (1959, p. 254) summarizes "the epistemological dogma" that Emerson defended in *Nature* in his attempt to find a harmony between his own philosophy and Plato's theory of ideas:

1. Words are signs of natural facts.
2. Particular natural facts are symbols of particular spiritual facts.
3. Nature is the symbol of spirit.

Nevertheless, a comparative reading of these passages suggests that Thoreau breaks away from, and goes beyond, the epistemological dogma of Emerson and the other members of the Transcendental Club. Thoreau's writings show a sustained effort to overcome and update the metaphysical postulates of Transcendentalism in his own original quest. In particular, Thoreau dismisses statements (2) and (3) and defends a negative theology rooted in the longing to see, hear, and experience God in Nature with "new organs" and "renewed senses." Thoreau endorses a philosophy of life characterized as a cosmological panentheism rooted in three basic statements:

(i) The wild natural world is a spiritually significant world.
(ii) "Is not Nature, rightly read, that of which she is commonly taken to be the symbol merely?"
(iii) "To find God *in* Nature".

The wild natural world is a spiritually significant world. There is a natural world to which humans and non-humans are inextricably bound, and this world is connected in some mysterious way with God. Thoreau stresses that a sense of unity, entanglement, and interdependence runs through all life. In Thoreau's vision, sacred, beautiful, divine, good, elevated, and other similar terms are not exclusive attributes of God, but properties that we are effectively discovering rooted in the world, and that constitute the textures, smells, and modulations of wild nature. Wild nature is spiritually significant in yet another sense: it is through it that we humans constitute a sense of home, dwelling, and of having a "seat" (*sedes*) in-the-world. In an instructive passage in *Walden*, Thoreau tells us (1971, p. 81):

> Wherever I sat, there I might live, and the landscape radiated from me accordingly. What is a house but a *sedes*, a seat? —better if a country seat. I discovered many a site for a house not likely to be soon improved, which some might have thought too far from the village, but to my eyes the village was too far from it. Well, there I might live, I said; and there I did live, for an hour, a summer and a winter life; saw how I could let the years run off, buffet the winter through, and see the spring come in.

This vision of conscious dwelling in Thoreau is far removed from our modern vision of welfare rooted in materialistic conceptions of progress, comfort, and a high standard of living. The first two lines of the quote have a strong eco-theological connotation: "Wherever I sat, there I might live." As Schneider atates (1995, p. 100): "Thoreau shows his fellow travelers through life that they have choices (freedom) about how to use their time of which they are unaware. This freedom, if used wisely, can allow them to seek their full spiritual rather than monetary potential."

Consciously dwelling in a place means having a "site;" being linked to a place which is "one's own" seat. The first "seat" and the first "home" in which human beings dwell is the body; from there, we build our sense of self as embodied and vulnerable beings. The second sentence is also especially illuminating: "the landscape radiated from me accordingly," since it refers not only to the landscape that we find around us, to the geological aspects of the concrete environment, but, particularly, to the topology of the self as rooted in the place. The landscape is neither interior nor exterior; instead, it crosses and connects the self and the world. In Thoreau's thought, as Cafaro stresses, a utilitarian and excessively economic view of nature can erode the depth and inherent value of wild nature (Cafaro 2012, p. 85). This is remarkably clear in *Walden*, where Thoreau voices powerful criticism of "Farmer Flint": "Flint's Pond! Such is the poverty of our nomenclature." He adds: "I respect not his labors, his farm where everything has its price; who would carry the landscape, who would carry his God, to market" (Thoreau 1971, pp. 175–176). Farmer Flint is not only ignorant of the deeper value and meaning of nature, including the multi-species relationships among plants and animals that inhabit ponds, rivers, and grasslands, but he is also totally unaware of ecological successions, and how his history is "interwoven" with the history of nature:

> I go not there to see him nor to hear of him; who never saw it, who never bathed in it, who never loved it, who never protected it, who never spoke a good word for it, nor thanked God that he had made it. Rather let it be named from the fishes that swim in it, the wild fowl or quadrupeds which frequent it, the wild flowers which grow by its shores, or some wild man or child the thread of whose history is interwoven with its own (Thoreau 1971, p. 196).

Thoreau emphasizes that the landscapes and places that we inhabit have an intrinsic value, which exists beyond economic considerations; and that value is tied not only to the dynamic relationships between human and non-human life, but also to the story that we create about ourselves as beings entangled with place. In other words, the inherent value of the landscapes that we inhabit is linked to the natural history which is part of a broader network embedded within the history of civilizations. Nevertheless, it should be noted that: "Natural histories describe the natural world yet may fail to appeal to the imagination wherein our emotions, values, and beliefs are forged" (Spratt 2012, p. 158). In short, each little part of the wild world has its own value because it carries a deep truth that echoes through the ages.

Thoreau suggests that, with enough cultivation and a deliberate focus on developing ecological virtues, humans could unravel on their own this truth: that nature is also part of their own history and their own web of life. These truths ultimately show the "essential facts of life," both human and non-human (Sandler 2007; Cafro and Sandler 2010). As Furtak (2023, p. 4) argues: "Thoreau does not introduce an artificial distinction between facts and values, or between primary and secondary qualities. When we perceive sights, sounds, and textures, we are not standing as disembodied consciousness apart from a world of inanimate mechanisms; rather, we are sentient beings immersed in the sensory world, learning the 'essential facts of life' only through 'the perpetual instilling and drenching of the reality that surrounds us.'" (Thoreau 1971, II).

"Is not Nature, rightly read, that of which she is commonly taken to be the symbol merely?" This statement not only undermines any attempt to read Thoreau exclusively as a symbolist, but also highlights how his natural philosophy and his spiritual search can plausibly be interpreted as panentheism:

> A distinction needs to be made between true pantheism, which equates the divine with the totality of the natural world and panentheism which accepts that idea but adds that the divine is both immanent in the world yet still somehow transcendent in some manner. This view allows some "otherness" to God beyond his or her presence within the natural world. In panentheism, God suffuses the world but is not exhausted by the immanence in the world. There is a sense in which the fullness of God goes beyond the material world of nature (Larson 2010, p. 1283).

Thoreau's position here is based on two related axioms: first, God is immanent in the world but not exhausted by it, transcending the natural world in different ways. At the same time, natural facts can only be properly understood by stepping outside the material world, in other words by using things to see what *is* and not what *appears to be* (Thoreau 1971, p. 96). And second, God transcends the world insofar as he or she is not confused with wild nature or exhausted in the world of material facts. However, Thoreau believes that God is only realized "in the present moment," and at the same time he stresses the notion that there is a further goodness and a sacredness in human dwelling in nature (Schneider 2016). In other words, it is not necessary to go to any other place or await the "right time" to see and learn what is noble, beautiful, and good.

Thoreau recognizes that there is something precious and sublime in eternity, but that this eternity is only materialized in the fragility of the present, in the *here* and the *now*: "In eternity there is indeed something true and sublime. But all these times and places and occasions are now and here" (Thoreau 1971, p. 97). And he adds: "God himself culminates in the present moment, and will never be more divine in the lapse of all the ages. And we are enabled to apprehend at all what is sublime and noble only by the perpetual instilling and drenching of the reality which surrounds us" (Thoreau 1971, p. 97). This is the view of God as immanent in the world and at the same time able to transcend it. "As with Emerson, it is the immediacy of the moment that most fully reveals the sacred, and Thoreau understands this not primarily in ethical terms, but in terms of its felt aesthetics. For Thoreau, this attitude opens into a direct sense of presence or beingness, especially in the stillness and power of nature" (Hunt 2003, p. 154).

Thoreau struggled tirelessly throughout his life to find God in nature and to discover a sense of entanglement and interconnection among God, nature, and the self, but his attempt to overcome transcendental symbolism often fails: "In fact, Thoreau does not show a decrease in imagination or abandon metaphor. Instead of using nature as a metaphor for human concerns, in his late 'nature' writings he frequently reverses the process by using human metaphors to apply to nature" (Schneider 2016, p. 17). As Walls (2017, p. 273) emphasizes:

> In *Nature*, Emerson announced that "nature is the symbol of spirit," and natural facts are the materialization of preexisting "Ideas in the mind of God." When Thoreau first read these words, they cracked the shell of Harvard open: every least object in nature signified a hidden life and a final cause. For years it was all he needed. But now, at the end of *A Week*, he turned on Emerson with an anguished plea: "May we not see God? Are we to be put off and amused in this life, as it were with a mere allegory? Is not Nature, rightly read, that of which she is commonly taken to be the symbol merely? [...]. What is it, then, to educate but to develop these divine germs called the senses?"

15.2.1 "To Find God in Nature"

Thoreau struggled assiduously to give narrative form to his experiences of entanglement and emotional connection with the wild nature to which he attributed a spiritual connotation and restorative power. In this part I will return to Thoreau's central tenet of seeking "God in nature," as cited above: "My profession is to be always on the alert to find God in nature, to know his lurkingplaces, to attend all the oratorios, the operas, in nature" (*J*, 4. 53–55). In what sense did Thoreau love and seek God in nature?

One of the central tenets of Transcendentalism was that wild nature "played a central role in religion and human life in general, that nature could be 'read' spiritually" (Botkin 2001, p. 2). Nevertheless, Thoreau's vision of Nature was astonishingly original and followed a different philosophical strand. In Thoreau's vision, the search for God is open to human understanding thanks to the fact that both Nature and God have a certain *sonorousness*. In his *Journal* he states: "Nature always possesses a certain sonorousness, as in the hum of insects, the booming of ice, the crowing of cocks in the morning, and the barking of dogs in the night, which indicates her sound state. God's voice is but a clear bell sound. I drink in a wonderful health, a cordial sound" (Thoreau 1961, p. 24). The prior training required to be able to hear the sonorousness of God and wild nature was to purify the senses and cleanse the soul of mundane distractions, and ensure that the bodily self could get in touch with that other reality that is only accessible to renewed eyes and ears. Thoreau was convinced that unity between God and the world was much closer than Emerson and the other transcendentalists believed (Moore 1932). Ultimately, he devoted himself to showing that "the realm of spirit is the physical world, which has a sacred meaning that can be directly perceived" (Furtak 2023, p. 5). In another entry in his *Journal*, Thoreau (1961, p. 65) insists that his task is to "state facts that they shall be significant." The world is rich in values and meanings that are not human-centered but emerge as we broaden our perspectives and become "part and parcel of the natural world".

Thoreau insistently calls for an effort to renew and expand the self, which implies abandoning the *precarious existence* of "quiet desperation" (Thoreau 1971, p. 8) – in which most people waste their lives – and struggling for *living deliberately life* in *voluntary simplicity*. He tries to discern between those experiences that feed and enliven our existence and those which make it a burden, in order to achieve a deep and meaningful life (Furtak et al. 2012). Thoreau stresses that, to accede to this higher nature suffused by God, what we really need is to make a determined, self-conscious effort to educate our own senses. In a way, Thoreau suggests that we could see God *in* and *within* nature if our senses and instincts were sufficiently pure and healthy to perceive the plenitude and the sonorousness of eternity in the present moment:

> Now or never! You must live in the present, launch yourself on every wave, find your eternity in each moment. Fools stand on their island opportunities and look toward another land. There is no other land; there is no other life but this (Thoreau 2017).

This panentheistic search to discover 'God in nature' was linked to the desire to attain "a purely sensuous life" removed from both sensualism and narrow hedonism:

> We need pray for no higher heaven than the pure senses can furnish, a purely sensuous life. Our present senses are but the rudiments of what they are destined to become. We are comparatively deaf and dumb and blind, and without smell or taste or feeling (Thoreau 1900, p. 425, 387).

Thoreau stresses that "is only necessary a moment's sanity and sound senses, to teach us that there is a nature behind the ordinary" (Thoreau 1900, p. 347, 389). And it is this other life hidden behind things that reflects an indissoluble bond among God, wild nature, and the self. Nevertheless, it is precisely this wounded cry of Thoreau's in *A Week* that reflects a decisive break in his eco-theological search for unity. Thoreau feels "at home" in Walden Pond, but when he finds himself driven to leave the forests, lakes, and paths with which he is familiar, he encounters *another* nature that is much more evasive and unshakable. It is on the Ktaadn excursion that Thoreau finds himself forced to recognize that he has lost the connection with place, leading him to admit that a manifestation of wild nature exists that exceeds his most intimate postulates as to the unity of the Self, nature and God. The view of the expanded self, in connection with place, falls apart, as does the attempt to find God in nature. As Cramer (2009, p. xv) comments in the Introduction to *The Maine Woods*:

Thoreau made his Ktaadn excursion during his second year at Walden Pond, despite his statement at the end of *Walden* that his "second year was similar" to the first. [...]. During this time Thoreau made one of the few early ascents of Mount Katahdin by a non-Native American, finding a primordial landscape in which he felt like an intruder. "For what canst thou pray here," he wrote in his journal, "but to be delivered from here."

15.3 The Dis-connection Between Nature and Self: Ktaadn and the Emergence of Thinking *in* the Place

Thoreau saw nature as a continuous revelation of God; but at the same time, his experiences on Ktaadn reflect the break between God, nature, and the self. The nature that Thoreau found was simultaneously "something savage and awful, though beautiful;" and this second nature was something terrible, portentous, and violent. As he would have to admit: "It was not lawn, nor pasture, nor mead, nor woodland, nor lea, nor arable, nor waste-land. It was the fresh and natural surface of the planet Earth" (Thoreau 2009, p. 63). Certainly this was not the familiar, pristine nature that was called to be our seat and final destination, 'the resting-place for our bones;' what was there, simply was: "It was Matter, vast, terrific,—not his Mother Earth that we have heard of." It was a *terra incognita:* "It was a place for heathenism and superstitious rites, —to be inhabited by men nearer of kin to the rocks and to wild animals than we" (Thoreau 2009, p. 63). Here Thoreau's cosmos becomes disjointed and the lights of *kaos* break up any attempt to seek and preserve unity:

> I stand in awe of my body, this matter to which I am bound has become so strange to me. I fear not spirits, ghosts, of which I am one, —*that* my body might, —but I fear bodies, I tremble to meet them. What is this Titan that has possession of me? Talk of mysteries!—Think of our life in nature,—daily to be shown matter, to come in contact with it,—rocks, trees, wind on our cheeks! the *solid* earth! the *actual* world! the *common sense! Contact! Contact! Who* are we? *where* are we? (Thoreau 2009)

Thoreau is overwhelmed by a nature that presents properties and powers that fill him with bewilderment and grief. On Ktaadn, Thoreau no longer finds this coincidence between the self and wild nature that he discovered in Walden Pond; he becomes aware of the break and the lack of continuity between the self and "the actual natural world." On Ktaadn, Thoreau was forced to reassess whether man really is "part and parcel of wild nature." The attempt to find unity in nature fails in the face of what Newman calls the "irreducible facticity" of the natural world. The fact is that the world of natural processes, as it is presented to us, shows an *excess of potentiality, violence*, and *sudden changes* that shatter human capacity to comprehend and narrate this world in terms of an unbroken unity. As Newman (2005, p. 162) comments:

> Thoreau's retrospective outburst about climbing Ktaadn is so extravagantly fractured that its argument can get lost. It begins with relaxed contemplation of the central Romantic idea, "our life in nature," and then descends rhythmically from the abstract to the concrete until it grounds that idea in the irreducible facticity, the thingness, of the planet. Having found hard ground, it rises again, the rhythm more insistent and confident now, opening outward from the actuality of the nonhuman "earth," to the materiality of the complete ecosocial

> "world," and finally to the patterns of human understanding that bind its communities. The climactic ejaculation —"Contact! Contact!"— occurs only now, with the thought that we live together not only with nature but also with each other, that we, people and the earth, are in material fact a "we."

The description of wild nature in Thoreau's writings is dual. On the one hand, he seeks refuge in a *wild nature* invested with a strong religious signification, to which he has constant recourse to keep himself sane and away from the anxieties of the big cities. At the same time, he defends an *actual nature*, much more fluid and processual, the value of which resides in the actuality, potentiality, and physicality of the natural world "rather than in metaphors for human experience" (Johnson 2009, p. 20). And it is precisely this experience of an absence of correspondence and continuity among God, the self, and nature that triggers what I have called 'the emergence of thought *in* the place.' This is thought not obsessed with the abstract, but based on sensitivity and the ecological consciousness linked to the experience of the self *in* the place.

It is true that "Thoreau never lost sight of the power of poetic language generally, and metaphor specifically, to reveal nature's spiritual truths and its material" (Spratt 2012, p. 159). However, it is also true that Thoreau's experiences of immersion in *altered* environments – that have suffered the *intervention* of human activities, culture, and technology – reflect a thought that places a positive value on disconnection and the need to construct a sense of belonging to and links with the landscapes and places that we inhabit. From the experience of disconnection and his recognition of the break between the self and the natural world grows a new sense of "we" and of the search for new types of *belonging*, *connection*, and *entanglement* between our physical, personal narrative experience in the world and the concrete places and environments that surround us (Schneider 2000; Sellers 1999). In his ascent of Ktaadn, Thoreau corroborates the existence of *another* nature: "Vast, Titanic, inhuman Nature has got him at disadvantage, caught him alone, and pilfers him of some of his divine faculty. She does not smile on him as in the plains" (Thoreau 2009, p. 57). This nature makes manifest the dual experience of a break, but also of a search for new forms of belonging, unity, and connection among God, nature, and the self *in* the place (Pratt 2022).

15.4 Final Remarks: What Kind of Pantheism Did Thoreau Endorse?

In this final section I present additional arguments to defend the position that Thoreau's philosophy can better be interpreted from a panentheist perspective than from a superficial pantheism (Cranwell 2010; Dombrowski 1988; Emerick 2011; Henriksen 2017; Lamm 2010; Levine 1994; Levine 1992a, b; Meister 2017; Murphy 2013; Sprigge 1997; Steinhart 2004; Valera and Vidal 2022). The first idea that I want to stress is that Thoreau's panentheist position is linked to the metaphysic of the extended self, linked to place. Firstly, Thoreau's view was more radical than Emerson's, both in the search for unity among God, nature, and the self, and in his political philosophy. Thoreau argues for a radical change in the self-constitution of the self. In his writings and experiences, he does not simply call for a passing, external change, but seeks to decipher the ultimate, sacred reality manifested here and now. His own life can be seen as an example of corporal-spiritual renewal and of eco-theological search of the self in the place. "But Thoreau's importance as a social reformer and an early spokesman of environmentalism resides more deeply on his metaphysics of the self" (Tauber 2003, p. 98).

Secondly, Thoreau believed firmly in a "divine unifying principle" and in the existence of certain "higher laws" that explained the beauty and unity of the cosmos. If any one part of nature was valuable and sacred, then there must exist a principle of commonality, interdependence, and interconnectedness among God, self, and nature capable of embracing the whole of life, both human and nonhuman.

Thirdly, Thoreau stressed that we habitually assign a value to the wild life around us from a self-centered, egoistical perspective. However, as Arne Naess (1977) noted, a pantheistic environmental ethic such as that professed by Spinoza requires a solid ontological and normative foundation to go beyond narrow anthropocentrism:

> The pantheist's ethics— environmental ethic and ethics more generally—is meta-physically grounded in the divine unity, the unifying principle that accounts for our commonality with other living and nonliving things and the grounds for extending our notion of moral community to those other living and nonliving things. Everything that is part of the divine unity (as everything is) can be thought of as also part of the moral community (Levine 2009, p. 140).

There is a paradox in Thoreau's corporal ecology and his view of the self-linked to the place. In some passages in *Walden* he exalts the sensitive, corporal connection with the places and landscapes in which the embodied self is evidenced, *entangled in* and *nurtured by* the place. In others passages in *Ktaadn*, in contrast, he feels uncomfortable, and in some sense removed from his corporality; there is a sense of "estrangement," of rupture between the self and the wild nature to which he sought to unite himself. As Hodder (2001, p. 305) remarks:

> Indeed, it is just this sense of disjunction—perceptual, imaginative, and spiritual—that distinguishes so much of his art. Whether in the delicate conceit of witnessing himself in the transparent surface of Walden Pond, the unsettling construction of himself in *Walden* as the 'spectator' of his life's dramatic scenes, or his preoccupation with self-transparency in the late journal, this awareness of self as in some sense separate from thought and activity governs his overall religious and artistic outlook.

Analysis of several fragments of Thoreau's *Journal, Walden,* and other writings suggests that it is plausible to interpret his views as a form of panentheism. Panentheism is the view that wild nature and the all-encompassing world are an extension of God or divinity, and most traditional positions within pantheism agree that there is an identity between God, Nature, and the universe. Thoreau, on the other hand, appears to hold that God is more than just the universe. "Panentheism (meaning 'all is in

God'), in contrast, advances a strong connection between God and world but with the important ontological distinction that God transcends the world; that is, the world does not exhaust the divine being" (Emerick 2011, p. 1755). In general terms it is indeed possible to interpret Thoreau's view within the naturalistic pantheism which holds that the entire universe, with all its natural processes, is identical with God or divinity. This position does not necessarily imply belief in a personal God or a supernatural world beyond the physical world of "material facts." It is true that in some passages of *Walden* and the *Journals* Thoreau seems to concede that nature is the "ultimate source of value and meaning" in the universe (Forrest 2016). As Leftow (2016, p. 66) asserts: "Pantheism which eliminates everything distinctively theistic is naturalist. For naturalistic pantheism [...] the universe at its basic level consists entirely of the basic entities of physics, and its laws are only those of physics." Nevertheless, this also does not fit well with Thoreau's vision because his position with respect to the science of his time was clearly critical (Metzger 2006; Walls 2000).

For Thoreau, "spiritual facts" were hierarchically superior to "natural facts;" although they intersect at some point, neither finally excludes the other. The ontological difference endures: "It is more proper for a spiritual fact to have suggested an analogous natural one, than for the natural fact to have preceded the spiritual in our minds" (Thoreau 1961, p. 18). In a broader reading, Thoreau's positions on nature and the divine could be interpreted as panentheism; for example, Thoreau often wrote about the spiritual significance of nature and believed that the natural world could provide spiritual sustenance or nourishment. He stated that "I found in myself, and still find, an instinct toward a higher, or, as it is named, spiritual life, as do most men, and another toward a primitive rank and savage one, and I reverence them both. I love the wild not less than the good" (Thoreau 1971, p. 210). The world of "natural facts" is rich in textures, smells, and sonorousness, and "whatever we have perceived to be in the slightest degree beautiful is of infinitely more value to us than what we have only as yet discovered to be useful and to serve our purpose" (Thoreau 1993, p. 144). According to Corrington (2016, p. 2),

> For the panentheist, the divine retains its ontological perch by being both in and beyond the worlds of nature but in different respects. Panentheism refuses to collapse the distinction between god and nature into one identity, whereas classical theism insists on an absolute ontological difference between creator and creation.

Finally, Thoreau's panentheist position can be posited as a search for unity based principally on a belief in a spontaneous order from which derive certain principles that unify reality and life on planet Earth as a single whole (Peyser 2017). In *A Week*, Thoreau (1900, p. 104, 68) develops an exhaustive critique of Christian theology, which "treats of man and man's so-called spiritual affairs too exclusively." Here he claims that he is not interested "solely in man's religious" or moral issues, or "in man," but in a much deeper search for God and the sacred that can be found in the actual world by a concrete, eco-theological search. God, in Thoreau's view, is immanent in the world and at the same time able to transcend it. Thoreau rejects systematic metaphysics and theology, which he considers too formalist and anthropocentric. At the same time, instead of renouncing religious search, he undertakes an eco-theological investigation based on *ecstatic experience* and with the aspiration of discovering God in nature. As Hodder (2001, p. 132) points out:

> His own 'theology,' if we can use this term for such a subversive religious thinker, was rhetorically invested, pointedly critical, deliberately unsystematic, and experientially fueled. It offered a vision of the world beholden to no tribunal but nature and his own experience. It was, in effect, a kind of negative theology, or what, in keeping with its experiential character, we might call a theology of ecstatic experience.

According to Thoreau, God is not exhausted in wild nature, and is not an instance of the Sacred that we can discover within ourselves; but neither is God exhausted by nor reducible to the human language with which we try to characterize our experience in the world. Thoreau's panentheism is expressed in a "thirst for communion" and "Contact!" to bring God, nature, and the self ever closer to one another; but this religious search is also vulnerable to "fracture" and what we might call the "breakdown" of the self in the world. Ultimately, Thoreau's God always *exceeds* and *surpasses* human reason and understanding. As Hodder (1993, p. 96) says:

> Thoreau was no stranger to religious feeling, nor of religious faith per se; what he opposed was sectarian exclusivism and, to him, the arrogant claim that human language, texts, or institutions could contain ever adequately the splendor of the universe or the mystery of God. […] There is little evidence in his writings that he conceived the sacred in any personal sense, and his overall critique of the limitations of theological language led him to espouse a kind of negative theology grounded in an almost instinctive conviction of the absolute ineffability of divine reality.

One of the vital aspects of Thoreau's panentheism can perhaps be summarized in three simple ideas found in *Walking*, which continue to resonate down to his last writings: "How near to good is what is *wild*! Life consists with wildness. The most alive is the wildest" (Thoreau 1977, p. 22). Wildness appears as a resource whose value is linked not only to human welfare and survival, but to the health of planet Earth as a whole. Thoreau finds a point of connection among human health, the health of wild animals, and the places that we inhabit (Rodríguez 2021). As Thoreau asserts: "A man's health requires as many acres of meadow," […] adding: "To preserve wild animals implies generally the creation of a forest for them to dwell in or resort to. So, it is with man." Every one of these passages "illustrates Thoreau's belief that we rather than nature benefit from our seeking affiliation with it; nature does not need us, but we need it, not only for our material but for our spiritual well-being. (Spratt 2012, p. 162).

Nevertheless, in these and other fragments we find that Thoreau's ecotheological search shows a counterpoint between the two poles of human health and the health of wild life. This meeting point is the perspective of a *hybrid nature*, constructed and modified by human action. Thoreau maintained that there is a space of indecision between the "objective" and the "subjective," between the "inner life" and "external reality;" it is this fold, this *between* that allows a connection between the two forms of care (Specq et al. 2013). On the one hand, we humans need healthy forests and meadows for a broad, deep, significant view of our own existence as vulnerable beings rooted in place. On the other, Thoreau recognizes that we care for and strengthen our health by increasing our affiliation and connection to nature; also that we, like wild animals, need to "create a forest," to construct healthy environments and to preserve the health of Planet Earth to preserve our own health. "Why, 'nature' is but another name for health."

References

Botkin D (2001) No Man's garden: Thoreau and a new vision for civilization and nature. Island Press, Washington D.C.

Cafaro P (2004) Thoreau's living ethics. Walden and the pursuit of virtue. The University of Georgia Press, Athens

Cafaro P (2012) In Wildness is the preservation of the world. Thoreau's environmental ethics. In: Furtak RA, Ellsworth J, Reid JD (eds) Thoreau's importance for philosophy. Fordham University Press, New York, pp 68–90

Cafaro P, Sandler R (2010) Virtue ethics and the environment (Cafaro P, Sandler R, eds.). Springer, New York

Corrington R (2016) Deep pantheism: toward a new transcendentalism. Lexington Books, Lanham

Cranwell C (2010) Embracing thanatos-in-eros: evolutionary ecology and panentheism. Sophia 49(2):271–283

Dewey J (1882) The pantheism of Spinoza. J Specul Philos 16(3):249–257

Dombrowski D (1988) McFarland, pantheism and panentheism. Hist Eur Ideas 9(5):569–582

Dustin C, Ziegler JE (2005) Walking: Thoreau's prepared vision. In: Dustin C, Ziegler JE (eds) Practicing mortality: art, philosophy, and contemplative seeing. Palgrave Macmillan, New York, pp 17–48

Emerick C (2011) Pantheism and panentheism. In: Kurian GT (ed) The encyclopedia of Christian civilization. Wiley-Blackwell, Hoboken

Forrest P (2016) The personal pantheist conception of god. In: Nagasawa Y, Buckareff AA (eds) Alternative concepts of god. Oxford University Press, Oxford, pp 21–40

Furtak RA (2023) Henry David Thoreau. In: The Stanford encyclopedia of philosophy. Available at: https://plato.stanford.edu/archives/spr2023/entries/thoreau/

Furtak RA, Ellsworth J, Reid JD (2012) Thoreau' s importance for philosophy. Fordam University Press, New York

Henriksen J-O (2017) The experience of god and the world: Cristianity's reasons for considering panentheism a viable option. Zygon 52(4):1080–1097

Hodder A (1993) Thoreau's religious vision. Ultim Real Mean 26(2):88–108

Hodder A (2001) Thoreau's ecstatic witness. Yale University Press, New Haven

Hunt H (2003) Lives in Spirit: precursors and dilemmas of a secular Western mysticism. SUNY Press, Albany

Hunt L (2015) The philosophy of Henry Thoreau: ethics, politics, and nature. Bloomsbury Academic, London

Johnson R (2009) Passions for nature: nineteenth-century America's aesthetics of alienation. University of Georgia Press, Athens

Kopp CK (1963) The mysticism of Henry David Thoreau. Pennsylvania State University Press, State College

Lamm JA (2010) Romanticism and pantheism. In: Fergusson D (ed) The Blackwell companion to nineteenth-century theology. Blackwell Publishing, Hoboken, pp 165–186

Larson P (2010) Pantheism. In: Leeming DA, Madden K, Marlan S (eds) Encyclopedia of psychology and religion. Springer US, Boston, pp 663–664

Leftow B (2016) Naturalistic pantheism. In: Nagasawa Y, Buckareff AA (eds) Alternative concepts of god. Oxford University Press, Oxford, pp 64–88
Levine M (1992a) Pantheism, substance and Unity. Int J Philos Relig 32(1):1–23
Levine M (1992b) Transcendence in theism and pantheism. Sophia 31(3):89–123
Levine M (1994) Pantheism: a non-theistic concept of deity. Routledge, London
Levine M (2009) Pantheism. In: Callicott JB, Frodeman R (eds) Encyclopedia of environmental ethics and philosophy, vol 2. New Macmillan, New York, pp 138–142
Malpas J (2004) Place and experience. A philosophical topography. Cambridge University Press, Cambridge
Malpas J (2014) Rethinking dwelling: Heidegger and the question of place. Environ Architect Phenomenol 25(1):15–23
Meister C (2017) Ancient and contemporary expressions of panentheism. Philos Compass 12(9):1–12
Metzger C (2006) Thoreau on science. Ann Sci 3790(1956):206–211
Moldenhauer JJ (1959) Images of circularity in Thoreau's prose. Tex Stud Lit Lang 1(2):245–263
Moore JB (1932) Thoreau rejects Emerson. Amer Lit 4(3):241–256. https://www.jstor.org/stable/2919880
Murphy M (2013) The emergence of evolutionary Panentheism. In: Biernacki L, Clayton P (eds) Panentheism across the World's traditions. Oxford University Press, Oxford, pp 177–199
Naess A (1977) Spinoza and ecology. Philosophia 7(1):45–54
Newman L (2005) Our common dwelling. Henry Thoreau, trascendentalism, and the class politics of nature. Palgrave Macmillan, New York
Peyser T (2017) Henry David Thoreau on spontaneous order and its enemies. Mosaic Interdiscip Crit J 50(4):1–16
Pratt SL (2022) Lessons in place: Thoreau and indigenous philosophy. Metaphilosophy 53(4):371–384. https://doi.org/10.1111/meta.12563
Rodríguez J (2021) Interdependence and vulnerability: a contribution to the one-health approach from the Arne Naess's philosophy. Australian Catholic University. Available at: https://acure-searchbank.acu.edu.au/item/8y091/interdependence-andvulnerability-a-contribution-to-the-one-health-approach-from-the-arne-naess-s-philosophy
Sandler R (2007) Character and environment: a virtue oriented approach to environmental ethics. Columbia University Press, New York
Schneider RJ (1995) Walden. In: Myerson (ed) The Cambridge Companion to Henry David Thoreau. Cambridge University Press, pp. 92–106. https://doi.org/10.1017/CCOL0521440378
Schneider RJ (2000) Thoreau's sense of place. Essays in American environmental writing. University of Iowa Press, Iowa
Schneider RJ (2016) Civilizing Thoreau: human ecology and the emerging social sciences in the major works. Camden House, Rochester
Sellers C (1999) Thoreau's body: towards an embodied environmental history. Environ Hist 4(4):486–514
Smith L (1985) "Walking" from England to America: re-viewing Thoreau's romanticism. N Engl Q 58(2):221–241
Specq F, Dassow Walls L, Granger M (2013) Thoreauvian Modernities. Transatlantic Conversations on an American Icon. University of Georgia Press, Athens
Spratt S (2012) "To find god in nature": Thoreau's poetics of natural history. Mosaic: An Interdiscip Crit J 45(1):155–169
Spratt S (2018) Thoreau and the Capitalocene. Metacritic J Comp Stud Theory 4(2):20–34
Sprigge TLS (1997) Pantheism. Monist 80(2):191–217
Steinhart E (2004) Pantheism and current ontology. Relig Stud 40:63–80
Tauber AI (2003) The philosopher as prophet: the case of Emerson and Thoreau. Philos Contemp World 10(2):89–103
Thoreau H (1893) The Maine woods: the writing of Henry David Thoreau, vol III. The Riverside Press Cambridge, Cambridge

Thoreau H (1900) A week on the Concord and Merrimack Rivers. Thomas Y. Crowell & Company Publishers, New York

Thoreau HD (1984) The journal of Henry David Thoreau (Torrey B, Allen FH, eds). Peregrine Smith Books, Salt Lake City

Thoreau H (1961) The heart of Thoreau's journals. Dover Publications, New York

Thoreau H (1971) Walden. Princeton University Press, Princeton

Thoreau H (1977) Walking. In: The portable Thoreau. Penguin Random House, New York

Thoreau H (1993) Faith in a seed: the dispersion of seeds and other late natural history writings. Island Press, Washington D.C.

Thoreau H (2007) I to Myself: An Annotated Selection from the Journal of Henry D. Thoreau. Yale University Press, New Haven and London

Thoreau HD (2009) The Maine Woods (Cramer JS, ed.). Yale University Press, New Haven & London

Thoreau H (2017) Thoreau's journals by Henry David Thoreau. Delphi Classics, Hastings

Thorson R (2019) Thoreau's great insight for the Anthropocene: wildness is an attitude, not a place. The Conversation. Available at: https://theconversation.com/thoreaus-great-insight-for-the-anthropocene-wildness-is-an-attitude-not-a-place-113146

Valera L, Vidal G (2022) Pantheism, Panentheism, and Ecosophy: getting Back to Spinoza? Zygon 57:545–563

Walls LD (2000) Believing in nature: wilderness and wildness in Thoreauvian science. In: Schneider R (ed) Thoreau's sense of place: essays in American environmental writing. University of Iowa Press, Iowa, pp 15–27

Walls LD (2011) Henry David Thoreau: writing the cosmos. The Concord Saunterer 19(20):1–21

Walls LD (2017) Henry David Thoreau: a life. The University of Chicago Press, Chicago

Willsky-Ciollo L (2018) Apostles of wilderness: American Indians and Thoreau's theology of the wild. N Engl Q 91(4):551–591

Chapter 16
Genesis 1 as Ecosophy

Konrad Ott

In memoriam Christof Hardmeier

Abstract In the Apron-Model of Deep Ecology (Naess A. The deep ecological movement. Some philosophical aspects. In Sessions G (ed) Deep ecology for the twenty-first century. Shambhala, Boston, pp 445–453, 1995), the supreme layer is open for a multitude of "ecosophies." As Arne Næss has argued, such ecosophies must be different from scientific and technological outlooks on nature which are among the causes of our ecological crisis. Ecosophies can find roots in ontologies, but also in religious traditions. From the perspective of discourse ethics, Jürgen Habermas has argued that religions are storehouses of moral wisdom which should be translated into the secularized moral and political parlance of modern societies. The monograph "*Naturethik und biblische Schöpfungserzählung*" by Christof Hardmeier and Konrad Ott combines both ideas. It outlines an ecosophy being based on a rigorous reading of the narrative of Genesis 1 which is connected to the universe of discourse in environmental ethics. The reading of Genesis 1 in Hebrew language shows that Genesis 1 is not dominionistic at all. It tells that humans being blessed are entitled to take the earth under their feet as representatives of a divine spirit. Key of Genesis is "hinne!" which might be translated as "Look! How great!" which constitutes basic attitudes towards nature, as excited and enthusiastic admiration. From such attitude, specific "biophilic" outlooks on fertile earth, plants, and the animals of land, air, and sea can be derived. In this chapter, I present the core claims of such Biblical ecosophy being translated into environmental ethics. Finally, I propose to improve the apron-model.

Keywords Discourse ethics · Deep ecology · Hebrew bible environmental ethics · Ecosophy

---

K. Ott (✉)
Department of Philosophy, Christian-Albrechts Universität zu Kiel, Kiel, Germany
e-mail: ott@philsem.uni-kiel.de

L. Valera (ed.), *Pantheism and Ecology*, Ecology and Ethics 6,
https://doi.org/10.1007/978-3-031-40040-7_16

16.1 Introduction

This chapter is based on the previously published book by Hardmeier and Ott (2015, with further references) giving the essential findings and claims of the monograph in a nutshell. Methodologically, the monograph rests on Jürgen Habermas's proposal that secular and religious citizens should both engage in the practice of translating religious claims into discursive speech (Habermas 2005). The proposal supposes that much human wisdom has been stored in religious and theological traditions which might be of ethical relevance for modern and even post-modern humans. The burden of translation initially falls upon religious persons, but secular persons should be attentive to what religious persons have to say. Without such religious legacy being translated, even secular moral discourse may impoverish conceptually, since religious traditions include ideas about how to live a mortal life, how to approach nature as a more-than-human world, which attitudes to adopt as virtues and standards of decency, and so on. The methodological idea of discursive translation has been applied to the narrative of creation in the Hebrew Bible: *Genesis 1*. Such translation has two dimensions: (a) a close reading of this entire narrative, and (b) bridging such reading to environmental ethics. Such re-reading of Genesis refers to the general environmental discourse within civil society, but it also refers to the particular discourse on ecosophies within the deep ecology movement.

From the perspective of deep ecology, the new reading of Genesis 1 is presented as candidate for an ecosophy to be placed at the supreme layer within the apron-scheme. As Næss has clearly stated there can and should be a plurality of ecosophies at this layer beside Næss's own "ecosophy T." From the philosophical traditions, Spinoza, Schelling, Whitehead, and late Heidegger may be regarded founding fathers of ecosophies, while Buddhism, Jainism, Daoism, and variants of shamanism are "ecosophical" religious traditions. I take the term "ecosophy" as denoting non-scientific and non-technological approaches to nature. The supreme layer of the apron-scheme is, however, not the supreme layer of human thought. Thus, a proposal for an amendment is made in Sect. 16.5 of this chapter.

Right from the scratch, it must be noted that there is no direct competition between Genesis and Darwinism. Genesis 1 is a utopian outlook on nature "as (if it were)" very good creation. Such outlook provides a specific role to the human being (Sect. 16.3). Darwinism and modern genetics refer to processes of gene-driven evolution which were beyond the minds of archaic Hebrew humans. It is, however, a simplification to suggest that one must choose between the Bible and Darwinism. One can have it both.

16.2 The Dominionistic Reading

From the Biblical tradition, the elder "paradise" narrative of creation looks more appropriate than Genesis 1 because it seems to constitute a theocentric stewardship-approach while Genesis 1 seems to demand that humans should "subdue" nature.

To Tirosh-Samuelson (2017, p. 62), humans should "till and protect" the garden "not as controlling managers, but as loving gardeners." To many scholars, it looks counterintuitive to interpret Genesis 1 as an ecosophy. Hardmeier and Ott (2015) adopt this burden of proof. They presume not just to give a "green" reading of Genesis but to come close to this narrative in its original meaning. This presumption was based on the expertise of Christof Hardmeier with archaic Hebrew language. We argue that the original Hebrew meanings have been distorted by Latin metaphysics and Roman imperial mindsets (Groh 2003).

We concede that the intellectual history of Genesis 1 was about "*dominium terrae*." The standard "imperial" reading starts with the Latin tradition of Christianity. There were three core assumptions (White 1967).[1]

(i) Only humans are created in the image of "God." Thus, they are superior to all other creatures as they participate in God's spiritual, intellectual, and creative properties. From the Hebrew-Christian tradition, a denial of human superiority (Taylor 1986) seems impossible. Thus, the modern secular anthropo-centrism is grounded in Biblical theo-centrism.
(ii) The superiority-status is connected to the imperative to "reign" and "subdue" nature, which is not just a permission, but a divine command. This command is repeated at the end of the story about Noah.
(iii) Nature is not sacred to the Bible. There are no sacred mountains, sacred groves, or sacred rivers etc. All divinity vanishes from nature within monotheism.

These core assumptions are secularized in modern philosophy, as in Descartes, Bacon, Kant, and Marx. Here, the divide between culture and nature originates. Thus, modern secular thinking operates in the intellectual horizon of the "imperial" reading of Genesis 1. Worster (2017, p. 353) quotes John Winthrop who in 1630 legitimizes the conquest of North America with Genesis 1 and the narrative of Noah.

Hardmeier and Ott (2015) argue that a stewardship-approach is misleading if it is based on the paradise-narrative. The divine demand that Adam should cultivate and sustain the fruit-tree garden is restricted to the innocent life in "paradise" (garden with fruit trees). After the so-called "fall" the human couple is expelled from the garden and nature is *cursed*. Out of Eden humans have to work hard in the fields, have to combat pests and weeds, and nourish themselves by cereals until they die. Giving birth is painful to females. The original stewardship approach vaporizes within a struggle against a hostile nature. Fall, curse, and hard life constitutes the human predicament. The decisive difference between Genesis 1 and Genesis 2 and 3 is the following: While in the paradise-story nature is cursed, it is blessed in the Genesis 1. Let's take a closer look.

[1] This standard reading was, from the perspective of intellectual history, presented by Lynn White (1967) in his seminal essay published in the scientific journal *Science*. It is also notable that German philosopher Ludwig Klages (1913) blamed Christianity for the destruction of planetary nature before the first world war.

16.3 Coming Close to Genesis

The first line of Genesis 1 is a headline over a complex narrative: "In the beginning God created the heaven and the Earth." The narrative follows this headline.

The original state of the world is maximum entropy: no order, no distinct entities, no structures, no colors. The Greek word "chaos" refers to the even more radical *tohuwabohou* in the Hebrew Bible. It means an abyss of nothing-ness, empty-ness as in a final triumph of the second law of thermodynamics. Such state was referred to as *ex nihilo* in Latin. "And the Earth was without form, and void, and darkness was upon the face of the deep" (Gen 1:2).

Then, however, a difference occurs. Neither reason nor explanation is given why a divine spirit ("God"[2]) moved upon the abyss of nothingness. The divine spirit is a force calling entities into existence by speech acts: "Let there be x." By divine speech acts an ordered lifeworld[3] emerges stepwise. The basic orderings are full of goodness. All acts of God are affirmed as "being good:" "it *was* good" (Gen 1:9, 12, 18, 25). Goodness within creation is, in modern parlance, ontological. The divine spirit is not just pleased by his doings but made them "good." It takes no wonder that "Biblical" inspired philosophers as Hans Jonas (1973) and Holmes Rolston (1988) both wished to defend an ontological axiology against secular individualistic axiology, based on preferences.

The term "day" means "period" and the six periods of creation form the narrative. Day and night, dry land and seas, sun and moon are more physical orders. The archaic cosmology of a division of waters is left aside. From physical orders the divine spirit moves to the bio-spherical ordering. Note that the divine spirit does not create plants him/her/itself. "Let the earth bring forth grass" (Gen 1:11). A fertile earth brings forth the spheres of the plants which, according to the Hebrew thought, are not living beings yet. Plants cover the dry land and prevent it from becoming a desert. A productive earth and its plants constitute habitats for living beings. Three realms (spheres) of life come into being: waters, air, and dry land. Like the air, the waters "bring forth" abundant marine life. Fishes and other marine animals are creatures "that hath life" (Gen 1:20). To animal life, the capacity of active movement is essential. The divine spirit blesses water-borne life: "Be fruitful, and multiply, and fill the waters in the sea" (Gen 1:22). Marine life is seen as abundant. "To fill" means that there is no sea where marine life is absent. Together with sea-life "winged fowl" appear which fill the airy realm of the skies. "Winged fowl" fill the air but multiply on earth, nesting. It can be abundant if the air is full of birds.

Not only the sea and the sky, but also dry land ("earth") brings fourth living creatures "after his kind." Terrestrial animal life is divided in three large categories: cattle, beasts, and creeping things. It seems strange, that domesticated animals ("cattle") are created before humans are present. The difference between beasts and

[2] "God" is neither a concept nor a name. It is an "instance" that reveals as such, as in a burning shrub (Exodus 3:2, 14) or in a still small voice (1 Kings 19:11–12).

[3] "Lifeword" understand as nested spaces that enable animal and human life: livable spaces.

cattle is of relevance to the archaic mindsets of hunters and herdsmen. The category of creeping things entails insects, spiders, bees, dragonflies, worms, and other groups of species ("after his kind"). Species are seen as natural kinds.

There are special motives in the Genesis narrative, as abundance and fertility of living beings. The periods of creation bring about a biosphere which intrinsically runs counter to the Second Law of Thermodynamics. Life is fighting entropy and it has done so for billions of years. Here, we can see that the environmental ethics given by Holmes Rolston (1988) comes very close to a Biblical environmental philosophy. To Rolston, there is "objective" ontological value in nature. Moreover, nature is "pro-jective," being full of wonderful projects. To Rolston, fertility is a basic ontological-dispositional goodness. Fertility and diversity are two sides of the same coin.

After having made animals, the divine spirit talks to itself. "Let us make man in our image, after our likeness, and let them have dominion" over all other animals. Human beings were created male and female which indicates original equality between male and female members of humankind.[4] There is a great difference between both narratives of creation, since in the paradise-narrative the female stems from the skeleton of the male and is made in order to assist the male. Since, according to Genesis 1, females are created in the image of divine spirit, it is beyond body, sex, and gender. The archaic patriarchic mindset regards God as a "He," but "he" is not male.

Without doubt, the Hebrew Bible gives exceptional status to the human being (Psalm 8). Hebrew thought does not deny human exceptionalism, but takes it seriously. To be created in the image of God means that all humans have to behave as representatives of "God" on Earth. If there is ontological goodness in "creation," a representative of the original creator won't be disrespectful against such goodness. To be created in the image of God means to act as if to show reverence, give testimony to, and praise such ontological goodness. Humanity, dwelling on Earth, should be grateful to live within such goodness.

The second blessing refers to the human beings, but, perhaps, it also refers to the fertility of animal life. In such blessing the order is given: "Be fruitful, and multiply, and replenish the earth, and subdue it, and have dominion over the fish of the sea, and over the fowl of the air, and over every living thing that moveth upon the earth" (Gen 1:28). Here, we touch the core of the dominionistic reading of Genesis 1 which has been pervasive in the history of Christianity since Roman times.

The two Hebrew verbs *kabash* (subdue) and *radah* (dominate) have, as most Hebrew words, different meanings according to different spheres of human life. Verbal meaning must be contextualized (Neumann-Gorsolke 2004). *Radah* means reigning and denotes archaic ideals of "good" kingship. A "good" reigning authority takes care of its subjects, it protects and assists them. The image of the good shepherd (Psalm 23) also applies. Sometimes, the reigning authority acts on behalf

[4] Our current debates on LGTBIQ are beyond the horizon of the Hebrew modes of thought. In Hebrew thought there is the binary condition between males and females of all species, as the story of Noah shows.

of its subjects, but it also keeps them in order and controls them. Humans are entitled to combat beasts as they intrude into the human dominion (gardens, fields, cattle). The human way of life is seen as terrestrial life, mainly farming and herding.

The decisive word *kabash* has a violent meaning in the context of warfare. In this context it means to trample down a beaten enemy and step with one's foot upon his crest. Therefore, *kabash* was translated into Latin as "trampling down." In the context of vine-making, however, *kabash* means to treat harvested berries with bare feet in order to make the juice run. Both meaning refer to the basic core meaning of *kabash* which refers to the noun *kav* which is "foot." Here, it is of importance to understand properly the context of a blessing. To be blessed means to the Hebrew mind "to have life at fullest degree" and "to be flourishing alive." A blessing is a firm promise for a "good" life at "good places" and on "good routes" (Psalm 23). Here, Hardmeier and Ott (2015) assume that many correspondences hold between Genesis and the Book of Psalms, especially Psalms 8, 23, 90, and 104. In the context of such divine blessing, humans are entitled to *take the earth under their feet*. That's exactly what humans did, as they moved into all continents except Antarctica. *Kabash* refers to nomadic ways of life, but also to practices of settling. To take the earth under one's feet in the image of the divine spirit ("God") means that at each and any place at which humans beings arrive they *should act as genuine representatives of the original "creator."* If so, God itself would be re-presented inside creation. To act as spiritual representatives is incompatible with "subduing" in the dominionistic reading. If one wishes to keep the concept of stewardship, it should be based on the role of blessed representatives feeling responsible for original goodness.

Such human arrival on many places requires proliferation. From the meaning of "replenish" all our current ideas of "overpopulation" must be abstracted away. At the times, when Genesis was composed in Babylonian exile, world population was roughly 100 million humans (Scott 2017), most of them living in the fertile crescent, territories surrounding the Mediterranean, Indus valley, and parts of China. Large areas were highly sparsely populated or completely devoid of humans. Replenishment means that humans should settle down at many places *as* representatives of the divine spirit. If so, the Earth would be filled with living representatives of a divine creator. This is a spiritual image full of plenty, beauty, worship, praise, and joy.

Verse Gen 1:29 denotes at a more plant-based diet of the human being. The divine spirit tolerates meat eating but does not recommend it. In archaic times, cattle were both wealth and meat. Only occasionally, as in sacrifices, at festivities, or to honor guests, cattle (mainly sheep and goat) are slaughtered and consumed. In archaic realities, daily foraging mainly refers to plants. A plant-based diet includes olives, fruits, and vine (Psalm 104). Beside nourishment, there is oil to balm the body and skin. Humans become strong via cereals, cheerful via vine, and beautiful via oil (Psalm 104).

Gen 1:31 is crucial: "And God saw everything that he had made, and *behold*, it was very good." At the end of the sixth period of creation, the divine spirit looks back. But the "behold" indicates that the judgement of goodness is different from the statements before "and God saw that it was good." In the Hebrew, the "behold" is a specific marker: *hinneh*. This marker indicates: "Look! Be attentive!" There are reasons to interpret that the *hinneh* is not a speech act performed by the divine spirit, but by a human being which looks at the results of the six periods of creation with God's eyes, taking the role of a representative. The *hinneh* indicates a fusion of horizons: The many manifestations of ontological goodness are confirmed with high enthusiasm by the human being who is entitled to have a blessed life within such goodness. Such confirmation takes the performative status of praise. The meaning of "behold" is: "*Just look! How very good! How wonder-full! Simply great! Incredible!*"[5]

Finally, Hardmeier and Ott press their reading into the following phrasing: "Humans being blessed are entitled to take the earth under their feet as representatives of 'God,' being fully attentive to the many ontological goodness of life on earth." A world which is praiseworthy without end should not be impaired, devastated, polluted, spoiled by representatives.

The seventh day is blessed and sanctified in its own right. From a lifeworld temporal perspective, it constitutes a "week" which is divided 6:1. The human being should take a break to be used for praising, relaxing, making love, spiritual recovery, and embodied recreating. Humans who honor the sabbath, will be rewarded by doing so: They live a 1:6-fraction of their life like in paradise. Thus, Hebrew wisdom runs counter to the many tendencies to remove the 6:1-structure and replace it by 7/24. From a spiritual perspective, however, the divine spirit *retires*. The project which was started in the initial move upon the abyss of nothingness has been completed at the end of the sixth day. There is nothing to do any more. The plants will grow, the animal will move and proliferate (and evolve). This retirement was speculated as *Zimzum* in the Jewish Kabbalah (Schulte 2014) and has been adopted by Hans Jonas (1992). To Jonas, God has retired completely and, by doing so, has passed on the predicament of God's work to the representatives. By retirement, God becomes absent. God is present in absence only. Since then, "we" humans are fully responsible for the fate of land and sea, water and air, plants and animals, and, last but not least, for the fate of humankind itself. If humans act as representatives they can make an absent spirit present within nature. By doing so, humans continue "creation." If they don't act as representatives there will bring about much evil.

16.4 Secular Environmental Ethics

If such reading of Genesis 1 can be defended within the discourse on the Hebrew Bible, this narrative resonates with many motives of contemporary environmental ethics. It denies the denial of human superiority (Taylor 1986) since it sees humans

[5] It makes sense to compare this marker with the brief sentences which, according to Næss, open the system of norms and hypotheses (Næss 1989, pp. 196–210). To Næss, the sentence "Let there be light!" is an archetype of such sentences.

as, indeed, somewhat exceptional beings which are morally responsible for their actions in ways other beings are not. Humans are "logical" beings which are able to act according to reasons being given in discourse. Plants are able to exchange information by molecules, animals are able to communicate, but only humans can exchange speech acts and argue for or against validity claims. This is true as well for discourses about how humans should and should not interfere into nature, how they should treat (or abolish) domesticated animals, and how they should take care for future generations, being the offspring of procreation. Discourse ethics and the Hebrew Bible share human exceptionalism, but not human destructiveness.

Meanwhile, environmental ethics is an established field in practical philosophy. I distinguish four approaches in environmental ethics. The first "classical" approach searches for the "correct" solution of the demarcation problem and it builds obligations and virtues on this solution (Taylor 1986). The second approach is environmental pragmatism that relies on ideas about sustainability and pleas for reforms in our practical interactions with nature (Norton 2005). The third approach is "deep ecology" (Næss 1989). The fourth approach is postmodern environmental philosophy. Prominent proponents of this fourth approach are Morton (2016) and Haraway (2016). Biblical thinking is an uncommon combination of pragmatism and "deep ecology." It recognizes nature as praiseworthy wonderland but it also recognizes that humans must interfere into nature and struggle for life by foraging.

After decades of environmental ethics discourse, some essential patterns of reasoning have been identified and mapped (Rolston 1988; Kellert 1997; Krebs 1999; Ott 2010; Muraca 2011). My original idea is to reconcile discourse ethics (Habermas 1981) and environmental ethics (Hendlin and Ott 2016; Ott 2020). The core of environmental ethics is a logical-discursive space of reasoning which can be analyzed as specific normative discourse (Taylor 1961). Such space is mainly secular, but it can be opened for "religious" reasons if such reasons have passed a Habermasian translation and won't simply reenchant nature. The main lines of environmental reasoning are given in a nutshell.

16.4.1 Metabolic and Reliant Values

The *first* pattern argues that human systems profoundly rely on natural systems which provide many different resources, goods, and services. Reliance upon nature, however, differs according to spatial scales and degrees of substitutability. In (post-) industrial and urbanized societies, reliance is often overlooked. Environmental ethics is critical of such forgetfulness. Humans have no alternative but to organize their metabolism with nature. The global societal metabolism has intensified since the Neolithic age until the great acceleration of the present age. Basic Neolithic achievements paved the way towards the full-blown Anthropocene. Such achievements have been urban settlement, agriculture, plows, nets, techniques of domestication and breeding, storage, crafts, and medicine. The modern age can be seen as a dialectical turn from qualitative to quantitative achievements. It is clever to catch fish with nets, but now the nets have become massive in size, catching species ever further down the marine food web, influencing the evolution of fish species.

The dialectics of metabolic values is the tension between expansion and restriction. How can we (and who are "we?") downsize the excessive industrial metabolism ("degrowth") without impairing the entitlements and aspirations of humans to live better lives? Most environmental ethicists claim that sufficiency in lifestyles is key. Only through the majority of humankind living more sustainable lifestyles will we succeed in keeping within planetary boundaries. If so, a voluntary sufficient lifestyle is a mandatory moral aspiration. To the Hebrew Bible, modest and even frugal lifestyles are rewarded by much embodied spiritual goodness in life.

16.4.2 Eudemonic Values

The *second* line of reasoning in environmental ethics is related to the human capacity for pleasure, joy, amazement, bonding, and even bliss and awe. These are the so-called *eudemonic* values, our capacity to enjoy nature's beauty, a (deeper) sense of place (*Heimat*), recreation, joyful embodied practices, biophilic sensations, and spiritual encounters with nature. Here, nature is valued as a substantial and essential dimension of a good, flourishing, and meaningful human life.

Eudemonic values constitute a "deep" anthropocentric environmental ethics (Ott 2016), but they also open the gate to the phenomenology of nature and anthropology. Phenomenology sharpens our embodied senses and attentiveness. Eudemonic values also refer to outdoor activities that humans perform for their own sake, like hiking, sailing, diving, climbing, and even hunting. If one goes hiking for the sake of hiking, one gives intrinsic eudemonic value to this activity. While hiking, the trail itself matters. I leave it as an open question whether the biophilia hypothesis (Wilson 1984; Kellert 1997) may serve as an evolutionary explanation of this sphere of values. The hegemonic doctrines of modernity have concealed our biophilic dispositions, while an ecological civilization will liberate them. From an environmental ethics perspective, the age called the Anthropocene *should* be an age of attentive biophilic humans wishing to restore nature as joyful focal practice (Borgmann 1984) or even restore nature for its own sake. They *should* also bring up children so that they become adults who appreciate nature's contributions to a flourishing human life.

16.4.3 Future Ethics

The *third* pattern of reasoning is related to the responsibility towards future generations with respect to *both* metabolic and eudemonic values. This pattern of reasoning is critical against the promise of a growth-driven economy claiming that future generations will be better off because of GDP (Gross Domestic Product) growth and technological innovations. Future ethics is about fair legacies on different scales.

If the chain of generations implies a basic egalitarianism between generations, one may adopt a *comparative* standard. Such standard demands that members of future generations should roughly have the same living conditions as the current generations. Since contemporary humans are highly unequal in many respects, the comparative standard becomes almost impossible to apply on a global scale. Rather, one should adopt an absolute standard as threshold of a dignified human life. On particular scales, however, political communities (states, nations) should pursue the strategy of bequeathing legacies based on comparative standards to its future members. Conservation, preservation, and restoration of nature are never completely universal in scope but will always remain particular, located in specific sites and territories. Meeting an ambitious comparative standard is, however, constrained by moral commitments not to live at the expense of others, especially not at the expense of the poor and vulnerable ones. Yet, another conflict-laden dialectic occurs. If intertemporal justice based on absolute standards can't be restricted to future people, ignoring contemporary poverty and misery, and if some states safeguard a high comparative environmental standard for their own current and future populations exclusively, then the demands of global justice will, necessarily, put high pressure on such comparative standards. This moral dialectic will also affect immigration policies because wealthy states with ambitious environmental (and social) policies will become attractive destinations to migrants.

This line of reasoning implies "strong sustainability" (Daly 1996; Ott 2014) in conjunction with a Leopoldian land ethics. Daly (1996, ch. 14) tried to ground his concept of strong sustainability in a Biblical ethics, but remained at the surface of both secular ethics and Biblical theology. One can do better. Leopold's idea of an "ethics of land" was not, as Callicott (1980) argued, ecocentric. The famous statement, that a thing is right, "if it tends to preserve the integrity, stability, and beauty of the biotic community" (Leopold 1949) can be upgraded into contemporary parlance and taken as yardstick of dealing with land: "Interfere with land and sea only in such ways as to preserve and restore fertility, resilience, and diversity of ecological systems." This yardstick can be specified to a system of rules, including a "constant natural capital rule" (Ott and Döring 2011), a rule to invest in degraded stocks of natural capital (restoration), rules for maintaining all kinds of ecosystem services (Ott and Reinmuth 2021), and even more specific concepts for fields of environmental policy making.

16.4.4 Inherent Moral Value

The *fourth* pattern of reasoning is about *inherent* moral values (IMV), as attributed to natural beings. IMV belongs to the dimension of deontology. Often, environmental ethics was conceived as a two-step enterprise: first present a convincing solution to the IMV problem, and then derive a theory of nature conservation. This approach is "monistic" since it supposes that there must be the "true" (or "best") solution to the demarcation problem. This creates a dialectics where, on the one hand, the concept of a problem is pragmatic. In ethics, we must suppose that moral problems can be resolved as other problems: by way of solution.

The scope of presumptive solutions to the IMV problem can be determined as follows: (a) sentientism, (b) zoocentrism, (c) biocentrism, (d) ecocentrism, (e) gene-based approaches, and (f) (pluralistic) holism. As I have argued elsewhere (Ott 2008), the demarcation problem requires identifying morally relevant features (properties) and attributing them to natural beings. By doing so, one doesn't fall prey to the naturalistic fallacy. Morally relevant features may be sentience, the ability to communicate, and openness to a world "outside." To discourse ethics, the ability to communicate deserves close attention (Hendlin and Ott 2016). There are various levels of such abilities (Ott 2015): exchange of information (plants), expression and articulation (mammals and birds), speech acts, and, finally, discourse (humans). Plants do not communicate *with* each other, but they spread signals that are decoded by other plants. This is different from vocal gestures that dogs interact through, which again differs from lingual interaction between chimps and humans. If animals can give voice to their mental state, we can and should interpret and translate such voices into human discourse. Humans have, for a long time, underrated nature's capacity for communication, and they have wrongfully silenced nature. There is a dialectics of recognition in this respect: The more we recognize communication within nature, the more we can and should admire the exclusive human faculty of arguing in discourse.

A being capable of discourse, then, would not be on a par with a being only capable of exchanging information via biochemical signals. Giving IMV to sentient and communicative beings coheres with giving dignity only to humans. If IMV rests on the assumption of morally relevant properties, and if the most relevant properties (sentience, communication) come in degrees, then it might be permissible to grade IMV. The cattle and the beasts of the sixth day are sentient creatures, while the "creeping things," as spiders, dragonflies, bees, snails, etc. may "prehend" an outside world. A principle of species-adequacy combined with the conjoined criteria of sentience, faculty of communication, and biological strategies (r-strategy versus K-strategy) may provide a solid ground for gradually overcoming anthropocentrism.

From a Biblical perspective, we should respect the creatures of land, water, and air as what they are. We should respect fish as fish, fowl as fowl, beasts as beasts, and creeping things as creeping things. Such respect "as such" implies due respect to "God" who created different kinds of animals differently. Thus, the Hebrew Bible allows grading according to respect "as such." It recognizes that humans must protect cattle against predators, must protect orchards against beasts, must combat creeping things etc.

16.4.5 Ecosophies

Within the *fifth* dimension of environmental ethics, we find outlooks on nature based on a non-scientific concept of nature. Many environmental philosophers are critical of a hard-core interpretation of the project of modernity. In this interpretation, nature is seen as nothing but (a) an object of scientific explanation, (b) a storehouse of resources, and (c) an inimical force in a world where humans long for safety, health, and convenience. Science, seeing nature as objectivity under general laws does not allot nature any value. Natural beings are objects of manipulation and control, mere inputs into the fabrics of commodities.

The idea of deep ecology, as conceived by Arne Næss (1989), was to replace such scientific ontologies and attitudes by "ecosophies" as to be found in late Heidegger (Foltz 1995), Whitehead's process ontology, Spinoza's *Ethics*, Schelling's *natura naturans*, Rolston's "projective nature" (Rolston 1988), and "green" religious doctrines (Buddhism, Taoism, Vedic wisdom). Ecosophies presume that nature can reveal itself in modes beyond scientific observation, data mining, and causal explanation. Ecosophies can bypass the naturalistic fallacy because it is only fallacious if a scientific concept of nature is presupposed. If there is goodness within nature, one doesn't commit an inferential fallacy if one derives obligations and commitments from such ontological goodness.

16.4.6 Environmental Virtue Ethics

Environmental virtue ethics (EVE) can be regarded as sixth line of reasoning, if one assumes that virtues can be argued. EVE should not rest on arbitrary traits of individual characters but should be grounded in the patterns of reasoning. Thus, EVE is derivative to the previous lines of reasoning. The virtue of sufficiency is, for instance, grounded in reasons against the consumerist excessiveness. Many (biophilic) virtues are grounded in eudemonic values. Eudemonic values may also have a transformative effect upon virtues (Norton 1988). Future ethics and strong sustainability require the prudent virtues of foresight and precaution (Klauer et al. 2016). They also imply being aware of one's own finitude and mortality. IMV requires the virtues of restraint, respect, and care against sentient and prehensive animals. It may also ground the existential virtuous attitude of reverence for life (Schweitzer 1923). Ecosophies may ignite the virtues of humility, gratitude, restraint, care, and generosity. Environmental virtues can also be grounded in phenomenological accounts, making one sensitive to the many values of nature.

Given these six broad lines of reasoning, the reading of Genesis 1 has, first, a direct connection to the ecosophical line. Moreover, it is connected to environmental virtue ethics. As Hardmeier and Ott (2015) argue, religious ethics are more about attitudes toward life and death than about obligations (deontology) and consequences (utilitarianism). A crucial Biblical virtue is gratitude because humans are gifted by many natural goods (Primavesi 2007).

16.5 Conclusions: Discourse, Genesis, and Deep Ecology

As it has been shown in Sect. 16.3, the narrative of Genesis 1 suggests a general attitude with respect to nature: To take the earth under one's feet in the image of the divine spirit ("God") means that at each and any place of the Earth at which humans beings arrive they should act as representatives of the original "creator." *In their beings and doings, humans should re-present the Genesis narrative as a whole.* They should do so keeping always in mind the *hinneh*: being fully attentive of an Earth which is praiseworthy without end. From the retirement of "God" at the seventh day, however, a sharp sense of responsibility shall emerge. "God" remains present in absence only.

In principle, humans can adopt the basic self-understanding as "being" representatives of divine spirit on a very good earth. Such "being in the world" comes close, as Psalm 8 states, to divine modes of existence. If humans act as representatives they will be rewarded with experiences of nature at good places and on good routes.

Hardmeier and Ott devote several subsections of the final chapters to sensual encounters with nature. Humans are open to having "deep" experiences with nature. Original Biblical thinking is not devoted to transcendence, but fully worldly. Human senses can be exalted, as they look, listen, smell, touch, and taste. Sensual perception can be enthusiastic. Phenomenology of nature is not alien to Hebrew thinking. In some blessed moments, humans may perceive nature as if they may look at, listen to, or smell "creation" as such.

Human existence can be glorious and gorgeous, but it comes at a price: The higher one can climb, the harder one can fall. Humans feel desires and wishes whose fulfillment is not in line with the status of a representative (greed, selfishness, arrogance, hate). Here, moral ambivalence occurs. Humans are made out of crocked wood, as Kant says. To the Bible, humans are fragile beings. They have a moral sense, but can fail on moral grounds. The first human who failed killed his brother. God became angry about humans. They should act as representatives, but they mostly act otherwise. "The wickedness of man was great on earth" (Gen 1: 6). At the end of the Noah narrative, God promises not to destroy the earth again. A covenant holds and history begins. Quite soon, however, humans wish to build a tower whose top may come close to heaven. From early times on, human history is full of corruption, failure, and guilt ("sin"). Humans can act as "mis-representatives" and they often do so. Genesis remains an original utopia, but human history seems to deny it almost always. Humans do their worst as to make "a wasteland of what was creation" (Hans Jonas in his last public speech 1993, quoted in Troster 2007, p. 352).

Ironically, humans misread the narrative of Genesis 1. Such misreading has been secularized. Therefore, the project of Western modernity was successful in a one-sided manner. It made nature a set of objects under general laws. It became technological, since objects can be manipulated according to human ends. It became economical, extractive, and expansive. For long, this project has been forgetful about the many benefits, blessings, and gifts of nature. Environmental ethics wishes to correct the one-sidedness of modern mindsets on secular grounds.

The lines of reasoning from the previous section are non-exclusive and can be combined. I dub my approach in combining those lines of reasoning *deep anthropocentrism plus X* (Ott 2020). I rely on Hargrove (1992). "Plus X" means that different solutions of the demarcation problem are reasonable, as (gradual) sentientism or (gradual) zoocentrism. Contents of the Biblical narrative and lines of reasoning resonate with each other. The reading of Genesis 1 and the lines of reasoning are correlative and reinforcing, as chapters 5 and 6 in Hardmeier and Ott (2015) show in detail.

Finally, I wish to present a proposal of how the apron-model might be improved. There should be a discursive-logical layer on top of the model. All ecosophies are thoughtful, and thoughtfulness is essential reasonable, discursive, and "logical" (in a Hegelian sense). Arne Næss was trained in the spirit of the Vienna Circle and its epistemological focus on truth-semantic and science-based epistemology. Næss himself remained skeptical against normative ethics since he adopted an imperative metaethics from Carnap (Ott 2006). If one broadens the realm of discourse in a, say, Hegelian-Habermasian spirit with reference to different modes of validity, such discursive-logical layer should precede any substantial ecosophy. An ecosophical layer should be endorsed from a discourse-ethical perspective. Since translations between secular and religious ways of reasoning are demanded by discourse ethics (Habermas 2005), and since the outcomes of such translations will, in the field of environmental philosophy, be (in some sense) "ecosophical," discourse ethics requires conceptual space for ecosophical dialogues. If so, ecosophical dialogues are *required* by discourse ethics.

The second layer of the "classical" apron-model entails eight principles or prescriptive statements (Ott 2023). The first principle (statement) is about the demarcation problem. The second is about the intrinsic values of diversity of life. It opens a vast array of problems which an ethics of biodiversity has to address. The third statement is a rigid version of sufficiency. The fourth statement demands a substantial reduction of human population. The fifth statement sees human impacts in nature as "inappropriate" and states that the situation is constantly worsening. It should be seen as place-holder for empirical analysis about how environmental states change over time. The second part of the statement denies success of environmental reforms dogmatically. The sixth statement demands profound change in political and economic institutions. The seventh statement distinguishes between standard of living and quality of life. It comes close to proposals for (physical and economic) degrowth. The final statement commits persons who agrees to the seven statements to engage accordingly. These platform-principles are not firm principles but they constitute topics of debate in environmental philosophy. I propose to replace these prescriptive statements by the six lines of reasoning in environmental ethics (Sect. 16.4).

The third layer of the classical apron-model is about policy making and political engagement. Here, the concept of strong sustainability becomes a platform for environmental policy-making on different fields: climate change, biodiversity, forestry, agriculture, ocean etc. The fourth layer is about choices, actions, and lifestyles. It gives room for cultural innovation, new imaginaries, and new ways of living in communities.

To Næss, the logical relation between those four apron-layers is "loose derivation" (inferential reasoning) if it runs top-down from ecosophy to lifestyles. If the relation is "bottom-up," it is dubbed "deep questioning" (asking for reasons). One may and should ask which convictions make oneself an environmentalist, campaigner, vegetarian etc. Starting deep questioning will end up in the field of ecosophies. Both "loose derivation" and "deep questioning" presuppose a logical-discursive layer which is required by both logical operations. Thus, an improved apron-model with specific content looks like this:

1. Discursive logic (Habermas 1981; Taylor 1961; Brandom 1998).
2. Ecosophies (*specified to* Genesis 1, see Hardmeier and Ott 2015).
3. Reasoning Space of Environmental Ethics (*specified to* "deep anthropocentrism plus x", see Ott 2020).
4. Environmental Politics (*specified to* strong sustainability, see Ott 2014).
5. Lifestyles, daily practices, community life, political engagement.

This 5-layer model keeps the kind of inquiry at layer "1" which Næss performed in the 1950s and 1960s. There is no reason to abstract such inquiries away from environmental philosophy. The replacement of 8 "principles" by lines of reasoning looks far more attractive than the principles or even the simplification to six postulates being made by Sessions (1990). Strong sustainability is compatible with deliberative environmental democracy (Mason 1999) and with the spirit of reforms in the tradition of environmental pragmatism (Minteer 2006). The new model integrates discursive logic, spirituality, ethics, political pragmatism, and wisdom. If so, it isn't shallow.

There should be wisdom in ecosophies. At the level of lifestyle, Biblical thinking gratefully accepts the predicament of having a mortal life. As Psalm 90 states, human life is finite. If it can be lived until old age, the mere number of years doesn't matter much.

> So teach us to number our days, that we may apply our hearts unto wisdom. […] O, satisfy us early with your mercy; that we may rejoice and be glad all our days. Make us glad according to the days wherein thou hast afflicted us, and the years wherein we have seen evil (Psalm 90).

One should not pray for a very long life, but, more wisely, for lifelong joyful and grateful rejoicing as spiritual representative on planet Earth. Living a human life, *as if* one could make "God" feel satisfied with human existence ("finding grace"). Perhaps, even secular persons might find such way of life charming.

References

Borgmann A (1984) Technology and the character of contemporary life. University of Chicago Press, Chicago

Brandom R (1998) Making it explicit. Reasoning, representing and discursive commitment. Harvard University Press, Harvard

Callicott JB (1980) Animal liberation: a triangular affair. Environ Eth 2(4):311–338
Daly H (1996) Beyond growth. Beacon Press, Boulder
Foltz B (1995) Inhabiting the earth. Humanities Press, Atlantic Highland
Groh D (2003) Schöpfung im Widerspruch. Suhrkamp, Frankfurt
Habermas J (1981) Theorie des kommunikativen Handelns. Suhrkamp, Frankfurt
Habermas J (2005) Religion in der Öffentlichkeit. In: Habermas J (ed) Zwischen Naturalismus und Religion. Suhrkamp, Frankfurt/M, pp 119–154
Haraway D (2016) Staying with the trouble. Duke University Press, Durham
Hardmeier C, Ott K (2015) Naturethik und biblische Schöpfungserzählung. Kohlhammer, Stuttgart
Hargrove E (1992) Weak anthropocentric intrinsic value. Monist 75(2):183–207
Hendlin Y, Ott K (2016) Habermas on nature: a postnormal reading between moral intuitions and theoretical restrictiveness. Environ Eth 38(2):183–208
Jonas H (1973) Organismus und Freiheit. Vandenhoeck, Göttingen
Jonas H (1992) Der Gottesbegriff nach Auschwitz. In: Jonas H (ed) Philosophische Untersuchungen und metaphysische Vermutungen. Insel, Frankfurt, pp 190–208
Kellert S (1997) Kinship to mastery. Island Press, Washington, DC
Klages L (1913) Mensch und Erde. Kröner, Stuttgart
Klauer B, Manstetten R, Peters T, Schiller J (2016) Sustainability and the art of long-term thinking. Routledge, London
Krebs A (1999) Ethics of nature. De Gruyter, Berlin
Leopold A (1949) A Sand County Almanac. Oxford University Press, Oxford
Mason M (1999) Environmental democracy. Earthscan, London
Minteer BA (2006) The landscape of reform. MIT Press, Cambridge
Morton T (2016) Dark ecology. Columbia University Press, New York
Muraca B (2011) The map of moral significance. Environ Values 20(3):375–396
Næss A (1989) Ecology, community and lifestyle. Cambridge University Press, Cambridge
Naess A (1995) The deep ecological movement: some philosophical aspects. In: Sessions G (ed) Deep ecology for the twenty-first century. Boston, pp. 445–453.
Neumann-Gorsolke U (2004) Herrschen in den Grenzen der Schöpfung. Neukirchener Verlag, Neukirchen-Vluyn
Norton B (1988) Why preserve natural variety? Princeton University Press, Princeton
Norton B (2005) Sustainability. Universe of Chicago Press, Chicago
Ott K (2006) Zur Bedeutung Spinozas für die Tiefenökologie. Studia Spinozana 15:153–176
Ott K (2008) A modest proposal of how to proceed in order to solve the problem of inherent moral value in nature. In: Westra L, Bosselmann K, Westra R (eds) Reconciling human existence with ecological integrity. Earthscan, London, pp 39–60
Ott K (2010) Umweltethik zur Einführung. Hamburg, Junius.
Ott K (2014) Institutionalizing strong sustainability: a Rawlsian perspective. Sustainability 6(2):894–912
Ott K (2015) Kommunikation, Sprache und das Inklusionsproblem der Umweltethik. Zeitschrift für Semiotik 37(3–4):151–169
Ott K (2016) On the meaning of eudemonic arguments for a deep anthropocentric environmental ethics. New German Crit 43:105–126
Ott K (2020) Mapping, arguing, and reflecting environmental values. In: Lysaker O (ed) Between closeness and evil. A Festschrift for Arne Johan Vetlesen. Scandinavian Academic Press, Oslo, pp 263–292
Ott K (2023) Deep ecology. In: Wallenhorst N (ed) Handbook of the Anthropocene. Springer, Cham
Ott K, Döring R (2011) Theorie und Praxis starker Nachhaltigkeit. Metropolis, Marburg
Ott K, Reinmuth KC (2021) Integrating environmental value systems: a proposal for synthesis. In: Beckmann V (ed) Transitioning to sustainable life on land. MDPI, Basel, pp 43–76
Primavesi A (2007) The preoriginal gift –and our response to it. In: Kearns L, Keller C (eds) Ecospirit. Fordham University Press, New York, pp 217–232
Rolston H (1988) Environmental ethics. Temple University Press, Philadelphia

Schulte C (2014) Zimzum. Gott und Weltursprung. Jüdischer, Berlin
Schweitzer A (1923) Kultur und Ethik. Beck, München
Scott J (2017) Against the grain. Yale University Press, New Haven
Sessions G (1990) Deep ecology and new age. In: Mesch H (ed) Ecoresistance/Ökowiderstand. Argument, Hamburg, pp 96–105
Taylor P (1961) Normative discourse. Prentice-Hall, Englewood Cliffs
Taylor P (1986) Respect to nature. Princeton University Press, Princeton
Tirosh-Samuelson H (2017) Judaism. In: Jenkins W, Tucker ME, Grim J (eds) Routledge handbook of religion and ecology. Routledge, London, pp 60–69
Troster L (2007) Hearing the outcry of mute things: toward a Jewish creation theology. In: Kearns L, Keller C (eds) Ecospirit. Fordham University Press, New York, pp 337–352
White L (1967) The historical roots of our ecological crisis. Science 155:1203–1207.
Wilson E (1984) Biophilia. Oxford University Press, Oxford
Worster D (2017) History. In: Jenkins W, Tucker ME, Grim J (eds) Routledge handbook of religion and ecology. Routledge, London, pp 347–354

Chapter 17
Panentheism in Christian Ecotheology

Guillermo Kerber

Abstract Since early nineteenth century the notion of panentheism has been used in Christian theology as a way to deal with the presence of God in the universe. In the second half of the twentieth century the concept has been reframed considering new developments in cosmology and ecology. In Latin America, an author like Brazilian Leonardo Boff, uses this concept to articulate his proposal for an eco-theology. He discusses the difference between panentheism and pantheism and expresses the challenges panentheism poses to various domains of theology. An interesting development of the concept has been done by ecofeminist theologians, such as American Sallie McFague, who reflects on the world or, more widely, the universe, as the body of God. Together with building new bridges between theology and science or philosophy, panentheism, as understood in Christian theology, has contributed to a common interfaith approach and action regarding the ecological and climate crises the world live in.

Keywords Ecology · Theology · Panentheism · Ecofeminism · Ethics

17.1 Introduction

The notion of pantheism has long been rejected by Christian theology. Classical Christian theology and normative instances such as ecumenical councils and the hierarchy of the Churches have opposed and condemned pantheism and its consequences and stressed the difference between God and nature, the world or, in Christian terminology, creation. Over the centuries, Christian theologians have, however, proposed formulations very similar to pantheism. This is the case for instance of Johannes Scotus Eriugena, a ninth century Irish theologian who affirmed that God has created the world out of his own being and saw creation as the

G. Kerber (✉)
Adult Faith Formation, Roman Catholic Church, Geneva, Switzerland
e-mail: guillermo.kerber@cath-ge.ch

L. Valera (ed.), *Pantheism and Ecology*, Ecology and Ethics 6,
https://doi.org/10.1007/978-3-031-40040-7_17

self-manifestation of God. His doctrine has been later labeled as "pandeism" or even "pantheism" and criticized because he didn't acknowledge the ontological difference between God and creation (Moran 1990).

In the twentieth century, Christian theological efforts to reflect on the relationship between God and creation gave origin to "panentheism." This notion which is composed by the Greek words *pan* = all, *en* = in and *Theos* = God, would mean "all is in God" or "God is in all" and has had a strong development in Christian ecotheology; this is theology that tries to include the challenges posed by ecology. The latter is at the same time understood as a new science and as a perspective to understand the state of nature, including human beings on Earth.

In this chapter I will develop the implications of panentheism for Christian theology, emphasizing various expressions of contemporary Christian theology, among them ecofeminism and Latin American liberation theology.

17.2 The Notion of Panentheism

Contemporary American philosopher and theologian Jay McDaniel recalls that the concept of panentheism was coined in the nineteenth century by German philosopher Karl Christian Friedrich Krause (1781–1832) and literally means "everything in God" (McDaniel 1993, 1995). For McDaniel, however, in the last 50 years in the context of the current interest in ecology, the term has undergone a change of emphasis. In his words:

> [*P*]*anentheism* implies an ecological way of thinking about God in which, even as God and creation are distinguished, God is understood to be intimately connected to creation, and vice-versa... What makes God 'God' is not that God is least connected to the universe, but rather that God is most connected (McDaniel 1995, p. 97).

In these sentences McDaniel offers a perspective on panentheism making a differentiation in two ways. On the one hand, it is distinguished from strict pantheism, which implies an absolute equation of God and creation. On the other hand, it is also distinguished from strict dualism, which implies an absolute gap between God and creation. Panentheism is thus the view that the creation and its processes are somehow "in God," even though God is "more than" creation. If God is the Sacred Whole, then that Whole is indeed more than the sum of its parts.

This offers a new perspective for Christian theology. By stressing God's presence *in* creation, panentheism can affirm images of the Divine that are not only personal but also transpersonal suggesting that Christian ecotheology can use a variety of images. "We can meaningfully speak of the Divine as 'the Mystery' and 'the Adventure of the Universe as One' and 'the Ultimate Context' and 'the cosmic Ocean', but also as 'Mother' and 'Father' and 'Bear' and 'Coyote Trickster,' as do some Indigenous groups in North America" (McDaniel 1995, p. 99).

God therefore can be not only a 'he' but also a 'she' and a neutral entity expressed by different concepts. Consequently, instead of fixating on particular images, a variety of images should be accepted as ways to image the Divine. From a Christian understanding, since God can also be envisioned in personal terms, panentheism coincides with biblical points of view. For panentheists, as for many biblical authors, God is "something like" a person, albeit not located in space or time. This personal and transpersonal perspective highlighted by panentheists has been expressed by Christian mystics over the centuries. It is one of the expressions of what has been called apophatic or negative theology. Cataphatic of positive theology would affirm, for instance, that God is Father. Apophatic theology would say that God is so different from what we can imagine that we cannot affirm anything about the Divine because God is totally beyond what we can imagine. This "totally beyond" is what theologians have called the transcendence of God which should be considered together with the immanence of God, as I will develop later.

Christian panentheists would take the personalist line of thought a step further. Not only is the Divine Life "something like" a person, but it is also something like Jesus of Nazareth. This would not mean that Jesus is the only way to God, but it does mean that in the attitudes we see in Jesus, amplified into the figure of Christ we glimpse the very heart of the Divine. If the word "Christ" can be used as a name for God, the body of Christ would be not simply the Christian Church as expressed in Paul's letters in the New Testament (Colossians 1:18) but the entire universe.

The notion of pantheism has long been rejected by Christian theology. Classical Christian theology and normative instances such as ecumenical councils and the hierarchy of the Churches have opposed and condemned pantheism and its consequences and stressed the difference between God and nature, the world or, in Christian terminology, creation.

17.3 Panentheism for Ecofeminist Theologians: The World as the Body of God

The world or, more widely, the universe, as the body of God has been one of the themes developed by Christian eco-feminist theologians. For eco-feminists, a critical perspective is needed to evaluate the heritage of Western Christian culture. The former requires a new awareness, a new cultural symbolism, and a new spirituality. In all three panentheism has a role to play (Ruether 1993). Together with stressing that the world can be seen as the body of God (McFague 1993; Dietrich 1996). Sallie McFague (1933–2019) gives complementary images of God for the present time, which she characterizes as the ecological or nuclear age, such as God as mother, God as lover, God as friend (McFague 1987, pp. 97–180). McFague sees these images as models or metaphors which means that God cannot be directly identified with those models, but they do help us to understand other aspects of God that have been obscured by traditional concepts.

The model of the world as the body of God presented by McFague has considerable implications. On the one hand, it is sharply distinguished from the model of

God as king or monarch. In these images God is separate from the world, above it and ruling over it with authority. On the other hand, the model stresses the panentheistic perspective as God takes on fresh connotations in this metaphor. God cares for the world, but at the same time God is exposed to risk. God loves bodies (McFague 1987, pp. 59–90). God suffers in the world, thus the relevance of earth healing as the prime aim of ecofeminism (Ruether 1993).

American ecofeminist Eleanor Rae (1934), for her part, links the world as the body of God. Her view converges with German theologian Jürgen Moltmann's (1926) idea of the Shekinah.

> While Jürgen Moltmann does not speak of creation as the body of the Divinity, his theology could be developed in that direction through his understanding of the Creation as the presence of God. He sees in the *Shekhina*, the feminine presence of the Divinity in the Kabbala, 'the direct presence of God in the whole material world and in every individual thing within it' (Rae 1994, p. 73).

17.4 "God in Creation:" Bases for a Panentheistic Ecotheology

The reference to Moltmann is relevant for our chapter as he entitles to one of his books *God in Creation* (Moltmann 1985). In the book he develops the idea of God's immanence in the world and the principle of mutual interpenetration which are at the basis of panentheism.

Moltmann reacts to those ideas of God and of God's relation with the world which have contributed to strengthening the idea of humankind as ruler. A transcendent God separates God from the world, and consequently humankind from the world and from God. Humans then seek to be like God by distancing themselves from the earth and appointing themselves as their lords and rulers. "Through the monotheism of the absolute subject, God was increasingly stripped of his connection with the world, and the world was increasingly secularized" (Moltmann 1985, p. 1). Over against this concept, Moltmann stresses the trinitarian nature of God, a key component of Christian faith which proclaims God as Father, Son, and Holy Spirit. In his words:

> If we cease to understand God monotheistically as the one, absolute subject, but instead see him in a trinitarian sense as the unity of the Father, the Son and the Spirit, we can then no longer conceive his relationship to the world he has created as a one-sided relationship of domination. We are bound to understand it as an intricate relationship of community – many-layered, many-faceted and at many levels (Moltmann 1985, p. 2).

Moreover, within this trinitarian conception of creation, in which traditional theology has placed almost exclusive emphasis on God the Father as creator, Moltmann has developed the concept of "Creation in the Spirit" (Moltmann 1985, pp. 9–19). This perspective enables us to regard God as immanent in creation, through the energies and possibilities of the Spirit. "The Holy Spirit… poured out on the whole creation… creates the community of all created things with God and with each other, making it that fellowship of creation in which all created things communicate with one another and with God, each in its own way" (Moltmann 1985, p. 11). This theological perspective of creation, Moltmann argues, is supported by new concepts emerging in cosmology, ecology, and physics. Quoting Capra's *The Tao of Physics*, Moltmann (1985, p. 322) affirms that: "In the new world view, the universe is seen as a dynamic web of interrelated events. None of the properties of any part of this web is fundamental; they all follow from the properties of the other parts, and the overall consistency of their mutual interrelations determines the structure of the entire web". This mutual interrelationship of the whole universe has been highlighted by Pope Francis' ecological encyclical *Laudato si'* when he uses the affirmation that "everything is connected, interrelated, interconnected" as a leitmotif all over the document (Pope Francis 2015, §70, 92, 117, 120, 137, 142).

As I said earlier, from the Christian theological perspective, *God's immanence* in the world (*God within*) must be understood together with *God's transcendence* (*God beyond*). This produces a tension, which is not only a tension in theological discourse, but is also "an immanent *tension* in God himself: God creates the world, and at the same time enters into it… The God who is transcendent in relation to the world, and the God who is immanent in that world are one and the same God" (Moltmann 1985, p. 15). In order to understand this tension Moltmann has recourse to two concepts: the rabbinic and kabbalistic doctrine of the Shekinah, as Eleanor Rae highlighted, and the Christian doctrine of the Trinity. In the first, the emphasis is on the indwelling of God in the creation of his love. God gives Godself away to God's people and suffers with their sufferings. In the second, the eternal God goes out of Godself and makes a creation. Father, Son, and Holy Spirit are involved in creating. God creates, reconciles, and redeems creation through the Son, who is the eternal counterpart within God. In the power of the Spirit, God is present in creation. Creation, therefore, is not only something that happened millions of years ago (*creatio prima*) but something that is happening now (*creatio continua*) through the Spirit. The outward movement of the Trinity in creation reflects the dialectical movement within the Godhead itself. The mutual indwellings and the eternal community of the Father, the Son, and the Holy Spirit are manifested in their mutual interpenetration or perichoresis (Moltmann 1985, pp. 16–17). Perichoresis is a Christian theological term used since the early Church to refer to the relationship of the three persons of the Trinity, the Father, the Son, and the Holy Spirit as a mutual indwelling.

If Eleanore Rae and others refer to Jürgen Moltmann as a source to understand the implications of panentheism, other authors, especially North American, would also quote American ecotheologian Thomas Berry (1914–2009). To go beyond the anthropomorphic tendency in Christian theology, Berry believes that even animals, plants, cells and protons are imbued with interiority. In this sense at least, panentheism is not anthropomorphic but rather biomorphic or even cosmomorphic. For McDaniel (1995, pp. 103–104), Thomas Berry adds a story and a scope to the panentheistic perspective.

For Berry (1988, pp. 44–46), humanity is now entering the ecological age after having lived for two centuries the technological age. The ecological age "fosters the deep awareness of the sacred presence within each reality of the universe". A follower of French Jesuit priest and palaeontologist Pierre Teilhard de Chardin (1881–1955), Berry proposes to overcome the immanence and the transcendence of God with the notion of diaphany or transparency because "even our sense of divine immanence tends to draw us away from the sacred dimension of the earth in itself… We go too quickly from the merely physical order of things to the divine presence in things. While this is important, it is also important that we develop a sense of the reality and nobility of the natural world in itself" which leads us "to speak of the diaphanous quality of matter" (Berry 1988, pp. 81–88). Teilhard presented transparency or diaphany in the following way: "If we may slightly alter a hallowed expression, we could say that the great mystery of Christianity is not exactly the appearance, but the transparency of God in the universe. Yes, Lord, not only the ray that strikes the surface, but the ray that penetrates, not only your Epiphany, Jesus, but your Diaphany" (Teilhard de Chardin 1965, p. 131).

For Berry, entering the new ecological age includes recovering the religious and sacred dimension of nature. This is the "ultimate lesson in physics, biology, and all the sciences, as it is the ultimate wisdom of tribal peoples and the fundamental teaching of the great civilizations." This recovery implies an ethical imperative as "the earth is mandating that the human community assume a responsibility never assigned to any previous generation" (Berry 1988, pp. 47–49).

17.5 Panentheism in Leonardo Boff's Ecotheology

Panentheism has not only been a concept developed by North American theologians quoting sometimes Europeans, like Moltmann. As I have argued elsewhere (Kerber 2006), I believe panentheism is the concept which might encapsulates Leonardo Boff's theological thought. Leonardo Boff was born in 1938, and is a well-known Brazilian liberation theologian. In his book *Ecology and liberation*, as a sub-title to the section on ecology and theology, we find "Christian panentheism," and under that title he deals with issues such as Christianity's co-responsibility for the ecological crisis, the liberation of creation theology, the Trinity, and the Holy Spirit, to arrive at Christian pan-en-theism (Boff 1995, pp. 43–51).

For Boff, the concept of panentheism arises from the new insights provided by contemporary cosmology. "The ecological view of the cosmos emphasizes God's immanence. God is seen to be involved in all processes without being merged into them… God does not figure merely as Creator, but as Spirit of the world" (Boff 1997, pp. 152–153). For Boff, there are various routes in our cosmological story that are a preparation for "God's advent in consciousness: quantum reality, the cosmic evolutionary process, the process and eschatological character of nature, the sacramentality of all things, and panentheism" (Boff 1997, p. 142). What is the latter for Boff?

> God is present in the cosmos and the cosmos is present in God. Theology in the early centuries expressed this mutual interpenetration with the concept of perichoresis. Modern theology has coined another expression: *panentheism* (Greek: *pan*=all; *en*=in; *theos*=God); that is, God in all and all in God. This term was first proposed by Karl Christian Frederick Krause (1781–1832), who was fascinated by the divine splendour of the universe (Boff 1997, p. 153).

Boff insists that panentheism must be distinguished from pantheism, which does not accept any difference, and where everything is God. For Boff, this lack of difference easily leads to indifference. Panentheism respects such differences, while pantheism denies them. All is not God. But God is in all and all is in God, by reason of the creation by which God is permanently present in the creature. The creature always depends on God and carries God within it. While God and world are different, they are not separated or closed but open to one another. They are always intertwined with one another. As he put it in another book, "Panentheism… starts from the distinction between God and the creature, yet always maintains the relation between them. The one is not the other. Each of them has his/her/its own relative autonomy yet is always related" (Boff 1993, p. 52).

As Berry, Boff affirms that the intimate relationship between God and creatures underlies the transparency of God in the created world. Boff here takes up Teilhard de Chardin's ideas on the transparency, or diaphany, of God, that I mentioned earlier. He says:

> this mutual presence means that simple transcendence and simple immanence are overcome. There emerges an intermediate category, transparency, which is precisely the presence of transcendence within immanence. When this happens, reality becomes transparent. God and the world are therefore mutually transparent (Boff 1997, p. 153).

If panentheism is the inevitable consequence of incorporating the new cosmology into theology, then theology in its turn will see some of its features reformulated because of this panentheistic emphasis. This is the case, for instance for theology of creation and the doctrine of the Trinity.

For Boff, the panentheistic perspective helps us to understand differently the theological notion of creation. To say that we are created means affirming that we have come from God, that we bear in ourselves the marks of God, and that we are travelling toward God (Boff 1995, pp. 45–48). As ecofeminists quoted earlier, but looking at the Latin American reality, Boff stresses that theology of creation has been subjugated to the dominant theology, which has concentrated much more on the mystery of redemption rather than on creation.

An ecotheology of creation, based on the new cosmology, emphasizes that human beings are not above but within creation. The world does not belong to humans, it belongs to God, its creator. But the world is given to humans as a garden to cultivate and tend, and they thus have a relationship of responsibility towards it. In my view, this calls for an ethics of care expanded to the whole creation understood as our common home, as the subtitle of Pope Francis' encyclical evokes. Human beings can only be human and fulfil themselves as they bring the world to fulfilment by becoming involved in it by way of work and by care for it. By contrast with a destructive, dominating involvement in the Anthropocene era, a panentheistic perspective directs to maintaining the balance of creation by taking care (i.e., doing justice) of the most vulnerable species (including humans) and environments.

Panentheism is, secondly, as regards the Trinity, "the interplay of perichoretic relationships" (Boff 1997, p. 154). For Boff, the ecological discourse makes it possible to speak of God as a trinity of persons, because ecology stresses the study of relationships and, in classic Christian teaching the three Persons of the Trinity are constituted by their relationship. In fact, when Ernst Haeckel coined the word Ecology he said: "By ecology, we mean the whole science of the relations of the organism to the environment" (Haeckel 1866, p. 286). Boff points out that the Trinitarian God is not an exclusively Christian doctrine. There is a trinitarian thread running through other great religious traditions of humankind.

If God is communion and relationships, then everything in the universe lives in relationship and everything is in communion with everything else. Thus, the Trinity emerges as one of the most suitable representations of the mystery of the universe. And perichoresis, a concept applied in the first place to God's essential being, as the relationship between Father, Son, and the Holy Spirit, can also be applied to the mutual relationship between God and creation. Yet further, the world is complex, diverse and one, intertwined and interconnected, because it is a mirror of the Trinity.

17.6 Conclusion

Christian theologians and more recently ecotheologians have used the concept of panentheism to reflect on the relationship between God and the world, the universe. This notion is different from pantheism as developed by some philosophers like Spinoza and by other religions. The use of panentheism makes theologians highlight the limits of the transcendence of God in Christian theology and essays to understand the relationship between God and the universe, understood as creation in Christian theology, considering the new cosmological visions that emerged in the last century. Various approaches to ecology have shaped what theologians have said about the relationship between God and the world. In various authors the reference to Teilhard de Chardin's perspective on the diaphany of God in the universe is a key to understand what Christian panentheism is. This new perspective of God in creation has had ethical implications which have been highlighted by ecofeminist and Latin American theologians.

References

Berry T (1988) The dream of the earth. Sierra Club Books, San Francisco
Boff L (1993) Ecología, mondialização, espiritualidade. Atica, São Paulo
Boff L (1995) Ecology & Liberation; a new paradigm. Orbis, New York
Boff L (1997) Cry of the earth, cry of the poor. Orbis, New York

Dietrich G (1996) The world as the body of god. In: Ruether RR (ed) Women healing earth: third-world women on ecology, feminism, and religion. Orbis-Maryknoll, New York
Francis (2015) Laudato si'. Encyclical letter on care for our common home. Libreria Editrice Vaticana, Vatican City
Haeckel E (1866) Generelle Morphologie der Organismen. Georg Reimer, Berlin
Kerber G (2006) O Ecológico e a Teologia Latino-americana. Sulina, Porto Alegre
McDaniel JB (1993) Emerging options in ecological Christianity. SUNY Press, New York
McDaniel JB (1995) With roots and wings: Christianity in an age of ecology and dialogue. Orbis, New York
McFague S (1987) Models of god. Theology for an ecological, nuclear age. Fortress, New York
McFague S (1993) The body of god. Fortress Press, Philadelphia
Moltmann J (1985) God in creation. An ecological doctrine of creation. SCM Press, London
Moran D (1990) Pantheism from John Scottus Eriugena to Nicholas of Cusa. Am Cathol Philos Q 64(1):131–152
Rae E (1994) Women, the earth, the divine. Orbis-Maryknoll, New York
Ruether RR (1993) Gaia and god: an ecofeminist theology of earth healing. SCM Press, London
Teilhard de Chardin P (1965) The divine milieu. Harper and Row, New York

Chapter 18
Theism Versus Pancomprehensions

Sixto Castro

Abstract In this chapter, I briefly outline the standard characterizations of pantheism and panentheism –which I call pancomprehensions– and their relation to theism. I point out that from the point of view of the thought underlying theism the panentheistic proposal is more suggestive, and that it has on many occasions infiltrated the theistic understanding. Second, I compare the pancomprehensions with the theistic proposal, especially by bringing to the fore the divine properties that both approaches assume and the philosophical conception of the world that underlies them. I point out that the key idea that divides pancomprehensions and theism is the idea of *creatio ex nihilo*. Finally, I refer to some possible models for explaining the God-world relationship.

Keywords Theism · Pantheism · Panentheism · Aquinas · Analogy

18.1 Introduction

Many of the issues dealt with by the philosophy of religion are done from a theistic perspective, which inevitably leads to determined solutions, paradoxes, or dilemmas. Despite this, we find global frameworks, such as the pantheist and the panentheist, which have a stronger presence in contemporary debate. This is something that can be regarded as enormously enriching due to the theoretical proposals that have arisen from these alternative frameworks. These proposals are relevant insofar as they are in keep with the spirit of our times and have the capacity to adopt religious interpretations of contemporary concerns like the environment and question of sustainability. It is easy to see how the latter can be examined within a religious framework, given the sacralization of nature due to its either partial or total identification with the divine, respectively.

S. Castro (✉)
Department of Philosophy, Universidad de Valladolid, Valladolid, Spain
e-mail: sixto.castro@uva.es

L. Valera (ed.), *Pantheism and Ecology*, Ecology and Ethics 6,
https://doi.org/10.1007/978-3-031-40040-7_18

The issue at hand when positing the role of theism in the development of these perspectives is not merely any God-world connections,[1] but the actualized modelling of our contemporary debates around theological questions. In this essay I will defend, as Richard Swinburne puts it, *the coherence of theism* and its theoretical superiority over its "rivals" –pantheism and panentheism– which I will refer to conjointly as "pancomprehensions."

18.2 Pantheism and Panentheism: Pancomprehensions

It is not easy to find a single definition or description of pantheism. The word "pantheism" is ambiguous, to say the least (Mander 2022). To begin with, one would have to define what "pan" means in relation to God. However, in an effort to avoid becoming entangled in philological debates, we can set down a maxim, Spinoza's *Deus sive natura* ("God or Nature") for example, as a paradigmatic form of pantheism.[2] Spinozian reality is composed of a sole substance –one with its own particular attributes, modes, etc.; however, it remains fundamentally unique. The *sive* in Spinoza's formulation, is unequivocal: the terms that flank it may be substituted perfectly by each other *salva veritate*. It would seem, therefore, that both the former and the latter point to the same reference in such a way that there is a total identification of God with the world.

And so it is that pantheism remains a popular way to present religious convictions in keeping with the spirit of our time, a time in which nature is seen both as divine and as an incarnation of the sublime. However, "sublimity" has proven mutable, in that it heretofore was traditionally limited to an association with beauty and with its manifestations. Yet, today it takes up that Platonic and Neoplatonic "revelatory" potential once reserved for beauty. Sublime nature can thus be said to be a theoretical celebration of the identification of God with the world (Castro 2021). It must be said, however, that pantheism cannot be considered philosophically articulated, a conclusion reached by century nineteenth-century thinkers who perceived that the difference between pantheism and atheism was not one of degree but of education. Upon infusing the world, nature, and the cosmos with divinity, pantheism becomes what Schopenhauer aptly termed "polite atheism," because referring to the world as God is basically just assigning the term "world" a superfluous synonym. Similarly, Feuerbach made quick work of the notion, concluding

[1] In what follows I use indistinctly nature, world, or universe to refer to whatever is not God, or which is practically or completely identified with God, without going into details about possible differences that may be suggested between those notions.

[2] However, in his opinion, there is always someone who thinks they represent panentheism. In his *Ethics*, he states: "Whatsoever is, *is in God*, and without God nothing can be, or be conceived" ("quicquid est, *in Deo est*, et nihil sine Deo ese neque concipi potest" [Spinoza 1914, I, XV]). Spinoza still thinks in terms of *natura naturans* (God*) and natura naturata* (God's modes).

that pantheism is simply a theologized atheism that sanctions the conversion of matter –the theological opposite of God– into an attribute of the divine essence. Thus, pantheism and atheism play toward the same end. With subtle difference in means, the former's divinization of everything leads to a theistic error, for if everything is divine, nothing is divine.

Theoretically speaking, the panentheistic proposal is far more interesting, particularly from a theistic point of view, because it touches on its very definition. Historically, panentheism, which states that nothing exists outside of God, has been confused with the pantheist idea that God and the world are one and the same. In fact, though similar on the surface perhaps, the theses are quite different. The panentheistic thesis came into its own with the coining of the term "panentheism" in the nineteenth century, Thus, both definition and reference were granted this conception of the Divine, which somehow seeks to mediate between theism and pantheism, and which up until then had been often confused with either or both. As opposed to pantheism, panentheism does not identify God with the world. On the contrary, it sustains that God is more than the world but that somehow, God is *in* the world and/or the world is *in* God (Culp 2021; Meister 2017). Moreover, panentheism's central tenet, as opposed to theism, sustains that God and the world cannot be split apart. Panentheism holds that even if, to arrive at a final explanation, we establish an epistemological distinction between God and the world, there can never be any substantial ontological distinction between them. At best, an ontic difference can be espied in the core essence of the divine being in order to attest that basic tenet that reality is in God, but that God cannot be reduced to being mere reality (Göcke 2017).

This "intermediate" approach has enjoyed various degrees of success and application in philosophical and theological debate because it is unifying while it remains differentiating. For example, Newton's notion of space and time as *sensoria Dei* fits perfectly into this schema, as does the Fichtean identification of God with moral conscience. The latter considers an infinite yet personal God to be contradictory, as does Schelling, who considers the real "being in God himself, is not God himself" (Pannenberg 1996). In general, nineteenth-century philosophers of the Absolute worked within a pantheistic framework, as process theologians did in the twentieth century. The latter labored under the influence of the theistic metaphysics of Alfred North Whitehead and Charles Hartsthorne, which are founded upon the notion that change and becoming are constitutive of both the world and God. Based on this tenet, which serves as an alternative to the notion of creation *ex nihilo,* they envisage a universe undergoing continuous creation and existing while God exists. Over and against the traditional theistic view of God as "pure act," they propose a universe and a God constantly going from potentiality to actuality. Thus, in a mysteriously intimate way, God and the world share a need to exist, which can be better understood as becoming.

18.3 Theism in the Face of Pancomprehensions

Classical theism is defined by Richard Swinburne (1993, p. 1) in the following terms:

> By "theism" I understand the doctrine that there is a God in the sense of a being with most of the following properties: being a person without a body (that is, a spirit), present everywhere (that is, omnipotent), the creator of the universe, perfectly free, able to do anything (that is, omnipotent), knowing all things (that is, omniscient), perfectly good, a source of moral obligation, eternal, a necessary being, holy, and worthy of worship.

As defined by Swinburne, God is a person with certain properties that derive from the fact that God is the *raison d'etre* of reality. Thus, theism admits the presence of God in the world with no confusion between these realities. It acknowledges God's personal presence in history, but not the presence of the world in God. This would make of God a reality that depends on the becoming of the world, either because the world and God are the same (pantheism), or because the changes suffered by the world in becoming affect God's very nature (panentheism); in other words, it affects what theism holds are constitutive properties of God.

Both pantheism and panentheism, each one in its own way, explain and limit the scope of God's properties. Among these properties are the omnipresence of God, God's eternity (if the universe is deemed eternal), or at least God's everlastingness (if this is what is meant by eternity). There is also God's necessity (connected and/or dependent upon the possible necessary nature of the universe) which is linked to God's role of creator (not Creation *ex nihilo,* but Creation of the emanationist or kenotic sort, etc.). However, none of the other properties of the theist God are applicable. For example, neither God's spiritual nature (excepting some emergentist idea of the divine), nor God's freedom, omnipotence and omniscience are possible, since God is always subject to the world's becoming. "Moral" properties attributed to God, such as goodness and sanctity, as well as God being considered the source of moral duty or God's deserving adoration are entirely absent in the pancomprehensions. For theism it is the total opposite: God is a reality worthy of adoration for the mere fact of being the source of all sanctity. Moreover, the latter worldview holds that the divinization of the world, of "everything in existence," is an act of human will.

Pancomprehensions require that "everything in existence" be an essential constituent part of the concept and the very reality of God. This implies that worshipping God is the same as worshipping the world and "everything in existence." In other words, in worshipping God, one accepts the *fatum or factum* as divine. For theism, however, worship is not a one-way idolatrous act, but presupposes that God is somehow "conscious" of our worship for it to make sense to engage in it (Leftow 2016). This is why the God of theism expects engagement from those who worship, which is both vital and existential, while worshippers in their turn believe that God will always intervene in reality in such a way as to fulfil their hope in God. This is clearly excluded from pantheism, and would be very confusing in panentheism, because God, in pancomprehensions, largely depends on "everything that exists." Consequently, either God is not free to change what exists (because God is dependent upon the world), or God is mutable.

The basic theistic tenet with regard to this is that God and the world are neither of the same substance, nor even of a similar substance; they are instead two separate and very different substances (or realities) one of which (the world) finds its origin in another (God). God is not the whole either, or even part of the whole. In the second century AD, Athenagoras of Athens, in an early defence of the Christian faith against those who judged it a form of atheism, wrote:

> But surely it is ludicrous to describe us as 'atheists,' when we distinguish God from matter; and teach that matter is one thing and God another, and that they are separated by a wide interval (for that the Deity is uncreated and eternal, to be beheld by the understanding and reason alone, while matter is created and perishable) (Athenagoras, *Apology*, IV, 10; quoted in McGrath 2008, pp. 28–29).

Thus, God is the *explanans* of the world, which is the *explanandum* (Göcke 2017). Ontologically, the world depends on God and an explanation of the world must take that into account. In classical Christian theism, theological debate around everything substance-related has developed mostly in reflections on the persons of the Trinity (*homoousios/homoiousios)* whose substantial identity (or similarity, such as for example, Arianism would have it) is not questioned. However, though the nature of the relationship between the divine persons has been discussed and argued over, these arguments have never focused on the relationship between God and the world. Instead, God and the world figure as immeasurably removed from one another, so much so that no other than Thomas Aquinas, a conspicuous theist, states that God's essential nature is being, whereas the nature of the world is not being. In *De aeternitate mundi* he writes: "A creature does not have being, however, except from another, for, considered in itself, every creature is nothing, and thus, with respect to the creature, non-being is prior to being by nature" (Aquinas 1991, 1997). For the creature, as well as for reality as a whole, nothingness is more essential than being, because, as Aristotle affirms, there must be a principle whose essence is actuality for that which exists potentially to exist (*Metaph.* XII, 6, 1071b). From the Thomist point of view, the being of the creature cannot be understood outside of its being created. Aquinas compares created beings with light, which is nothing by itself and falls into nothingness if it distances itself from its origin. Therefore, in theism, God is not the world but the One who preserves the world, preventing it from returning to the nothingness which is characteristic of it.[3] Both of these divine actions are to be understood as indications of God's power and goodness. Needless to say, none of this would seem to have any place in pancomprehensions.

If God and the world are not comparable, because God is not just another extant being, it is not possible for God and the universe to equal two. God is the reason why there is something instead of nothing, the condition for the possibility of every entity. From this derives the central importance for theism of the doctrine of

[3] "Secondly, a thing is said to preserve another *per se* and directly, namely, when what is preserved depends on the preserver in such a way that it cannot exist without it. In this manner all creatures need to be preserved by God. For the being of every creature depends on God, so that not for a moment could it subsist, but would fall into nothingness were it not kept in being by the operation of the Divine power, as Gregory says" (*Summa Theol.* I, q.104 a.1, in Aquinas 1920).

the *creatio ex nihilo*, which is absent from pancomprehensions: God creates without anything existing prior to the act of creation except Godself. There is no "nothingness" from which God creates, but the existence of God does not depend on the existence of the world either. Moreover, God being the reason why things exist instead of nothing existing does not simply mean that God caused the universe to exist in the past, but that God continues to work in the present as well. *Creatio ex nihilo* is also expressed in the idea of *creatio continua*; they are not, in fact, competing doctrines and cannot stand alone. Peter Geach, following in Aquinas' wake, devises an analogy of God as creator less like the blacksmith who makes horseshoes and such, and more like the minstrel who makes music: "For the shoe is made out of pre-existing material, and, once made, goes on existing independently of the smith; whereas the minstrel did not make the music out of pre-existing sounds, and the music stops if he stops making it" (Anscombe and Geach 1961, p. 110). Theism holds that this way of understanding creation accounts for divine omnipresence. The fact that God is everywhere does not place God in different places in space or filling the cosmos as a body or *sensorium.* Rather, God is everywhere as the cause of the existence of all those places. Being that God is the First Cause, omnipresence is implied, which is not to say that God is neither one with the universe, nor with everything that exists. Therefore, creation *ex nihilo* marks the ontological difference between God and what is not God, a step that medieval thinkers like Duns Scotus will attempt to mitigate by turning God into the being that is univocally said to be with all things that are, although in a more perfect way, so that God will easily be understood as just another element of the world system. By contrast, Aquinas preferred a theoretical device, the analogy, that keeps the world a good distance from God while maintaining some form of relationship. Seeing the world as the work of God does not require it to be God, thus allowing for its existential autonomy. God is present in the world as being the cause of its perfections, but the world is in no way God because of this.[4] Moreover, the perfections present in nature reveal the perfection of its author, while its defects are proof that the world and the creator cannot be one and the same.

The *ex nihilo* creation thesis bears upon all the properties of the divine being. Classical theism holds that God, *suum esse,* is necessary and the world is contingent. The world was created as a result of a freely made decision and is not necessary. Pancomprehensions, even if they may admit that the world is the outcome of God's free will and therefore contingent in its origin, must somehow come to terms with the notion of the world as being necessary, given the intrinsic dependence or

[4]Aquinas view is that "God is in all things; not indeed as part of their essence, nor as an accident, but as an agent is present to that upon which it works. Now since God is very being by His own essence, created being must be His proper effect; as to ignite is the proper effect of fire. Now God causes this effect in things not only when they first begin to be, but as long as they are preserved in being; as light is caused in the air by the sun as long as the air remains illuminated. Therefore, as long as a thing has being, God must be present to it, according to its mode of being. But being is innermost in each thing and most fundamentally inherent in all things since it is formal in respect of everything found in a thing [...]. Hence it must be that God is in all things, and innermostly" (*Summa Theol.* I, q.8, a.1, in Aquinas 1920).

identity they propose between God and the world (unless they state God's self being contingent). In order to do this, pancomprehensions attribute to the existence of the world the divine characteristic *per essentiam suam*, which ultimately devolves into some sort of necessitarianism, determinism or fatalism that holds that whatever happens occurs because God needs it to happen. It is, thus, impossible to affirm any sort of contingent element that would necessitate the miracle, i.e., the free intervention of God in the world.

Classical theism has also reflected upon the idea of the world as being necessary, but it has done so in a much different way. Thomism, in the medieval nominalist debate around divine omnipotence, holds that, once God has created, he is committed to creation: creation is in no case or way creation necessary for God, as it is for pancomprehensions, instead any necessity is consequence of the divine "commitment" to creation. This is why God does not change the past, which is necessary not *in se*, but *post factum creationis*. A square circle is not contradictory before creation, but after God's creation, the impossibility for a circle to be square is actualized and therefore it becomes necessary for a circle not to be square, since the square circle is outside the category of being possible. For Aquinas "God is called omnipotent because He can do all things that are possible absolutely, which is the second way of saying a thing is possible. For a thing is said to be possible or impossible absolutely, according to the relation in which the very terms stand to one another, possible if the predicate is not incompatible with the subject" (*Summa Theol.* I, q.25, a.3, in Aquinas 1920). Medieval nominalism, however, takes the opposite tack in denying creation a sort of "solid" character so that it is always contingent. This is proposed in order to preserve the notion of divine omnipotence: God may alter essences in virtue of God's own absolute power and unassailable will. Aquinas, on the other hand, concludes that "it is clear that the omnipotence of God does not take away from things their impossibility and necessity" (*Summa Theol.* I, q.25, a.3, ad 4, in Aquinas 1920).

As with divine need, so with to the immutable nature of God, a key issue for theism, that is rejected both by pantheism and panentheism for obvious reasons. The world, as a locus of change, changes God, for their substances are confused and mixed given that God is just another natural entity. Theistic difficulty to relate divine immutability and eternity to change and time seems to have led to the conclusion that equating God to nature presents a solution to this problem. Some theistic authors see God as everlasting (Wolterstorff 2000), however, the central classical thesis posits that eternity is atemporal. Furthermore, many of the theistic theological reflections on divine passibility seem to work best within a panentheistic framework rather than within a theistic one, especially when such reflections do away with theoretical prudence and, for example, affirm that human suffering, acts, decision-making, and temporal events work in such a way as to change God. The traditional Thomistic theistic solution presents a God beyond the order of creatures, given that these are ordered to God, but unto which God is not ordered. When we speak of the connection of God to creatures, we do it in a temporal sense, e.g., God removed the Israelites from Egypt, God will come again to judge the living and the dead, etc. This is not due to any change in God, rather the change occurs in the creatures just as –and this is the example utilized by Thomas Aquinas– one might say that a pillar moves from my left to my right simply because I have moved (*Summa Theol.* I, q. 13, a. 7, in Aquinas 1920).

In the same way, it would seem that God's infinite nature would be diminished due to God's limitation by the world; the same is the case with God's simplicity which, according to one of its meanings, implies that God's essential nature and God's being are identical. This is not the case in panentheism, where God is described in composite terms, since the world is somehow included. The notion of divine perfection is affected in the same manner. In pantheism, if the universe is God, God cannot be considered a perfect being in any sense of the term because the universe is always perfectible (as proven by Gaunilo of Marmoutiers' island). Therefore, God cannot be a source of values beyond those that already exist: there is no other duty besides being in a naturalistic way. Thus, the theistic suspicion of pantheism is also connected to the traditional theistic suspicion around relating any entity, including material ones, with God. It also looks with askance at the pantheist auto-justification of reality by reality itself, identified as God.

Many so-called theistic reflections, however, can be considered, at the very least, borderline panentheistic, especially those of a Neoplatonic bent, which describe divine substance emanating from the world and returning to a world made of that same substance, and those that understand divinity to be the "soul" of the world. Interpretations and musings along this line, among other things, led to the condemnations of Scotus Eriugena and Meister Eckhart. This line of thought is also present, in some form or another, in the theoretical writings of Nicholas of Cusa, Giordano Bruno, and very clearly in the theological proposals of German idealism, which, of course, was strongly influenced in its origins by the *Pantheismusstreit* around Spinoza's thought. According to the latter, God includes the world, or the world is included in God in a dialectical way. The same can be said about readings inspired by the Kabbalistic doctrine of Isaac Luria of the *tzimtzum,* recovered in modern times by Jürgen Moltmann and his idea of creation as *kenosis*. Somehow God must empty Godself to make room for the world.

The relevant paradigmatic theistic doctrine can be found in the condemnations of pantheistic and panentheistic theses in the *Syllabus* and in the documents of the First Vatican Council. It is within this framework that we must theistically and with analogous prudence interpret the countless biblical, patristic, as well as theological references that seem to describe God being in things and/or things being within God. In all cases, theism applies the hermeneutical principle that God and the world are two separate substances, with diverse identities. That is why it remains necessary to narrow the debate relative to how and up to what point the world influences God. In other words, it is necessary to suggest models that may explain the relationship between the world and God whereby they can relate to each other without becoming confused one for another, relating to each other, yes, but as separate entities.

18.4 Possible Theistic Ways to Explain the God-World Relationship

Classical theological theistic tradition is rich in analyses of the category of relationship. The trinitarian theology of perichoresis (referring to the communication among three persons of the same nature, where the notion of relationship avoids confusion and preserves the difference) or theological reflections on the hypostatic union –that Jesus is fully God and fully human– are examples of detailed analyses of the relationship between realities that are not intermingled. Key to theism is keeping different realities separate, whereas pantheism and panentheism fall into the temptation of mixing them up to varying degrees.

In order to better ground this sort of dualism (God and whatever the universe is), theism may appeal to theoretical tools like those used by Richard Swinburne to defend dualism between mind and body, with no need to share his conclusion on the subject. Swinburne's successful theoretical point is to argue for the existence of physical substances and pure mental substances that may interact without losing their specificity (Swinburne 2019). In the same way, one might further reimagine the fundamental elements of Aquinas' traditional strategy: namely, to think of God in terms of being the agent or cause who influences created order, as we can see in three of his five proofs: Motion (I), Efficient Cause (II), and Design (V),[5] We must keep in mind, however, that any discourse which places God and the world on an even footing has an evident analogical component. This is the reason why Thomas Aquinas states that the idea that God "contains things" is proposed "by a certain similitude to corporeal things" (*Suma Theol.* I, q. 8, a. 1 ad 2, in Aquinas 1920), by analogy with the body as contained by the soul. This similarity is balanced out by the fact that things are far from God "by the unlikeness to Him in nature or grace as also He is above all by the excellence of His own nature" (*Suma Theol.* I, q. 8, a. 1 ad 3, in Aquinas 1920). The difference, the non-similarity of things to God, is due to the excellence of the nature of God. Pancomprehensions, by contrast, are univocal. In Thomist theism, excellence, which is not a question of degree but of nature, demarks a point of separation that can only be illustrated and held by analogy. This is the key to theism.

References

Anscombe GEM, Geach PT (1961) Three philosophers. Basil Blackwell, Oxford

Aquinas T (1920) The Summa Theologiæ of St. Thomas Aquinas. Second and Revised Edition Literally translated by Fathers of the English Dominican Province. Online Edition. Available at: https://www.newadvent.org/summa/

[5] "God is in all things by His power, inasmuch as all things are subject to His power; He is by His presence in all things, as all things are bare and open to His eyes; He is in all things by His essence, inasmuch as He is present to all as the cause of their being" (Summa Theol. I, q.8, a.3, in Aquinas 1920).

Aquinas T (1991, 1997) De aeternitate mundi. On the Eternity of the World (trans. Miller RT). Available at: http://jonhaines.com/thomas/DeEternitateMundi.htm#f1

Aristotle (1933) Metaphysics. In: Tredennick H (ed) Aristotle in 23 Volumes. Harvard University Press, Cambridge. Available at: http://www.perseus.tufts.edu/hopper/text?doc=urn:cts:greekLit:tlg0086.tlg025.perseus-eng1

Castro S (2021) La sublimación de la belleza. Alpha 2(53):89–100

Culp J (2021) Panentheism. The Stanford Encyclopedia of Philosophy. Available at: https://plato.stanford.edu/archives/win2021/entries/panentheism/

Göcke BP (2017) Concepts of god and models of the god–world relation. Philos Compass 12(2):1–15

Leftow B (2016) Naturalistic pantheism. In: Buckareff A, Nagasawa Y (eds) Alternative concepts of God. Oxford University Press, Oxford, pp 64–90

Mander W (2022) Pantheism. The Stanford Encyclopedia of Philosophy. https://plato.stanford.edu/archives/spr2022/entries/pantheism/

McGrath AE (2008) Theology. The basic readings. Blackwell, Malden/Oxford

Meister C (2017) Ancient and contemporary expressions of panentheism. Philos Compass 12(2):e12436

Pannenberg W (1996) Theologie und Philosophie. Ihr Verhältnis im Lichte ihrer gemeinsamen Geschichte. Vandenhoeck & Ruprecht, Göttingen

Spinoza B (1914) Opera quotquot reperta sunt. Martinum Nijhoff, Hagae

Swinburne R (1993) The coherence of theism, 2nd edn. Oxford University Press, Oxford

Swinburne R (2019) Are we bodies or souls? Oxford University Press, Oxford

Wolterstorff N (2000) God is “everlasting” not “eternal”. In: Davies B (ed) Philosophy of religion. A guide and anthology. Oxford University Press, Oxford, pp 485–504

Chapter 19
The Hidden Theology in the New Naturalisms

Alfredo Marcos and Moisés Pérez

Abstract In this chapter we place contemporary naturalism on its proper intellectual ground. We maintain that naturalism is not a scientific hypothesis, it says nothing original about nature. Nor is it a philosophical thesis, except as natural theology. In fact, under the naturalistic label a series of doctrines about the (non) reality of God are grouped. That is, the adequacy of naturalistic ideas must be discussed within the territory of theology, to which these doctrines belong. From here, we draw a classification of naturalisms according to whether their theological outlook is more or less liberal. The strictest versions admit only an atheistic theology. Others would also accept deism. Some, more liberal, are also compatible with pantheism or even with panentheism. Only an extremely liberal version of naturalism would be compatible also with a theistic theology, but it is worth questioning whether in this case we could properly continue speaking of naturalism.

Keywords Radical naturalism · Radical ecologism · Scientism · Anti-theism · Ananthropic universe

19.1 Introduction

Current science-based naturalism rarely contributes positive original ideas. Its several varieties only share one distinctive characteristic, namely, their refusal of theism. Consequently, in its contemporary renditions, naturalism becomes mainly a theological thesis. On the other hand, in its process of radicalization, the ethically-concerned ecological naturalism also tends towards theological stances, either pantheistic or animistic. Our purpose in this chapter is not so much to argue about

A. Marcos (✉)
Department of Philosophy, Universidad de Valladolid, Valladolid, Spain
e-mail: amarcos@uva.es

M. Pérez
Philosophy Department, Universidad Católica de Valencia, Valencia, Spain

L. Valera (ed.), *Pantheism and Ecology*, Ecology and Ethics 6,
https://doi.org/10.1007/978-3-031-40040-7_19

naturalistic positions as to place them in the field of research where they belong. They do not belong in the field of science and, in the field of philosophy, they contribute very little to ontology, epistemology, and ethics. In fact, they contain notions that should be discussed in the field of theology since, in the long run, they deal mostly with the existence of God, the possibility of knowing God and the relationship God may have with human beings and with the world. So much so, that it may be possible to suggest, as we will do further on, a classification of current naturalisms in view of their respective theological stances, depending on whether they seem to be liberal or if they only admit to an orthodoxy that may range from atheism down to animism going through agnosticism and pantheism. Later, within theological research, someone may wish to rate their merit. But this last step lies beyond the scope of this chapter.

19.2 Naturalism and Natural Sciences

Describing naturalism is not an easy task since, in fact, there is no clearly defined version of naturalism. There are several varieties of a doctrine whose borders are rather blurred. The very fact that a good part of contemporary philosophy has taken *naturalism* as a kind of inescapable tenet has fostered the multiplication of *naturalisms*. In other words, because naturalism is taken for granted within some philosophical circles, differences among philosophers in those circles –regardless of how deep they really are– end up appearing as varieties of the same doctrine. And as differences among philosophers are never few, varieties bloom everywhere. A century ago, the philosopher Roy Wood Sellars (1922, p. vii) stated that "we are all naturalists now." And he added immediately: "But, even so, this common naturalism is of a very vague and general sort, capable of covering an immense diversity of opinion" (Sellars 1922, p. vii). We might think that in our times, naturalism is still the main philosophical orthodoxy. But naturalism remains as popular as it is diffuse.

19.2.1 Everything That Exists Is Natural

As a first approach we may say that naturalism holds that *everything that exists is natural* (Pérez 2021, 2022a). Let us begin with an analytic approach to this central statement of naturalism: "everything that exists is natural," which implies that whatever exists is natural and only what is natural exists. The thesis has two parts: it affirms all what exists is natural and denies the existence of anything that might be extra-natural. So far, the simple semantics of the statement.

Further, if we add some of the historical and pragmatic context, naturalism strongly denies not merely the extra-natural, but very specially the supernatural. It is not worth devoting too much time to this point. It is obvious that the whole poetic folklore of fairies, dwarfs, and goblins, as well as the ridiculous game of flying teapots and spaghetti are not the real issue. In this discussion, they are mere decoys. Here we are dealing with the supernatural and, especially, with God.

What do naturalists say? Two things: that the natural exists (which is so obvious that nobody questions it), and that God does not exist. The statement about the existence of nature cannot be taken as a statement belonging exclusively to naturalism but is a commonplace where most philosophies agree (acosmism would count as the one exotic exception). So, exactly what is current naturalism about?

About denying the reality of God. Namely, we are talking about theology. In fact, the main obsession of strong naturalism is the denial of the reality of God, especially the denial of a God as creator and provident. Its main tenet, therefore, is a theological thesis: God does not exist and, if he did exist, he would have no connection with nature or with the human being. The tendency of strong naturalism to deny or devalue freedom and human subjectivity is derived from this theological thesis (Pérez 2018).

In general, naturalism has closer links to theology than to natural science, despite what naturalist authors themselves usually say. On one hand, sciences do not need naturalism and they derive no benefit from it. On the other, naturalism does not need sciences although at the moment it is trying to feed off their prestige by adopting scientific apparel. In other words, all things being equal, should naturalistic philosophy disappear scientific activity would not be concerned at all; there would still be scientific research of the same quality as current research. Natural sciences have no need at all of naturalistic philosophy. Besides, naturalistic philosophy does not even need the existence of natural sciences. It is true that in its scientist version it refers to them and, somehow, feed off their prestige. But well before modern science was born there were naturalistic thinkers, also called materialistic (Soler 2013). Should our civilization pay no heed to natural science, philosophers would still be able to promote naturalistic ideas in non-scientific versions. And if there ever was ancient science, it was not carried out by the most naturalistic thinkers. Rather, precisely the less naturalistic ones were involved in it. Greek astronomy has clear roots in Pythagoras and Plato, biology has Aristotelian roots, while atomists or sophists, closer to naturalism, made scarce contribution to natural science.

The argument may be applied to almost every period in history (Arana 2020, 2021, 2022). If Aristotle wasn't a radical naturalist, Galileo, Descartes, Kepler, Leibniz, or Newton were much less so. Neither Lavoisier, founder of the new chemistry, nor the creators of the synthetic theory of evolution would claim to subscribe to strong versions of naturalism. Neither were evolutionary biologists and geneticists Ronald Fisher, Theodosius Dobzhanky, nor Francisco Ayala. Much less so was the proponent of the cosmological theory of the Big Bang, Georges Lemaître. And coming to a more recent field, the past director of the Human Genome Project, Francis S. Collins, would not rate himself as a naturalist in a strong sense either. One might dodge our line of argument with the assertion that, in fact, these outstanding scientists –from Aristotle to Collins– did not fully understand the philosophical implications of the natural sciences they worked on and, in some cases, even founded. But it would be a hard stance to maintain because in most of the cases mentioned –if not all– we are talking about thinkers who reflected expressly on the philosophical consequences of their scientific findings. Of course, many other scientists, current or historical, would back naturalism in philosophy. Nevertheless, historical evidence proves that natural sciences and naturalistic philosophy are mutually independent.

Such independence does not apply regarding naturalism and theology. In this case, connections seem to be strict. In fact, the theological condition of naturalism allows us to work out some gradation of its variables. We might list naturalisms according to how liberal they are as regards theology. We would have a distribution of the different versions of naturalism: some strict or orthodox, others more liberal regarding theology.

To begin with, there is a peculiar version of naturalism that denies the existence of nature itself. We might say: *naturalism without nature.* It is wonderfully expressed in the verses of Alberto Caeiro, one of the heteronyms of the Portuguese poet Fernando Pessoa (2010, p. XLVII): "*Vi que nao há Natureza, que Natureza não existe, que há montes, vales, planicies, que há árvores, flores, ervas…*" ("I saw there was no Nature, that Nature does not exist, that there are hills, vales, plains, that there are trees, flowers, herbs…"). There are natural things, in the sense that they are born and die, but they do not come from the same hand, they do not make up a whole, they do not form a system, nothing unites them, not even supposedly natural laws. If we insist on seeing them as a real and true whole it is simply due to some kind of "*doença das nossas ideias*" ("Disease of our ideas," Pessoa 2010, p. XLVII). This nihilistic version of naturalism requires atheism as its theological counterpart. It seems quite clear that it is not compatible with theism in its traditional format, but neither with pantheism. In both cases, some kind of unity and order of the world are presumed. In the case of the former, that unity and its order correspond to creation and providence. In the latter, the unity of the world and its order are due to the identification of nature with God, a God that is strongly rational at times (such as in the pantheisms of Spinoza or Einstein). But in the words of Pessoa (2010, p. XXVII), "*Só a Natureza é divina, e ela não é divina…*" ("Only Nature is divine, and it is not divine…"). It is questionable if this version of naturalism is compatible with polytheism and up to what point, but it seems clear that it does not fit with deism: the most important "task" of the deistic God is to order the world in a particular way. And the only way to know him is through rational knowledge of that way (it is a God of reason, of strictly natural theology, not revelation). On the other hand, it seems clear that such a version of naturalism is incompatible with natural science itself – a science that tries to obtain a scheme, either of the order which things work out among themselves, or else of the invariant relationships among them that allow us to speak about the existence of laws. Briefly, this nihilist-style naturalism would be strictly related to atheism as a theological stance.

19.2.2 Strong Ontological Naturalism

A second version of naturalism would be *strong ontological naturalism.* It is scarcely liberal as regards theology as well, though slightly more open than the former. Ontological naturalism says that only the natural exists; therefore, it denies the existence of a transcendent God. Despite which it leaves open a space for the tenet of pantheism. There is no God beyond this nature that accounts for it or explains it, but what if God is nature? What if we attribute some of the traits of divinity to nature? This type of naturalism seems to be compatible both with atheism and with pantheism, although obviously not with theism, deism, or panentheism (it seems unlikely that an ontological naturalist would accept a transcendent dimension such as that stated by panentheism).

19.2.3 Epistemological Universal Naturalism

The theological consequences of a third version of naturalism, based on epistemology, are like those of ontological naturalism. We mean *epistemological universal naturalism*: it holds that there is no knowledge besides that of the natural sciences. This type of naturalism is equivalent to *scientism*. As we said above, its consequences are like those of ontological naturalism: we only know through the instruments of science, so there is nothing like a transcendent God. But there is no objection to equating all that knowledge with God. Therefore, we may think that universal epistemological naturalism, compatible with atheism, also opens the door to pantheism, but it would hardly be compatible with panentheism, let alone with theism (there is nothing in science like "the transcendent").

19.2.4 Epistemological Naturalism

A fourth kind of naturalism, that we may name *local epistemological naturalism*, states that in some places -as science or philosophy- we must stick to the epistemological methods of natural science and only to those, because they are sufficient to provide a full explanation. If I am devising an experiment with a particle accelerator, for example, it makes no sense to address God at all. The theological consequences of local epistemological naturalism are different to those described above because, insofar as it professes some ontological agnosticism, it is far more liberal. We may consider it is compatible with atheism, agnosticism, pantheism, and deism. But it is unthinkable that this naturalism may be compatible with panentheism or with theism. It is not compatible with panentheism, because the knowledge of the transcendent dimension of nature is ruled out by science's methods. And it is incompatible with theism because, at its core, theism holds that God is the cause that explains reality. It is not that a theist wishes to bring God into the experiment of the particle accelerator but as, from God's perspective, the natural world has been caused by God and is providentially governed by God, a "full" explanation of the world would require, in the long run, an appeal to God. It is not that God must be introduced into science but the scientific explanation of nature, if it sticks to local epistemological naturalism, cannot be complete in some rational, relevant, or ultimate sense. For example, we have performed an experiment in the accelerator with no need to appeal to God, but then we might still ask: which conditions of reality have permitted to carry out this experiment, where does the order that makes it possible come from, why are human beings, in some way or another, able to understand or know reality at such a profound level as the one we find in the particle accelerator, why can mathematical tools devised beforehand by humans describe reality so precisely? All these questions surpass the realm of science, but they are also senseless in a philosophy that follows this type of naturalism.

19.2.5 Heterodox Naturalism

Lastly, we can talk about a fifth kind of naturalism. It would be more liberal than the previous one and, maybe, some may not even rate it as an authentic naturalism. In any case and in view of the current lines of the debate, it might be seen as a *heterodox naturalism*. It would be compatible with any idea of God, from atheism through theism. This kind of lenient naturalism limits itself to state the reality of nature and the importance of natural sciences, as well as the notorious interest of their results for philosophical research. It would consider that there can be no good philosophy if the results of natural sciences are ignored, let alone ignoring natural reality. It must be clear that this naturalism does not appeal to or necessarily end in the theistic idea of God, but it does not deny it *a priori*. It is as liberal as possible because it does not prejudge, rather it leaves it an open issue. In view of this, we might call it a-theological naturalism. Not because it denies theology; rather, it does not claim to have a theological solution from the start; that is, it is not essentially a theological thesis, as seems to be the case –somehow or other– with the other naturalisms. This open version of naturalism is attractive since it acknowledges the dynamics typical of nature, as well as the importance of natural sciences. And it achieves this with no need to make any theological commitment.

Briefly, we have a kind of nihilistic unscientific naturalism that leans towards atheism. Classifying it under naturalism is almost an overstatement since while it states the reality of natural things, it denies nature itself. At the opposite end we have a naturalism that is compatible with science and totally liberal in the field of theology. We subscribe to it, but some would hesitate before classifying it as fully naturalistic because they hold that the adjective should be reserved for philosophical doctrines that include –explicitly or implicitly– a theological commitment. These are the blurred borders of the naturalistic palette. In the centre, in the range of the undoubtedly naturalistic, we would have the ontological and epistemological varieties, which open up the possibility of pantheism, despite being deeply refractory to all varieties of theism.

19.3 Naturalism and Ecological Consciousness

Besides reflection on natural sciences, the other great agent for the introduction of naturalism in contemporary thought is ecological consciousness. Following this line of development, naturalism seems to be a doctrine whose main component runs along theological lines –be it pantheist/panentheist (Valera and Vidal 2022) or animistic.

Ecological consciousness belongs within the scope of practical reason, beyond ontology and epistemology. It includes knowledge of the ecological situation of our planet, as well as about the impact of human actions on it. This consciousness functions prescriptively as well as descriptively. Ecological consciousness aims to tell us what we *must* do as to our relationship with nature. It soon becomes an environmental ethics.

There is no reason why environmental ethics should result in any theology in particular, but that is how it has happened. This is due to the initial view of the ecological issue as a problem with intrinsic value (inherent, or in itself).[1] There is an intention to base ecological ethics on the inherent value, not of the human being but of nature as a whole or on the value of some natural beings. Anthropocentrism was left out from the start as a possible ground of environmental ethics. Biocentrism or ecocentrism have occupied that place. In fact, many thinkers have held that the ecological crisis was caused precisely by the anthropocentric mentality. Anthropocentrism has been systematically accused of devaluing all other beings, up to the point of conceiving them as mere instruments in the service of humans. Most environmental thinkers have understood that it is necessary to change the focus of ethics, the grounds of value, and place it in nature. Nevertheless, between this move and the divinization of nature itself there only is a very small step. And that step seems compulsory if we expect to acknowledge moral proposals, without handing everything over to pure and irrational emotivism.

Insofar as environmental ethics may wish to displace human beings from the centre, it will be seized on by pantheistic theological stances. In other words, nature will have to become divine: *Deus sive natura.* Otherwise, where would we find moral guidelines? Obviously, not on the grounds of its usefulness for humans, or based on the feelings humans may have, or social agreements they may reach. Any of these would lead us to fall again into reviled anthropocentrism.

Some environmental thinkers are more explicit in their pantheism. Among others, it must be examined as an implicit tenet or modified in panentheistic shape. Atheism and theism are excluded from any non-anthropological thinking in environmental ethics. The former, because it would either cancel any foundation of value, or it would lead us back to anthropocentrism. The latter, in view of its immediate connection to the special dignity of human beings, created in the image of God and after God's likeness, and destined by God to keep and care for the rest of

[1] These words are not strictly synonyms. However, they can be read as such in our argument, with no specification of different hues.

creation. In other words, among all things created, theism attributes special value to human beings. Consequently, a firmly anti-anthropocentric environmental ethics must be anti-theistic as regards theology.

However, it must be noted that an anthropocentric environmental ethics is open to theism but not tied to it. If it were stated in an anthropocentric key, it would lack strict theological implications. It might find backing in usefulness, feelings, consensus, deontology or in the dignity of human beings as creatures of God. Utilitarian, emotivist, contractarian, deontological formulae have their own virtues and defects; they may prove more or less efficacious, more or less reasonable, but they lack direct theological implications; they are compatible with atheism and deism. Moreover, so long as they are not absolutized, they can ground their sensible statements in the theistic notion of God. Anthropocentrism is also compatible with theism, as we have seen. But anti-anthropocentric environmental ethics nails its colors to the mast of a pantheist theology. Thus, it becomes theology.

Something similar might be said about animalistic variants. In their most radical versions, they have already gone from saying that the human being is an animal – where they might easily coincide with Aristotle, Saint Thomas, and Darwin– up to stating that human beings are *nothing but* an animal. All three would probably disagree with such statement. It must be noted that the passage from the first to the second statement is a logical fallacy. C.S. Lewis is said to have named "nothing-butterism" this peculiar inclination to fall into this kind of fallacy. In fact, this inclusion of humans among all animals may have two very different practical readings. The first one is truly alarming and can therefore be left aside. While the second one, morally more acceptable, surely leads to animistic theological stances.

According to the first reading, human beings are animals and may be treated as such, because their value is relative and gradual, not absolute. There can be no reference to human dignity as thematized by the Universal Declaration of Human Rights (UDHR). Accordingly, some human beings would be more valuable than others, just as we admit that some animals are more valuable than others (in fact, nobody demands the same treatment to a dog and to a sponge). If a human being were nothing but an animal, we might scarcely refer to the dignity of the most vulnerable and dependent human beings. And claiming a species barrier would be inconsistent with anti-speciesism. We should simply acknowledge that some human beings are more valuable than others. That is what Peter Singer does at different points of his work (Singer 1993, pp. 169–173; Marcos and Pérez 2018).

Should many people find it difficult to acknowledge this harsh moral stance, alarming for most, we might choose the second, kinder, reading of course. If we assimilate humans to all other animals, we might think that there is some kind of absolute value among the latter, some kind of animal dignity; in other words, that their life is as sacred as human life. It is not easy to put this moral conviction into practice: it would lead us to offer equal treatment to all animals. On the other hand, the restriction of the dignity of the animal realm will soon be questioned, since there are good arguments to posit that there is intelligence and sensitivity in the realm of plants too (Mancuso and Viola 2015). However, it is not the intention of this work to criticize animalistic ideas, but to reveal their deep theology –in this case, as animistic, because they lead to sacralization of all animal life.

Might we still develop an environmental ethics completely liberal regarding theology? From our point of view, yes. It can be done, precisely, if an anthropocentric stance is acknowledged, which does not exclude –rather, it grounds– the intrinsic value of all natural beings. Another point: without some anthropocentrism, the inherent value of natural beings would either decay or would have to find some backing in pantheistic or animistic theologies.

Let us see what anthropocentrism has to say. The human being, not in general but in the actual person of each one of us, has some kind of special value which is incomparable, non-negotiable, immeasurable, absolute; in a word, what we (following Kant) call dignity. And what is it that anthropocentrism does not say or imply? It does not say that all other beings lack their own value; it does not state that all other beings are mere instruments in the service of human beings. How can we harmonise anthropocentrism with the inherent value of all other beings? Hans Jonas (1985) shows us the way. But let us begin by stating the complexity of the case before giving it an answer.

Robin Attfield (2018, pp. 48–50) connects the birth of environmental ethics with a thought-experiment proposed by Australian philosopher Richard Routley who in 1973 proposed *the last man thought experiment.* If only one human were left on Earth, aware that he himself will die soon and with him, the entire human family, he might arbitrarily damage all other living beings, even putting an end to them, with no human being hurt. Even so, according to a commonly shared moral intuition, the free harm caused to all other living beings would not be ethically correct. Why? It may be because we acknowledge intrinsic value in all other living beings, besides the instrumental value they may have for us. Environmental ethics developed from these grounds. Biocentric and ecocentric ethics appeared to challenge traditional anthropocentric ethics (Attfield 2018, pp. 74–76).

We begin to see the problem that many find in anthropocentric ethics. It is not a problem connected to ethics, but to anthropology. Anthropocentrism is only wrong when it is associated with an erroneous idea of human beings. What if, everything considered, human dignity was compatible with the inherent value of all other beings? And if it were precisely humans who confer value, if it were humans who illuminate all other beings, who transmit intrinsic value to all? Should we not then be anthropocentric even in environmental ethics?

Let us consider another thought experiment that both takes a step beyond the previous one and in fact gives it its sense (Marcos 2021). Let us name it *the ananthropic universe thought experiment.* Let us now eliminate the last person. There are no human beings in the universe, there never were and never will be, there is not even any chance of their ever existing. We are in an *ananthropic* universe. How would value be distributed in such a world? Is there anything wrong about a black hole absorbing a star or a complete galaxy, should such items exist in our fictitious universe? An *ananthropic* universe is a universe lacking any possibility of value. Except, of course, if the universe itself were God (pantheism) or if some of the existent nonhuman beings were divine (animism). Besides, this second mental experiment grants sense to the moral intuition that emerges in the first one. *The last human being should not damage their surroundings because those surroundings make possible the existence of human beings.*

If we wish to instil some value in the universe, without falling into animistic or pantheist stances, we should reckon with the (possible) presence of human beings and with an anthropology of each person's dignity. If we acknowledge the infinite value of every person, an absolute value, then we may acknowledge the intrinsic value of all other beings. In order to know how to go about it we will have to go back to Jonas (1985, p. 7): "Insofar as it is the fate of *man*, as affected by the condition of nature, which makes our concern about the preservation of nature a *moral* concern, such concern admittedly still retains the anthropocentric focus of all classical ethics."

Once the absolute value of human life has been established, however, all other beings do not become mere instruments. They also have their own value since the sheer possibility or capacity of value is a value in itself. It is a line of metaphysical grounding of value in being. Thus, Jonas clears the way that leads from *is* to *ought* with no need to go through the naturalist fallacy. In his own words: "The ontological idea generates a categorical, not a hypothetical, imperative" (Jonas 1985, p. 43). So,

> it is important to see that the mere fact of value (with its opposite) being *predicable* at all of anything in the world, whether of many things or a few, is enough to decide the superiority of being, which harbors that possibility within its manifold, over nothingness, of which nothing whatever, neither worth nor worthlessness, can be predicated. [...] The capacity of value (worth) is itself a value (Jonas 1985, pp. 48–49, italics in the original).

This last sentence is the keystone. Based on it, an anthropocentric ethics becomes perfectly compatible with acknowledging the inherent value of all beings. It actually grounds such acknowledgement.

If we admit the dignity of each person, we should acknowledge the value of natural beings which make the existence of humans on Earth possible. We inhabit an anthropic universe, hospitable to humans, to an infinitely valuable being. Therefore, the sheer possibility of this universe to host human beings is an intrinsic value of such a universe. And the same can be said about each being in that universe. Its value does not depend on their being instrumentally *necessary* for our life, but on the fact that they make it *possible*, and that possibility is already a value we cannot rate as instrumental. It is an intrinsic one, because it happens regardless of whether there are human beings *de facto* or not.

Now we might enquire about the grounds of human dignity. And the reply would follow the Kantian line of autonomy. It would remind us of our common belonging to the human family, along the line of UDHR or it would use notions such as "image and likeness." In other words, an anthropocentric grounding of the value of natural beings may also find several different groundings. It is perfectly compatible with a theistic vision of God, the universe, and the human being, but it does not strictly depend on it. Therefore, it may achieve social consensus over and above religious beliefs.

On the other hand, when we try to find a biocentric or ecocentric grounding one of the following may happen. Either the inherent value of natural beings remains hanging over, with no rational justification. Or the inherent value of natural beings is founded on the divine condition of the entire universe or some of its parts. Here we face a clearly theological grounding, which should not act as a political argument. It is not legitimate to impose some kind of State pantheism, nor, of course, to turn animism into an official religion. What we have just said is not political fiction. Some countries have already courted with imposing pantheism as the State religion. And in others there are political parties that experience such a temptation. Among the advantages of the anthropocentric stance defended in this paper, there is one which is non-trivial: it allows us to acknowledge the inherent value of natural beings with no need to combine religion and state.

19.4 Concluding Remarks

The project of naturalizing philosophy, so widely circulated currently, is mainly theological both in its scientist and its ecological versions. There is practically unanimous agreement about at least part of the naturalistic proposal. That consensus goes well beyond the boundaries of naturalistic schools. Consequently, it cannot be taken as its own and distinctive part. Practically every sensible person admits to the existence of natural beings, that they are valuable and that they consequently deserve some respect, that nature has its own dynamics, and that the findings of natural science must be earnestly considered both for philosophical research and in order to reach practical decisions. Up to this point, naturalisms agree with common sense and hold widely shared ideas. The distinctive and exclusive trend of naturalisms comes to surface when each adheres to a particular conception of divinity: from strict and practicing atheism up to animism, going through the different bypaths of agnosticism and pantheism. And all of them, as a group, reject theism. If we wished to summarise it in an oversimplified formula, leaving variables aside, we would state that *naturalism is the word currently used for anti-theistic theology*.

References

Arana J (2020) La cosmovisión de los grandes científicos del siglo XX. Tecnos, Madrid
Arana J (2021) La cosmovisión de los grandes científicos del siglo XX. Tecnos, Madrid
Arana J (2022) La cosmovisión de los grandes científicos del siglo XX. Tecnos, Madrid
Attfield R (2018) Environmental ethics. A very short introduction. Oxford University Press, Oxford
Jonas H (1985) The imperative of responsibility. The University of Chicago Press, Chicago
Mancuso S, Viola A (2015) Verde brillante: Sensibilità e intelligenza del mondo vegetale. Giunti, Milan
Marcos A (2021) Contra la ética ambiental. In: Rodríguez Valls F, Padial JJ (eds) Ciencia y filosofía, Estudios en homenaje a Juan Arana, vol II. Sevilla, Thémata, pp 669–675

Marcos A, Pérez M (2018) Meditación de la naturaleza humana. BAC, Madrid
Pérez M (2018) ¿El naturalismo es un antihumanismo? La ontología naturalista y la idea de persona humana. In: Amengual G, Arana J, Génova FJ, Pérez M (eds) El ser humano: más allá del animal y la máquina. Instituto Emmanuel Mounier, Salamanca, pp 125–148
Pérez M (2021) La cosmovisión naturalista. San Esteban Editorial, Salamanca
Pérez M (2022a) El naturalismo cientificista como pseudo-religión y anti-teología. Scientia et Fides 10(1):73–90
Pérez M (2022b) ¿Naturalismo versus Teísmo? La polémica Diéguez-Soler. Anuario Filosófico 55(2):265–296
Pessoa F (2010) O guardador de rebanhos (poemas de Alberto Caeiro). Best Books Brazil
Sellars RW (1922) Evolutionary naturalism. Open Court, Chicago
Singer P (1993) Practical ethics. Cambridge University Press, Cambridge
Soler F (2013) Mitología materialista de la ciencia. Encuentro, Madrid
Valera L, Vidal G (2022) Pantheism, panentheism and ecosophy: getting back to Spinoza? Zygon 57(3):545–563

Chapter 20
Towards a Speculative Ecology. Monads, Habits, and the Non-otherness of the World

Emma Lavinia Bon and Francesco Vitali Rosati

Abstract In this chapter, we discuss metaphysical issues activated by ecological discourse, in order to coherently outline a speculative philosophy which rejects both holism and reductionism. In doing so, we will shed new light on classic concepts of Latin pantheism, such as locus, habitus, and ipseity, which are helpful to frame the God-world chiasm as a process of "making place," free of any negative movement. The chapter is articulated on two levels: individual and cosmological. In the first section, we consider the notions of living and psychological individuality, arguing that the problem of alterity may be faced from a perspective of primitive, displaced unity. Nonetheless, we should consider the fabric of all beings as an ecology of ideas, developing the analogy between ecological networks and the mental process. In the second section, we explore the same analogy in the light of the pantheistic concept of God as cosmogony: their mutual immanence, just as that of the self and the otherness, appears to be an *in locu* genetic act, employing a logic of *complicatio* and *explicatio*. Finally, we argue that a correct representation of the problem may avoid the vicious dissonances of rigid dualities, implying the possibility of turning them into virtuous resonances.

Keywords Cusa · Ecology of mind · Monad · Locus · Pantheism

E. L. Bon (✉)
Department of Philosophy and Communication, University of Western Piedmont, Vercelli, Italy
e-mail: emmalavinia@gmail.com

F. Vitali Rosati
Department of Philosophy and Educational Sciences, Università di Torino, Turin, Italy
e-mail: francesco.vitalirosati@unito.it

L. Valera (ed.), *Pantheism and Ecology*, Ecology and Ethics 6,
https://doi.org/10.1007/978-3-031-40040-7_20

20.1 Introduction

Since its appearance in the twentieth century, deep ecology has set the philosophical task of repairing the broken mirror placed between the individual and their *milieu*, between humans and cosmos. Stimulated by the material provided by the positive sciences, the young and multifaceted discipline turned to the fundamental work of bridge-making to build transversal connections among fields of knowledge, cultures, and times (Deléage 1991). Accordingly, it gradually defined its onto-epistemic task of revitalizing nature, as well as our representations of it. In doing so, a number of panpsychists (Whitehead 1929), biopsychists (Bateson 1979; Ruyer 1952), cosmologists (Vernadsky 1926), and neo-animists (Stengers 2011; Viveiros de Castro 2009) accounts have resurfaced, reclaiming sometimes the power of "ideas of the past [...] to affect the present" (Stengers 2011). Some have summoned a new holistic spirit (Lovelock 1979; Bohm 1980), opposing the reductionist and mechanistic, yet effective methods of the natural sciences. Others have turned their attention to non-Western and non-modern ontologies (Descola 2005; Kohn 2021), in order to rediscover an original agency or even a certain degree of subjectivity of the non-human worlds (Haraway 2003): forests, fungi, animals, matter, machines, technical objects. Asserting the transcendental status of the material, biological and cultural networks –that which surrounds and grounds human experience– meant imagining a non-human fabric or field that includes us and replicates through us.

Within the undoubted richness of the debate, some very ancient questions have been reactivated: insofar as some kind of universal animation is claimed, the millenary aporias of the soul, previously articulated by the doctrines of Renaissance vitalism and Idealistic panpsychism, still threaten to confine us to two alternate outcomes: holistic monism and dynamical pluralism, since "the world soul presupposes both an all-embracing, continuous living principle found throughout the world (*innata toti mundo*) and a totality of endless centres of energy endowed with self-sufficiency" (Giglioni 1995, p. 32). Yet, our challenge shall be thinking *together* the consubstantiality of forms and their disseminating multitude as "immanent differences" (Bateson 1979, p. 210). In these terms, the conceptual apparatus employed by Renaissance theologians, and especially by German theologian Nicholas of Cusa, may provide a fundamental key to articulate the relationship between one and many, the whole and the individuals, without collapsing one term onto the other.[1]

However, it is a common misconception, inside ecological discourse, to ascribe to forms a secondary reality, even an epiphenomenal illusion of the primary process. This is especially the case of radically relational and neo-materialist ontologies (Barad 2007), which privilege one of the terms of the individual-relations dichotomy. Such solutions indeed seem to place all the emphasis on an original

[1] It has been argued that what is commonly considered "pantheism," from twelfth century heresies to Giordano Bruno and Baruch Spinoza, does not properly concern an identity between God and world; rather it is an attempt to overcome the dualisms produced by language and fallacious substantialist representations (Dattilo 2021).

indeterminacy that blindly creates and dissolves worlds, disavowing all common sense and losing grip on concrete life. Rather, the multiform experience that animates human and non-human worlds requires a perspective of *betweenness* that cannot be reduced to either monism or pluralism, holism or reductionism. It must explore the topological particularism of the web of life (Deleuze and Guattari 1980), in order to establish a reason capable of "thinking from within the milieu."

But as long as the monadic structure of experience holds us on to our *singular* nature, "the plural of which is unknown to us" (Schrödinger 1944), this singularity tells us something about a higher, inexhaustible reality which penetrates all life. Such a transversal "bond" shall root all planes of existence in the same enfolding (*complicatio*) and unfolding (*explicatio*) rhythm, the same cosmic motion of systole and diastole. This greater bond (*λόγος;* "logos"), following the famous Heraclitean warning (DK 1 and 50) must be *listened to*: tuning to the "music of life" (Noble 2008) is the main quest of a speculative ecology.

20.2 The Form and Its Double

Both philosophical and scientific discourses struggle with the notion of individuality: we still lack understanding living organisms, whenever we face those strange "agents capable of creating their own norms, continuously harmonising their ability to create novelty and stability, combining plasticity with robustness" (Soto et al. 2016, p. 1). Organic and psychical processes seem to challenge our mechanistic views, claiming their own conditions of intelligibility: thus, when we observe in awe the morphological unity of the living forms, their ability to preserve and transform, we are mirroring our own essentially ontogenetic nature. Life appears to us as an endless process of individuation in the form of continuous and perpetual criticality. But how does this process flow? What kind of individuality does it express?

Gilbert Simondon attempted to settle the question in his doctoral thesis *L'individuation à la lumière des notions de forme et d'information*. According to the French philosopher, individuals cannot be considered as mere fixed identities, or some sort of epiphenomenal accidents, but rather as continuously emerging forms, critical transitions from a pre-individual plane to an actual, incarnated reality. If the individual is never separable from the process that generated it, the process itself is in turn unthinkable without the resolutions it comes up with: *forma dat esse rei* ["the form that gives existence to a thing"], in so far as without single beings, as caustically pointed out by Cusa, the universe could not be one, nor a whole, nor can it be anything at all (*De doc. Ign*. II, p. 4).

Nevertheless, from Simondon's point of view, to understand the ontological consistency of forms it is necessary to trace their genetic dynamics. Looking closely, when a form is generated, we witness some sort of displacement or symmetry breaking within a system, emerging from a truly *primitive tropistic unity*: an original condition of co-presence and mutual penetration between the living unit and its *milieu*, where the latter figures as a complex of directions and polarities that situates the individual *in the middle*, being the intermediate point of an "indefinite dyad [...] that develops from the individual itself" (Simondon 2005, pp. 166–168). The dyad, here, names the differential tensions of the becoming, of which form is always a resolution. This tension is immanent to the primitive unity, somehow incarnated by the individual existence. Then, at the very heart of experience seems to lie this *more-than-unitary* (but less than dual) plane, which recursively splits into asymmetric pairs, always polarizing towards new forms and new points of resonance, through endless cycles of "transductive" amplification. But this process, Simondon argues, does not exist *before* the individuals, but only *through* them (Simondon 2005, p. 256). Far from abstract, such *in situ* dynamism holds the conditions of our entire existence. Every single form exists by means of the "charge" of pre-individual reality that *contains* and *brings forth*, expressing the wider *puissance* we carry with us. In other words, the individual embodies the pre-individual intensity that it uses to create and transform. The same intensive state which, internally, animates and transforms the metamorphic being, externally connects it to other members of a broader individuation, by generating a wider "resonance" of which it is an element, hence capable of composition or dissonance. Simondon's pre-individual is therefore a plastic flow, a morphogenetic frequency which informs the bodies in rhythmic concentrations, constituting at the same time their mutual, "trans-individual" sensibility, their capacity for contact and reaction.

The complex, original unity that entangles any form with its causal principle, rejects any logical relations of identity and alterity. Instead, it reclaims some sort of platonic "bastard reasoning:" something capable of simultaneously grasping the symmetrical unity of the pre-individual level and the asymmetrical diffraction of the individuals animated by that. But how do we experience pre-individual animation? How can we witness it?

In his compelling work *Streifzüge durch die Umwelten von Tieren und Menschen*, Baltic biologist Jakob von Uexküll had hinted at an answer. By describing organisms as perceptual and operational agents, von Uexküll noticed that each form of life behaves as a *self*, a subjective unit, a monad, coextensive with its own operative domain. By their actions, organisms unravel peculiar relational patterns, as if being driven by psychic *themes* (von Uexküll 2010): semiotically distinctive orientations, vectors of purpose. Addressing the *signifying* agency of any form of life means considering cognition and representation as inseparable from simple being. Every agent is *a sign* insofar as it makes bodies, it grows, and it extends itself in semiotic chains, within the wider "ecology of selves" that envelops us all (Kohn 2021).

Furthermore, this meaningful organization of experience shows some kind of "magical," non-causal traits (von Uexküll 2010, pp. 143–144), as it appears to be imbued with imaginary elements that orient the action, even in the absence of sensory stimuli. These invisible attractors are "themes" or "emotional tones" (*Stimmung*) that produce "paths unfolded with the intensity typical of a magical apparition" (von Uexküll 2010, p. 144). And yet, such paths are not properly something external to the organism: they are "externalized" signs of the subject, working as erotic pulses that lure us into life and action. They are the same desires, the same *guiding themes* that produce the external milieu *together with* the inner organization of our body; in this sense, their magical activity could be seen as the actualization of the pre-individual primitive plain, its mode of existence, its unfolding. Its "transit," then, shall be interpreted according to a logic of *complicatio-explicatio*, as Cusa suggested. As the intensive plane manifests as a constellation of purposes and values (von Uexküll 2010), transformative forces play the role of entelechies, the highest elements of experience: themes are the very incarnation of "ideas", supra-individual realities grounding any change and creation, or, the other way around, "thematicity" is the way we acknowledge the eidetic essence in glimmering moments of divination. The immanent presence of transformative ideas within phenomenal existence ultimately elevates subjectivity to a cosmological status: "the whole universe consists of elements disclosed in the experiences of the subjects" (Whitehead 1929, p. 166).

If we understand properly the metaphysic status of the thematic forces, we should speak of a "divine" self as the true entelechy of all experiences: the *idea* that is none-other than each single thing. The Platonic tradition, particularly vivid in Cusa, foreshadows such intuition when it speaks of *idea idearum* or essence of every single essence (*De vis.* IX, pp. 34–35), although it seems more correct to state, in Ruyer's words, that "*Dieu n'est pas l'Agent des agents, leur fabricant, il est l'Agent qui est dans tous les agents. Sa liberté ou sa science ne contredit pas la mienne: il est ma liberté et ma science*" ["God is not the Agent of agents, their maker, he is the Agent who is in all agents. His freedom or his science does not contradict mine: he is my freedom and my science"] (Ruyer 1952, p. 300).

Furthermore, when we name the plurality of beings with terms such as "agents," "ideas," or "forms," we are assuming not only the continuity, but the very unity of life and thought. As far as "life thinks, thoughts are alive" (Kohn 2021), we shall immediately acknowledge *how* it does that. We are surrounded by the overflowing proliferation of modes of being, forms of life, *habits* which appear to be centres of a multifocal cosmos, each one being a specific *locus* of meaning, intrinsically colored by "purposeness" (Kohn 2021). Whether we choose to align or dissociate from this manifold texture, our growth and our health essentially depend on how we *represent* it. Shall we regard the ecology of forms as a process of adapting, symbiotic and co-evolving ideas, or an *evolutio idearum* of dialogic nature (Bateson 1972)?

If living beings are ideas, the "global mind" (Bateson 1979) is not simply the sum of them, nor a mere container, but rather, according to Bateson, it is the connection of all connections, a pure *connectivity* that entangles every pattern of life, not merely a part of it. Indeed, this kind of mental process entails an immense *mise en abyme* ("put in the center")[2] in which each singular part is a living recursion of the wholeness *and* a particular idea of the one mind. As a pantheistic *Mens* ("mind"), Bateson's global mental process exhibits a fractal structure, which repeats itself by varying in each thought: its unity, so to speak, self-reflects in the increasing diversity of "differences that make differences," and vice-versa, every single form fulfils

[2] In the terminology of heraldry, the abyme is the center of a coat of arms. The term *mise en abyme* means "put/placed in the center" and alludes to a coat of arms that appears as a smaller shield in the center of a larger one.

and penetrates the Mind. As the latter thinks and feels, it immediately creates and acts: being and thinking, from Elea to the Amazon forest, are the same.

In his famous treatise *De visione Dei* written in 1453, and even more so in a previous dialogue written in 1450 and entitled *De mente*, Cusa had engaged with this kind of paradoxes (see Cusano 2017).[3] He powerfully compared the relationship between God and the world to the perceptive act that *assimilates* the subject to what it sees: in the perfect simultaneity of creation and cognition, all things are conceived through the "force of his living plasticity (*in vi suae flexibilitatis vivae*)", meaning that the multiplicity of things is the modality of thinking of the one mind ("*Pluralitatem rerum non esse nisi modum intelligendi divinae mentis*") (Cusano 2017, VI, p. 95). In the divine mind, Cusa states, conceiving and creating are one (Cusano 2017, *De mente* III, p. 72), to the point that the very nature of thought is not cognition, but *creation*. In other words, God thinks what God creates and creates what God thinks: the object of thinking is immediately an existing, real object. This reversibility between the speculative and the ontological dimension – or, just as the Ancients suspected, between psychology and cosmology – is reflected in the experience of any living idea. As we are one with our thoughts and feelings (they are, so to speak, us in our wholeness), the divine Mind is one with us.

This particular, psychological foundation of cosmology can be described as resting on both as a *displaced* unity, and a *sur place* dyad. In this regard, for instance, Cusa uses a very special kind of double negation (*non-aliud*) to preserve the elusive concept from any dialectical movement of alienation. More specifically, he is able to overcome both duality and self-referential identity, proposing the broader notion of a constitutive differential unity, being more than itself. The pantheistic notion of *ipseitas* precisely conveys a closed unity, paradoxically interpenetrated with the other, insofar as it does not conceive an actual alterity, but only different moments of one immanent, self-explicative reality. Indeed, God cannot, strictly speaking, *relate* to the world, as far as the world is *represented* as a creature, separated from the creator. This way, God will always remain other than the creature. Rather, one must think of the world as one with God: an explained, unfolded, contracted God. Only when the genesis of the world is placed *in no other* –not merely in itself, but in "non-otherness itself" (*ab ipso non-aliud*: *De non al.*, VI, p. 22)– is its intimacy with the divine restored.[4] Thus, Cusa masterfully solves a fundamental theological tension, which haunted the whole history of ideas, recurring again in the final series of paradoxes in Alfred North Whitehead's *Process and Reality* (1929).

---

[3] The works of Nicholas of Cusa (1401–1464), Niccolò Cusano in Italian, are included in the first Italian version of all of Cusano's 'philosophical, theological, and mathematical writings'. The Italian translation was published in 2017. The text of the critical edition was edited by the Heidelberg Academy of Sciences, and incorporated an extensive set of notes and a systematic commentary on the individual writings. This Italian version was used during the preparation of this chapter and is cited in the text as "Cusano 2017."

[4] S. L. Frank suggested that Cusa gets rid of logical negation by means of a second-degree self-affirmation, which no longer tolerates any alterity, any "aut-aut" disjunction. The non-aliud only reveals the "self-evidence of God's creativity" in every form: not a stable and obtuse identity, but as a "concrete manifestation of absolute novelty" (Frank 1939, p. 101, 128).

Similarly enough, in the famous conclusion of Gregory Bateson's (1979) *Mind and Nature*, the author sketches the powerful image of an infinite sphere to portrait the problem of self-referentiality and non-otherness in the form of magical and ritual dances. Action meant not to produce an extrinsic effect, but a "difference that makes a difference," an operation that only "affirms membership in what we may call the *ecological tautology*, the eternal verities of life and environment" (Bateson 1979, p. 209). And at the beginning of the conclusive section of Mind and Nature, he states that:

> The Creatura, the world of mental process, is both tautological and ecological. I mean that it is a slowly self-healing tautology. Any large piece of Creatura tends to settle toward tautology, that is, toward *internal consistency* of ideas and processes. But every now and then, the consistency gets torn; the tautology breaks up like the surface of a pond when a stone is thrown into it. Then the tautology slowly but immediately starts to heal [...]. A self-healing tautology is also a sphere, a multidimensional sphere (Bateson 1979, pp. 205–206).

20.3 Wholeness and the Logic of *Locus*: Cosmos as Theogony

The mutual immanence of thought and being, *mens* and *creatio* is the key to develop a speculative ecology which is careful to preserve the singularity of each particular habit. Again, with Cusa, we can say that the plurality of things is the spontaneous result of the fact that the divine mind conceives different things in different ways (Cusano 2017, *De mente* VI, p. 94). Difference, therefore, is not a posthumous consequence of the act of creation but is eternally present in the divine *mens* as *complicatio*. In its self-enjoyment, God's thinking unfolds in the form of the truly existing world. If in God all things are *unitas*, as soon as they are expressed, they differentiate. This act is both simultaneous and genetic: what in the realm of the world is a process which prolongs itself spatially and temporally, in God it is eternally implicated: creation is nothing but the other side of eternity. In the same immediate vision in which God sees God's own self, God sees all things, and thus produces them. Indeed, as suggested in the ninth century by Irish theologian John Scotus Eriugena, God is the one who *sees* all the realities which are in God's own self, without seeing anything but God's self, and the one who *runs* through each reality, filling them all (Eriugena 2013, *De div. nat.* I, p. 452b–d).[5]

Borrowing a concept developed by Ruyer, it can be said that God is the *absolute surface* that flies over (*survole*) itself without doubling or surpassing itself: it is

[5] Eriugena refers to a double pseudo-etymology of the Greek θεός: on the one hand, he claims, it derives from the verb θεωρῶ, and therefore should be interpreted as "the one who sees;" on the other hand, it would instead derive from the verb θέω. In this case the meaning would be "the one who runs." Eriugena's major work, the *Periphyseon*, also en *De divisione naturae* ("On the Division of Nature") was translated ito Italian and published with the title "Divisione della natura" in 2013. Hence his work is cited in the text of this chapter as "Eriugena 2013."

completely and simultaneously enveloped in every point of the field (Ruyer 1952, p. 112). God is not contained in space-time, but God overflights it, *explicating* it as its absolute *domain*. This macrocosmic structure is repeated and renewed at the microcosmic level. Each self is the *locus* of its overflight, producing the spatiality and temporality that it dominates. In other words, the self is nothing but a surface that explores itself in a single gaze without perspectival distance between the source and the object. The dynamics of the *survol absolu* ("absolute flyover"), Ruyer points out, "*est la clé, non seulement du problème de la conscience, mais du problème de la vie*" ("is the key, not only to the problem of consciousness, but to the problem of life") (Ruyer 1952, p. 112). Each living being, from humans to bacteria, from animals to trees first experiences the world and its own self as the undivided domain –or *locus*– of an immanent and immediate overflight. This movement of cosmic *self-enjoyment* is nothing but *life* itself.

In this sense, there is no contradiction in saying that God is all things (Cusano 2017, *De doc. Ign.* II, p. 2); God is, in each self, *non-other* than the self-overflight. In Cusa's words, God *enfolds* everything, everything *unfolds* God. As Eckhart observed, the divinity (*Gottheit*) resides in the innermost, empty depths of the soul. There is no separation between God and all things, because God is *in* all things; God is more intimate to them than they are to themselves (Eckhart 1985, p. 224). There is one and the same point at the heart of every atom in the world (Cusano 2017, *De mente* IX, p. 118), of which the world is the infinite development.

Thus, speaking of the divine means *ipso facto* speaking of the nature of life in its differential proliferation. God and the world are the poles of the same genetic process, they are produced through each other, to the point that, as Whitehead argues, "it is as true to say that God creates the World, as that the World creates God" (Whitehead 1929, p. 348). In this mutual *creatio continua*, God unceasingly generates himself without going out of himself. To use Simondon's lexicon, it can be stated that there is no dialectic leap between the *pre-individual* level and the level of individuation. The latter regenerates in the extensive dimension of one intensive and timeless genetic impulse. The pre-individual is always present in the individual as the virtuality (*virtus*) that moves it from within, not towards an extrinsic realization, but as an expression of the particular *theme* that drives it. As it is said to be *essentia essentiarum* or *locus locorum*, God is thus definable as the theme of all the themes, the habit of all the habits.

In pantheistic terms, this is the idea of the universe as a *theogony* –continuous self-unfolding of the Divine–, in which God is portrayed as both the entelechy and the global morphogenetic field. Or rather, as argued by Cusa, each singularity enfolds (*complicat*) the wholeness, while the cosmos unfolds (*explicat*) its plurality (Cusano 2017, *De doc. Ign.* II, p. 5).[6] The *divine ontogenesis* is univocally the same

[6] Physicist David Bohm used the terms enfolding and unfolding to describe what he believes to be the elementary structure of the universe, which he defines through the principle of implicate order: fundamentally, universe is for him a holomovement, "an unending flux of enfoldment and unfoldment" (Bohm 1980, p. 235). Intensively, the whole enfolds in itself all realities, and at the same time each reality unfolds it. Phenomena –i.e., that we can know and perceive– do not exhaust all of

as any living individuation: it is a process of self-revelation, excluding any dialectical movement of alienation. Indeed, theogony is not a transition from an embryonic or lacking stage to a more complete one, but rather it is the becoming of an actuality which is at any moment the totality of its own concretizations. The divine is inseparable from the momentum of its actualization: it *is* this motion, as one eternal embryogenesis (Ruyer 1952). Neither abstractly possible nor merely actual, God is an actual power and a potential reality, it is all that can be and it can be all that is: *Possest*, as Cusa defines it.

This theological structure spontaneously converts into an ecological vision in which all the single entities are valued in their peculiar integrity. Indeed, the risk underlying every dualism –but also every dialectic– is that of tacitly implying a derealization of reality itself. Instead, the notion of cosmos as theogony values the singular insistence –the *conatus*– of every node, organism, form of life, as the differential nature of the same creative force. Once again, each self is a *habit*, a mode of reality, an immanent form prolonging the chain of being: following only their own intrinsic normativity, each individual adheres to the flow of the whole, tuning to the cosmic frequency that penetrates everything and, in turn, resonates *through* everything. The *ecology of the selves* is nothing but this universal symphony, in which every single movement renews the whole composition.

Cosmos equals plenitude: the deity is wholly present in each life. Reversely, everything is in God, as its own *place* and end, while making place for God. Although differentiated in itself, this place is not a difference between given differences, but the transcendental field where differentiation takes place. This also means that differences cannot dissolve into an abysmal *Ungrund* by which they are emptied of their ontological consistency. Instead, they are enlivened in their singular occurrence. On the one hand God is the *maximum absolutum* (Cusano 2017, *De doc. ign*. II, p. 4), the actual and intensive infinite, on the other hand the world is the *maximum contractum* which, by contracting, is already multiple and differentiated. God, in other words, is but the cosmic dissemination of its singular contractions. God explicates Godself in the universe which, in turn, contracts itself into things, thus enhancing its continuous ontogenesis.

This game of reversibility between *complicatio* and *explicatio* reveals the deepest logic of the *locus*. As Eriugena has pointed out, the latter is both the field that "contains" all entities without being itself contained by anything, and the very *limit* and *definition* of every singularity. It is the *terminum* and the *finem* in which every individual is enclosed and without which it would flow indefinitely (Eriugena 2013, *Periphys*. I, p. 470c–d). The *locus* is indeed infinite and singular at the same time,

---

reality, just as Spinoza's substance does not exhaust itself in the attributes of thought and extension, even though it is all expressed in each one. In this book *Wholeness and the Implicate Order*, Bohm never mentions Bruno or Cusa (while he instead quotes Leibniz and Whitehead), however the similarities between their positions are many. Not only does Bohm employ Cusa's lexicon –that of *implicatio* and *explicatio*– but he also seems to recall the well-known principle of *quodlibet in quodlibet* when he writes that "in terms of the implicate order one may say that everything is enfolded into everything" (Bohm 1980, p. 225).

unlimited and circumscribed to each individuality as the circle (Eriugena 2013, *De div. nat.* I, p. 483c–d) which defines its peculiar form and transformations. As a specific *habit* or *locus*, embodied in space and time, every life form coincides with its own limit and circles in its tautological definition, composing its musical theme as its own immanent norm and measure.

Ecologically speaking, the logic of the *locus* is a deeply inclusive one: to be placed, differences do not have to exclude, negate, or overcome each other. *In locu*, the determination is not understood as the negation of something else, but as the singular and recursive expression of the whole itself. Every singularity –from the infinitesimal constituent of matter to the most complex living organisms– has its own place. Accordingly, God is nothing but the place of places (*locus locorum*). Like the Platonic *χώρα* ("Khôra"),[7] it places every singularity allowing its own peculiar form, without taking any form on its own.

The *locus* is thus nothing but this pure inherence of every single thing to itself, through which it is at the same time *more* than itself: if *in* "x" God is nothing but "x", *in* God every "x" is God,[8] and in the divine *locus* otherness is lived and felt as self-expression. Ethically speaking, the *locus* shall reflect the *virtus*, as virtuous life consists in no longer understanding the relationships with others, society or environment as proper *relationships*, but instead as mutual penetrations within the non-otherness. The principle of *non-aliud*, thus defined, grounds both the horizontal diversity of things and the vertical bond between each thing and the source. The same intensity folds in an infinity of singular variations. This does not mean reducing everything to a static tautology, but thinking of tautology itself as a creative and metamorphic process: recursive repetition means infinite variety. The more a form persists in its own *virtus*, the more it relaunches the intensity that sustains it. The more a thing inherits itself, the more it touches the infinite field within which everything is eternally given birth.

In this sense, it is possible to understand the logic of *non-aliud* as a kind of monadology, by acknowledging every place as a "perpetual living mirror of the universe" (Leibniz 2007, p. 56), where the whole is reflected, and the unceasing movement of this mirroring is nothing but life itself. The monads, Leibniz points out, are given birth by God in continuous and recursive fulguration of itself. Thus, God makes God's own *locus*, a peculiar point of view *of* the universe. Through the monad, the universe sees itself, as well as the monad sees the universe.

---

[7] *Khôra* is the term used by Plato in his dialogue *Timaeus* to designate a receptacle, a formless interval (alike to a non-being) in between which the "forms" were received from the intelligible realm (where they were originally held) and were "copied," shaping them into the transitory forms of the sensible realm.

[8] To clarify this logical step, Cusa makes use of images, such as those of the sky and the relationship between light and colors. Thus, it can be said that in the sky, God is none other than the sky, just as in color light is none other than color (De non-aliud VI, 22).

20.4 Conclusion

The outlined monadological structure suggests, once again, an ecology that enlivens the singular and the particular, not to legitimate a naïve discretization of the universe, but to enhance an *harmony* founded by the fact that each monad already expresses the totality of the relations –as in Cusa's *mirabilissimam connexionem* (Cusano 2017, *De doc. ign.* II, p. 5)– holding the universe together. Reasserted in terms of *non-aliud*, the dual[9] oscillation between the one and the many, identity and difference, the world and the self, becomes the ground of a cosmology in which the non-otherness of the world displays its fulfilment as singularity. As previously suggested, this does not mean hypostatizing individuals as identical substances, but rather understanding the cosmogonic and metamorphic nature of every place.

If forms are the centres of an inexhaustible stream of elements that flows from the one to the many, the eternal *sur place* cosmogony appears to be a continuous "reflux" between them (de Gandillac 1993, p. 101). This oscillatory, ontogenetic movement never takes a stable or definitive configuration, but rather it is the *locus* of a *metastable equilibrium*. The ancient opposition between being and becoming is virtually solved. Being is nothing but a metastable state which prolongs and conserves itself *through* becoming (Simondon 2005). Only in this fundamental oscillation can life prolong itself and explode into the infinite variety of forms that enhance it. The unity is not something given, but must be renewed every time, through each singular expression.

Thus conceived, the ecological problem ultimately does not concern the conservation of a static balance, or the impossible restoration of some previous conditions. Much like life itself, it requires creative variations on the invariant theme of the dyad –i.e., resolutions to the differential criticality of becoming. This is the eternal joy of living thought, or, according to Baruch Spinoza, beatitude as virtue and *amor Dei*. The nature of this problem is most likely neither logical nor ontological, but rather dramatic or musical: experiencing differences as *se ipsum*, each world tunes to God and plays its own life.

References

Barad K (2007) Meeting the universe halfway. Quantum physics and the entanglement of matter and meaning. Duke University Press, Durham

Bateson G (1972) Steps to an ecology of mind. Collected essays in anthropology, psychiatry, evolution, and epistemology. Chandler Publishing Company, San Francisco

Bateson G (1979) Mind and nature. A necessary unity. Dutton, New York

---

[9] As noticed by de Gandillac 1972, p. 361, the platonic dyad plays a fundamental role in the development of Cusa's thought; for example, in the paradigmatic figure of the two pyramids opposing and interpenetrating each other (De coniecturis I 9, pp. 40–42). One of the two pyramids represents God, unity and light, and the other one otherness, multiplicity, and darkness.

Bohm D (1980) Wholeness and the implicate order. Routledge, London
Cusano N (2017) Opere filosofiche, teologiche e matematiche. Bompiani, Milan
Dattilo E (2021) Il dio sensibile. Saggio sul panteismo. Neri Pozza, Venezia
De Gandillac M (1972) Les "conjectures" de Nicolas de Cues. Rev Metaphys Morale 3:356–364
De Gandillac M (1993) Explicatio-complicatio chez Nicolas de Cues. In: Piaia G (ed) Concordia discors. Studi su Niccolò Cusano e l'umanesimo europeo offerti a Giovanni Santiniello. Antenore, Padova
Deléage JP (1991) Histoire de l'écologie. La Découverte, Paris
Deleuze G, Guattari F (1980) Mille Plateaux. Éditions de Minuit, Paris
Descola P (2005) Par-délà nature et culture. Gallimard, Paris
Eckhart M (1985) Sermoni tedeschi. Adelphi, Milan
Eriugena GS (2013) Divisione della natura. Bompiani, Milano
Frank SL (1939) Непостижимое. Онтологическое введение в философию религии. Dom Knigi, Leningrad
Giglioni G (1995) Panpsychism versus hylozoism. An interpretation of some seventeenth-century doctrines of universal animation. Acta Comeniana 11:25–45
Haraway D (2003) The companion species manifesto: dogs, people, and significant otherness. Prickly Paradigm Press, Chicago
Kohn E (2021) Come pensano le foreste? Nottetempo, Roma
Leibniz G (2007) Monadologia. SE, Milano
Lovelock J (1979) Gaia: A new look at life on earth. Oxford University Press, Oxford
Noble D (2008) The music of life. Biology beyond genes. Oxford University Press, Oxford
Ruyer R (1952) Néo-finalisme. PUF, Paris
Schrödinger E (1944) What is life? The physical aspect of the living cell. Cambridge University Press, Cambridge
Simondon G (2005) L'individuation à la lumière des notions de forme et d'information. Millon, Grenoble
Soto AM, Longo G, Miquel P-A, Montevil M, Mossio M, Perret N, Pocheville A, Sonnenschein C (2016) Toward a theory of organisms: three founding principles in search of a useful integration. Prog Biophys Mol Biol 122(1):1–7
Stengers I (2011) Reclaiming animism. In: Franke A, Folie S (eds) Animism: modernity through the looking glass. Generali Foundation, Vienna
Vernadsky VI (1926) Биосфépa. Nauchno-Tekhn, Leningrad
Viveiros de Castro E (2009) Métaphysiques cannibales. PUF, Paris
Von Uexküll JJ (2010) Ambienti animali e ambienti umani. Quodlibet, Macerata
Whitehead AN (1929) Process and reality: an essay in cosmology. Macmillan & Co, New York

Chapter 21
Anthropocene Narratives and New Cosmologies

Lisa Sideris

Abstract This chapter explores the intersection of Anthropocene narratives with science-based cosmological creation stories. Grand narratives like the Universe Story offer a universal, global story that distills contemporary science into a true myth of cosmic evolution that unites humanity and bonds us with the natural world. In their diagnosis of our current global crisis, and in prophetic claims about Earth's future, cosmic narratives parallel Anthropocene visions of a future Earth wisely managed by humans. This alignment is no coincidence, given that the Anthropocene concept shares elements of its history and development with ideas and intellectual figures that are foundational to the Universe Story. On both accounts, human evolution culminates in a cosmic turning point, when humans begin to take conscious control of planetary processes. This chapter assesses whether such narratives provide useful guidance in the current crisis, with emphasis on the problem of human-centeredness and the ethical potential of wonder.

Keywords Anthropocene · Universe story · Cosmology · Mythmaking · Wonder

21.1 Introduction

Humans are often depicted as story-telling creatures par excellence. Among all the animals, we humans are said to create our worlds through narrative acts that give meaning and purpose to individual lives and unite communities with a common sense of purpose (Gottschall 2012). According to some scholars, however, humanity as a whole has recently entered a crisis phase of storylessness, a condition of amythia (Rue 2004). Traditional stories, such as the myths of longstanding religious traditions, have become fragmented and ineffectual. In the face of cumulative

L. Sideris (✉)
Environmental Studies Department, University of California Santa Barbara,
Santa Barbara, CA, USA
e-mail: lsideris@ucsb.edu

L. Valera (ed.), *Pantheism and Ecology*, Ecology and Ethics 6,
https://doi.org/10.1007/978-3-031-40040-7_21

knowledge from the sciences, including alarming reports about the contemporary environmental crisis, the stories we have inherited look increasingly out of touch and implausible, on this account.

Meanwhile, a new Anthropocene epoch beckons (or threatens) with its all-encompassing vision of an Anthropos—the human species as a whole—that is collectively transforming the planet and determining its future course. In this new "Age of Humans," some argue, the discovery, or creation, of a grand unifying narrative to match this unified portrait of humanity is an urgently pressing matter (Swimme and Tucker 2011). If humans, collectively, are now charged with managing the entire planetary system, we must have a new myth to guide us into this perilous but also promising new territory. If no appropriate myth emerges organically from our common condition of crisis—and there is little evidence at the moment of its emergence—then a new myth must be consciously and intentionally crafted from the best ingredients we have at hand. Above all, those ingredients include the narrative elements, the mythic materials, provided by the sciences. The task before us consists of pulling together the various strands of scientific and interdisciplinary knowledge, weaving them together into a single coherent narrative of the cosmos and the place of humans within it. Once completed, the construction of this new myth—a uniquely *true* myth—will restore a sense of wonder to the broader culture, bringing zest to our daily lives and a common vision for humanity's future (Swimme and Tucker 2011).

But which humans should be entrusted with the authority to pull together these narrative strands and mythic materials? Who can claim to represent humanity writ large? Should we designate a select group of experts, a subset of humanity, to construct a common myth for the rest of the world and its diverse populations? A number of scholars and science popularizers who subscribe to a grand new myth, known variously as the New Story, the Universe Story or the Epic of Evolution, have volunteered their services as authors of common creation story for humanity, built upon the solid foundation of convergent scientific knowledge. These epic stories are inspired chiefly by the work of two seminal figures in the development of religion and ecology: The Jesuit priest and paleontologist Pierre Teilhard de Chardin (1881–1955), and the former Passionist priest, cultural historian, and "geologian" (as he dubbed himself) Thomas Berry (1914–2009). The insights of these key thinkers are celebrated and elaborated by a cadre of their devotees, a loose affiliation of scholars and science communicators who call for widespread adoption of the New Story. These "new cosmologists" include Mary Evelyn Tucker and John Grim, who laid the foundations for the study of religion and ecology across the major world traditions, the cosmologist and "evolutionary philosopher" Brian Swimme, as well as science popularizers and self-styled "Epic of Evolution" evangelists Connie Barlow and Michael Dowd, among others (Dowd 2007). My focus here will be largely on the Universe Story as narrated collaboratively by Swimme and Tucker in their multimedia project titled *Journey of the Universe*.

21.2 A Purpose-Driven Universe

For both Teilhard and Berry, the universe as a whole exhibits a psychic-physical character, a discernible evolutionary trend toward increasing complexity and consciousness, with humans representing consciousness evolved to its highest and most complex and self-reflexive form. The universe is seen to possess a kind of subjective interiority and purposiveness, even deep desires for relationality and bonding (Swimme and Tucker 2011). These cosmic features, as described by Swimme and Tucker in a public-facing book and film project *Journey of the Universe*, are presented not merely as poetic flourish. Rather, they reflect what these storytellers believe to be the empirical, observable reality of the universe. Modern science has revealed these qualities as integral to the universe and the evolution of all life. Similarly, on this account, the structure of the universe does not merely lend itself to narrative form. It *is* a story. The universe exhibits a storied nature, through and through. It unfolds according to a logic and beauty of its own, a trajectory built into the cosmos since its inception. It has a telos. "Our universe is a single immense event that began as a tiny speck that has unfolded over time to become galaxies and stars, palms and pelicans, the music of Bach and each of us alive today" (Swimme and Tucker 2011, p. 2). The evolution of consciousness is prefigured, foreordained, in the very processes that ignited a Big Bang. An overarching directionality can be discerned in the unfolding of the universe: cosmic processes move in the direction of, and find their fulfilment in, the emergence of highly complex and self-conscious creatures like ourselves.

Teilhard (1959) referred to this tendency of the universe to evolve toward greater cephalization—that is, development of a more complex nervous system—as "complexity-consciousness." At our present moment, where humans have come to know the details of cosmic evolution, and to grasp its all-encompassing story, we are also coming to know ourselves as an expression of the universe. Humans are the universe made conscious of itself, as new cosmologists like to say. And yet at this very moment in which we at last perceive ourselves as cosmic creatures—beings who embody and fulfill cosmic traits and tendencies—we simultaneously find ourselves in a time of extreme planetary peril. The evolutionary processes that gave rise to and shaped humankind are now threatened by the creature they generated. The direction of the planet's future evolution, if its evolution is to continue at all, will henceforth be determined by us. Our oversight of the planet must now assume a benign and beneficial form, a departure from the destructive and domineering control of nature that brought us to this crisis point. We must evolve to become wise planetary managers, turning our earth-shaping powers toward good.

But how will humanity acquire the wisdom of a benevolent overlord charged with steering an entire planet into a new era of mutual flourishing between it and ourselves? Universe story enthusiasts believe that the narrative itself, the new creation story of the cosmos and the discovery of our place within it, will inspire an awe-filled awakening. The story of the universe "changes everything," challenging our entire understanding of who we are and what role we play in the universe (Swimme and Tucker 2011, pp. 1–5). The wonder it produces will be our guide.

21.3 A Big Stories—Old and New?

Berry, like his supporters Swimme and Tucker, understood modern humans to be at a crossroads or chasm, where the traditional stories, represented by the great faith traditions, no longer provide sufficient or relevant guidance, and a new story has yet to arrive. In the late 1970s, Berry highlighted the need for what he called a "functional cosmology" that would more accurately reflect the broader cosmological context, as revealed by modern science, and humans' special role in the unfolding of the universe. "It's all a question of story," he famously opined. "We are in trouble just now because we do not have a good story" (Berry 1978, p. 1). The stories of the so-called Western religions are especially dysfunctional in the modern context, Berry believed. Now a "new creation story has evolved in the secular scientific community," he argued, "the equivalent in modern times to the creation stories in antiquity" (Berry 1978, p. 5). The new story seemed to him destined to become the universal story of all humanity. A new mythic age was dawning, and with it a new paradigm of the *human*, a new Anthropos.

For all its debt to Berry, the "New Story" in its latest incarnation, *Journey of the Universe*, goes a bit further than Berry himself did, staking a claim to an anthropic universe, a cosmos that is just right for the emergence of a species like our own. For example, Swimme and Tucker cite physicist Freeman Dyson's claim that the universe seems to have anticipated us, to have *known* we were coming. The anthropic principle has roots in Teilhardian cosmology. Teilhard himself subscribed to a strong version of this principle in his oft-cited idea of the Omega Point, which suggests a predestined and final state of material complexity and unified consciousness. The Omega Point had eschatological and Christological dimensions for Teilhard; secularized versions of this idea have penetrated the broader culture, inspiring space enthusiasts, transhumanists, and a variety of technophiles. John D. Barrow and Frank J. Tipler's watershed work, *The Anthropic Cosmological Principle* (1988) draws heavily on Teilhard in presenting a universe characterized by intelligence and information processing. A strong version of the principle understands the universe as compelled to bring life into being, suggesting the significance of humans for the universe. Belief in the strong anthropic principle is sometimes seen by skeptics and atheists as a covert effort to resuscitate an intelligent design-like belief that the universe is made for us. *Journey*'s embrace of this principle portrays the human as uniquely capable of piercing the cosmic veil, to find and tell a story that is ultimately telling *us*. With the emergence of humans, *Journey* explains, the universe reached a "new fever pitch," birthing a creature of "white-hot awareness" and "blazing imaginations" (Swimme and Tucker 2011, pp. 87–88).

Tucker, Grim, and Swimme have made careers of extending the reach of Berry's and Teilhard's ambitions. *Journey of the Universe* coheres with a mission that Berry called universe education, with the goal of bringing the universe story to all of humanity, so that they might understand their role in creating the next evolutionary stage. But *Journey* follows Teilhard's philosophy more closely than Berry's in articulating a mystical-scientific narrative over and above an ecological *praxis*. That is to say that Berry's environmental agenda seems somewhat eclipsed by excitement about the idea of a complexly conscious universe. The storyline's proximity to Teilhardian themes is especially evident in the anthropic plot of *Journey of the Universe*, which suggest the human as the microcosm of the macrocosm, something even more significant than the *center* of the cosmos.

21.4 New Cosmologies of and in the Anthropocene

Already, certain parallels with the Anthropocene storyline can be discerned in these storylines, notably, the idea of a new phase in the human-Earth relationship, inaugurated and directed by the human. Berry, in collaboration with Swimme, envisioned a dawning geological event—a geological era (longer than an epoch) to follow the Paleozoic, the Mesozoic, and the Cenozoic. He optimistically christened it the "Ecozoic," a period marked by "mutually enhancing human-Earth relations" and by a recognition that our species now occupies the geological driver's seat (Swimme and Berry 1992, p. 247). These ideas echo Teilhard's cosmology in which the development of the "noosphere," which constitutes a merger of the collective human mind with an emerging, global techno-sphere, begins to shift humanity's place within the larger cosmos, toward a directorial role and responsibility. To be sure, Berry and his protégé Swimme remain on guard against the extreme techno-optimism that characterizes Teilhard's thought. Yet even here, one detects a note of excitement in Berry's invocation of the Ecozoic, as a moment when humans take hold of the evolutionary reins. As we will see, there are many resonances between the noosphere and the Ecozoic, and between both concepts and what some have called a "good" Anthropocene inaugurated by wise human oversight of the planet.

Teilhard's vision of a storied universe alive with conscious intent and directionality, together with Berry's call for a new story grounded in the modern cosmological context, provide a blueprint for *Journey of the Universe*, a project built upon a book and film and supported by a barrage of public facing events. However, one feature of cosmic storytelling is made more explicit in *Journey*'s narrative than in Teilhard's or Berry's: the theme of wonder. The promissory note of *Journey* is its claim that a profound sense of wonder will result from understanding our place in the broader scheme of life. Wonder, inspired by our new myth, will orient humanity toward the necessary ecological awakening. This awakening is in part what Berry envisioned as an Ecozoic era. Swimme and Tucker's narrative amplifies wonder as a key affective and cognitive state, expressing an abiding trust that the same "guiding process" that awakened life in the universe and steered it through numerous plots twists and cosmic near-misses, will spark a new consciousness and sense of purpose. Wonder itself is "one of our most valuable guides on this journey into our future as full human beings" (Swimme and Tucker 2011, p. 113).

The refrain of wonder in the new cosmic story is both insistent and vague. At *what* should we wonder, exactly? Where might science-borne wonder lead, and why should we wish to go there? For all its amassing of revelatory scientific information, the narrative is often murky on these questions of great moral import and consequence. A closer look at the narrative's deep debt to Teilhard's cosmic philosophy and theology suggests some answers. As I will argue, the theme of wonder here remains stubbornly anthropocentric, positing humans as the center of the cosmos. More precisely, the storyline is grounded in an *anthropocosmic* vision of our species: that is, the human as the creature who completes the cosmos. To appreciate these points, we need to fill out the picture of *Journey*'s debt to Teilhard.

21.5 From Noosphere to Good Anthropocene

As noted previously, Grim and Tucker were both mentored by Berry, as was Swimme, who began an intense collaboration with Berry in the 1980s that resulted in the publication of *The Universe Story* in 1992, a forerunner to the *Journey of the Universe* project. All three of Berry's protégés, and many others influenced by him, have read deeply in the works of Teilhard. Swimme, for example, pinpoints the idea of complexification-consciousness as a centerpiece of Teilhard's thought: the whole process of the cosmos is "about complexifying and deepening intelligence or subjectivity" (Swimme in Bridle 2006, n.p.). Additionally, Swimme affirms Teilhard's understanding of the Earth system as comprised of developmental layers, living envelopes, such as the rocky lithosphere, the atmosphere, the hydrosphere, and the biosphere. According to this developmental theory, Earth is presently at a stage of its development where a radically new sphere or envelope is emerging, one with great portent for the future of the planet and its life: the noosphere. By noosphere, Teilhard meant a sphere of intellect or mind (Greek: *nous*), and its human-made technological creations as an outgrowth of mind. The noosphere, then, develops as a living, thinking layer of the planet itself, a merger of human and planetary dynamics, but with humans as the leading edge of Earth's evolution.

For Teilhard, the noosphere was the "ultimate and inevitable sphere" of an evolutionary process in which intelligent life—human life—takes on a wholly new form of existence, placing humans in "a superior position vis-à-vis the natural environment" (Samson and Pitt 1999, p. 3). Swimme cites with approval an idea Teilhard called "hominization," that is, "the way in which human thought transforms previously existing practices and functions of the Earth" (Swimme in Bridle 2006). The noosphere superposes on the biosphere as an agent of planetary transformation, resulting in greater hominization. To put it in plainer terms, the noosphere is reshaping the biosphere, to such an extent that our species will come to direct the course of evolutionary unfolding. Teilhard's work articulated a kind of philosophy of the Earth, a *prescriptive* geology (recall Berry's account of himself as a geologian, a term with normative and spiritual connotations compared to mere "geologist"), whereby planetary and cosmic developments give meaning and direction to the human species; humans, in turn, evolve to understand their role in these processes, eventually taking evolutionary forces into their own hands.

Now the similarities with Anthropocene narratives come into sharper focus. This is no coincidence, for Teilhard himself was among a set of key thinkers who popularized the noosphere concept that to this day strongly influences contemporary thinking about the Anthropocene and its normative import. Along with Russian biochemist Vladimir Vernadsky (1863–1945) and French philosopher and mathematician Edouard Le Roy (1870–1954), and others, Teilhard developed the concept in the 1920s. While there are differences between the Vernadsky and Teilhard's conceptions of the noosphere, their respective views are often conflated, particularly among supporters of Teilhard's work and environmental legacy (Samson and Pitt 1999). This conflation matters because Vernadsky's noosphere very clearly points not only to human transformation of the environment but to an idea that is clearly discernable in universe story narratives, that the accretion of knowledge enables our species to ascend to the role of planetary management (Samson and Pitt). Teilhard's followers in the lineage of Berry combine aspects of Vernadsky's and Teilhard's noosphere, such that this sphere of mind exists within the evolving material world. It is not merely a mystical and otherworldly phenomenon but something that takes place, physically, in what we would now call the Earth System. As human knowledge of the cosmos increases, the noosphere enfolds that knowledge into itself, including knowledge gained about the general direction of the cosmos. In this way, the noosphere acts as both a mirror and a directive agency of the cosmos. Humans are not the center of the universe, Teilhard believed, but "something much more wonderful—the arrow pointing the way to the final unification of the world" (Teilhard 1959, p. 226). Tucker, in an earlier essay, expressed a similar idea in claiming that "humans complete the natural and cosmic world by becoming participants in the dynamic transformative life processes" (Tucker 2003, p. 48). This is what it means to describe the new cosmology anthropocosmic.

These ideas find expression in the Universe Story which posits humans as fulfilling a role variously described as the heart, mind, or consciousness of the cosmos. Something like this completion and transformation of planetary and cosmic processes is also discernible in Berry's (and Swimme's) notion of an "Ecozoic" event: a dawning geological era, a higher stage of cosmic evolution, signified by intensified, but benevolent, human involvement and improvement of the planetary system. These ideas, across the work Teilhard, Berry, Swimme and Tucker, all resonate strongly with Anthropocene narratives, and specifically with narratives of the so-called "good Anthropocene," in which global-scale human impacts are seen to inaugurate an exciting phase of our species' conscious oversight of the planet (Revkin 2014). Here the Anthropocene is reframed as an auspicious sign of our species' promotion to a planetary-scale creature.

The idea of a good Anthropocene was promulgated in recent years by an association of scholars called the Ecomodernists in a widely read document, "The Ecomodernist Manifesto," published by the Breakthrough Institute, a neoliberal thinktank based in Oakland, California. The manifesto and the "new conservationism" it inspired scoffs at the idea of ecological limits, endorsing a radically human-centered vision, while dismissing as romantic and sentimental a whole host of environmental pioneers from Henry James Thoreau to Rachel Carson. Ecomodernists contend that the future belongs to those who embrace humans' expanding managerial control of the planet. Future challenges will be met with technological ingenuity and creativity. As with our cosmic storytellers, some ecomodernists take inspiration from concepts like the noosphere as developed by Teilhard and Vernadsky; science journalist and ecomodernist Andrew Revkin, for example, lauds the influence of these thinkers and similarly praises Berry's Ecozoic concept as pointing toward the exciting prospect of "taking full ownership" (in Revkin's words) of the Anthropocene (Revkin 2011). Revkin has thrown his support behind *Journey*, praising its upbeat storyline and disavowal of judgmental pessimism about human patterns of environmental harm (Gandy 2011).

While some scholars doubt that there are any authentic precursors to the Anthropocene concept, the noosphere and the Ecozoic appear as likely candidates. Clive Hamilton argues that the concept of the Anthropocene—which for him entails an unprecedented, radical rupture in the Earth system—could not have been grasped prior to the establishment of Earth System science. The development of Earth System science in the 1980s and 1990s marked a paradigm shift, a new meta-understanding of Earth as a unified totality, a complex system of intricately linked, interlocking spheres. The concept of the Anthropocene "would not have been possible without the emergence of Earth System science," Hamilton argues, "as a way of understanding the novel role of humankind in the Earth System, as distinct from the understanding embedded in traditional environmental science" (Hamilton 2017, pp. 17–18). He insists that scholars who identify precursors to the Anthropocene concept are referring to changes humans have wrought in particular landscape or in ecosystems, or to particular plant and animal species—but not a rupture to the *whole Earth*, the Earth system in its totality such as we now see with climate change and other systemic, planetary-scale disruptions. A rupture at on this scale could not be envisioned or measured without the relatively new understanding of Earth System science in place.

Part of Hamilton's concern in distinguishing the Anthropocene from various historical lookalikes (genuine or otherwise) is that these forerunners create a deflationary effect. That is, if the Anthropocene, as a concept or as a historical event in the human-nature relationship, lacks novelty—if it is not truly unprecedented as a concept and a real-world phenomenon—then our current global-scale impacts might merely be seen as a continuation or culmination of something humans have always thought or done. The Anthropocene loses its shock value in light of these purported precursor concepts. Perhaps, one might venture, we humans have *always* shaped and reshaped our world as a key strategy of survival. Perhaps it is even in our nature, or in the nature of our planet, that such a dynamic has unfolded. We need only learn to fulfill this longstanding role, armed with conscious intent and cutting-edge technology, and all will be well.

Hamilton's concern is well-taken, as some scholars (often those with a foot in the "ecomodernist" camp) see the Anthropocene not only as nothing new but as something to be *welcomed*. And indeed, as I have argued, there are hints of celebration in the cosmic stories that herald the arrival of the Anthropocene, or what Berry and Swimme would call the Ecozoic era. Again, it is important to note that however alarming our modern environmental crisis, our species' emergence as something that completes, transforms, or directs the planet (and perhaps the cosmos) is *not itself* seen as a troubling or lamentable outcome of evolutionary processes, for the creators *Journey*. It is a role foreordained for us in cosmic processes, something built into the irreversible progress of cosmic unfolding.

The danger of grand narratives like these lies in their power to normalize and naturalize events that are not, in fact, inevitable or irreversible. In suggesting their inevitability, and in linking humans' planet-shaping power to evolutionary ascent, these stories make it difficult to challenge or critique particular outcomes or to envision alternative paths. To his credit, Berry, as I have noted before, set out to temper some of Teilhard's excessive, human-centered optimism and faith regarding technological progress. He questioned whether technoscience was a benign, divine-like outgrowth of humans' creative evolution. Berry's Universe Story, co-written with Swimme, warns of a seductive but dangerous mystique of technology and affirms the Ecozoic against a possible Technozoic era that would lack the healing, mutually enhancing vision of the former. And yet, Swimme and Berry's Ecozoic welcomes pervasive human influence and intervention in nature. The Ecozoic still retains a powerful entrancement with the human, and with progressive evolution that positions humans "in the driver's seat of geological evolution, directing the course of Earth history out of the Cenozoic period, and into a new era whose features will be determined by humans" (Scharper 1998, p. 130).

The stories we tell in the Anthropocene are performative. They have the power to preclude or endorse certain kinds of action in response to the events they selectively narrate (Bonneuil 2015). The problem with the new cosmologies, then, is that they might normalize and justify an environmentally destructive status quo, potentially encouraging more of the same, in the form of intensified human "participation" in nature by the creature who completes the universe and represents its highest form of consciousness and complexity. Universe story narratives show troubling affinities with the main features of a "good" Anthropocene narrative that naturalizes human planetary management as an outgrowth of progressive natural and cultural processes, eclipsing both human culpability in creating our current crises and human agency to choose a different set of values and a different future. The "new" cosmic myths double down on human creativity and abiding trust in a cumulative, coherent body of scientific knowledge as leading the way to enlightenment. They mistake knowledge for wisdom. Ultimately wonder reveals itself to be a solipsistic form of awe at the human species and human knowledge, as the crowning achievement of a universe whose own internal logic impelled it to create exceptional beings like us. As such, the narrative fails to challenge longstanding ideas of human superiority and significance vis-à-vis myriad other living beings, and indeed it may further entrench these suspect values. The new story is not so new after all, then. Nor is the wonder it seeks to inspire.

References

Barrow JD, Tipler FJ (1988) The anthropic cosmological principle. Oxford University Press, Oxford

Berry T (1978) The new story: comments on the origin, identification, and transmission of values. Teilhard Studies 1, New York

Bonneuil C (2015) The geological turn: narratives of the Anthropocene. In: Bonneuil C, Hamilton C, Gemenne F (eds) The Anthropocene and the global environmental crisis: rethinking modernity in a new Epoch. Routledge, London

Bridle S (2006) The divinization of the cosmos: an interview with Brian Swimme on Pierre Teilhard de Chardin. What is enlightenment? https://enlightennext.org/magazine/j19/teilhard.asp

de Chardin PT (1959) The phenomenon of man (trans. Wall B). Harper and Row, New York

Dowd M (2007) Thank god for evolution: how the marriage of science and religion will transform your life and our world. Viking, New York

Gandy M (2011) Journey of the universe, the epic story of cosmic, earth and human transformation, premieres on PBS in December. KQED Pressroom. Available at: https://www.kqed.org/pressroom/220/journey-of-the-universe-the-epic-story-of-cosmic-earth-and-human-transformation-premieres-on-pbs-in-december

Gottschall J (2012) The storytelling animal: how stories make us human. Mariner, New York

Hamilton C (2017) Defiant earth: the fate of humans in the Anthropocene. Polity Press, Cambridge

Revkin AC (2011) Embracing the Anthropocene. New York Times. Available at: https://archive.nytimes.com/dotearth.blogs.nytimes.com/2011/05/20/embracing-the-anthropocene/

Revkin AC (2014) The good, the bad and the Anthropocene (age of us). New York Times Dot Earth Blog. Available at: https://archive.nytimes.com/dotearth.blogs.nytimes.com/2014/07/07/the-good-the-bad-and-the-anthropocene/

Rue L (2004) Amythia: crisis in the natural history of western culture. University of Alabama Press, Tuscaloosa

Samson PR, Pitt D (1999) The biosphere and the noosphere reader: global environment, society, and change. Routledge, New York

Scharper SB (1998) Redeeming the time: a political theology of the environment. Continuum, New York

Swimme B, Berry T (1992) The universe story: from the primordial flaring forth to the Ecozoic Era—A celebration of the unfolding of the cosmos. Harper, San Francisco

Swimme B, Tucker ME (2011) Journey of the universe. Yale University Press, New Haven

Tucker ME (2003) Worldly wonder: religions enter their ecological phase. Open Court, Chicago

Chapter 22
System as Paradigm for a New World View

Joseph Bracken

Abstract Alfred North Whitehead distinguished between quantitative wholes (in which the whole is nothing more than the sum of its parts) and qualitative wholes (in which the whole is other than the sum of its parts). This qualitative whole is its immaterial principle of self-organization that cannot be perceived by the senses but which controls the interrelation of the parts with one another and acts as a "constraint" on the existing structure of the whole. Not individual subjectivity from the top down but instead intersubjectivity from the bottom up governs the part-whole relationship.

Keywords Thomas Aquinas · Alfred North Whitehead · Karl Rahner · Jurgen Moltmann · Part-whole relations

22.1 Introduction

In 1982 the philosopher of science Stephen Toulmin published a book entitled *The Return to Cosmology: Postmodern Science and the Theology of Nature.* Therein he noted the difficulty currently experienced by academics in both the sciences and the humanities in finding a balance between absolutism and relativism in their publications as a result of the scathing critique of "meta-narratives" by postmodernists (Lyotard 1984, pp. xxiv–xxv). To deal with this issue, he proposed that "scientists, philosophers, and theologians […] sit down together and follow their joint discussion wherever it leads" in the hope that together they can come up with a new philosophical cosmology that would offer an "overall scheme of things" suitable to guide human beings in other more practical life-decisions (Toulmin 2022, pp. 272–274). But what kind of physical cosmology would be acceptable to all parties to the discussion? Scientists, for example, might be strongly tempted toward 'scientism',

J. Bracken (✉)
Department of Theology, Xavier University, Cincinnati, OH, USA
e-mail: bracken@xavier.edu

L. Valera (ed.), *Pantheism and Ecology*, Ecology and Ethics 6,
https://doi.org/10.1007/978-3-031-40040-7_22

namely, the belief that only scientific knowledge based on empirical evidence and deductive logic deserves to be taken seriously as genuine knowledge. Every other claim to truth, accordingly, would be seen as grounded in individual subjectivity or personal opinion. Yet further discussion with experts in the humanities might enable these scientists to acknowledge the limits of their own quest for absolute certitude. Likewise, after prolonged discussion with scientists, people in the humanities (especially those in philosophical theology) might see the advantages of integrating their religious beliefs within a scientifically grounded world view. Thomas Aquinas, for example, used the metaphysical principles of Aristotle for that purpose even though he likewise recognized the limitations of Aristotle's scheme in supporting his own religious beliefs (Jones 1969, pp. 211–212). But in today's pluralistic approach to reality, what other school of thought would equally well or even better serve as philosophical support for one's personal convictions (both scientific and religious)?

22.2 The Elusive Concept of Society in Whitehead's Cosmology

In this article I argue for the overall suitability of the philosophical cosmology of Alfred North Whitehead in meeting the needs of both scientists and theologians for a shared world view. But in my judgment Whitehead seriously misjudged the key role that his own concept of "society" should play in a world dominated by organized groups of entities or systems in dynamic interrelationship with one another and with the external environment. For, he likewise endorsed the belief that "the final real things of which the world is made up" are momentary subjects of experience or what he calls "actual entities" or "actual occasions" (Whitehead 1978, p. 18). But what is "a momentary subject of experience"? Is it in the first place an individual entity or Aristotelian "substance" (something existing here and now in its own right), or is it instead an underlying activity or metaphysical principle needed to sustain the ongoing existence and activity of individual entities? Whitehead studied the philosophical implications of Zeno's Paradox and eventually came to the conclusion that being an underlying activity and being an individual entity are really not contradictory concepts but instead strictly complementary or "non-dual" realities. Each needs the other simply to be itself and for the other (Whitehead 1978, p. 43). Yet Whitehead wanted to avoid even the appearance of endorsing the classical subject-predicate approach to reality. Thus, unlike Spinoza, Whitehead did not seek theoretical knowledge of God or the Absolute but instead knowledge of how the world itself worked in terms of a reciprocal relation between its component parts or members. But, given his commitment to philosophical atomism (Whitehead 1978, p. 35), he could not logically attribute agency or efficient causality to societies as such but only to each constituent actual occasion in its moment of self-definition (Whitehead 1978, p. 34). In his own words,

> The causal laws which dominate a social environment are the product of the defining characteristic of that society. But the society is only efficient through its individual members [actual occasions]. Thus in a society, the members can only exist by reason of the laws which dominate the society, and the laws only come into being by reason of the analogous characters of the members of the society acting in unison with one another (Whitehead 1978, pp. 90–91).

In Aristotelian metaphysics, of course, an immaterial substantial form regulates its "accidents" or material parts. In Whitehead's scheme, however, the "accidents" by acting in unison with one another co-determine the governing structure of the Whole. That is, the Whole is then capable of characterization only through its accidental embodiments, and apart from these accidents is devoid of actuality. In the philosophy of organism this ultimate is termed 'creativity'; and God is its primordial, non-temporal accident (Whitehead 1978, p. 7).

Quite apart from calling God an "accident" of the cosmic process, this can only mean that God and the World are the contrasted opposites in terms of which Creativity achieves its supreme task of transforming disjoined multiplicity, with its diversities in opposition [the World], into concrescent unity, with its diversities in contrast [God] (Whitehead 1978, p. 348). Whitehead thereby seems to forget or ignore the inevitable gulf between God and the World if they are polar opposites co-existing within the even broader reality of the God-World relation as a whole. He states:

> What is done in the world is transformed into a reality in heaven, and the reality in heaven passes back into the world. By reason of this reciprocal relation, the love in the world passes back into the love in heaven, and floods back again into the world (Whitehead 1978, p. 351).

But there is no reciprocal causal relation at work if either or both of these rival spheres of activity are unaffected by the exchange between them (Whitehead 1978, pp. 22–23). To be fair, classical metaphysical schemes like those of Aristotle in Antiquity and Aquinas in the Middle Ages were focused primarily on permanence of structure and design with corresponding neglect to the perception of the reality of never-ending change or evolution in the structure of reality. Likewise, early modern philosophers of science like Galileo and Descartes were more interested in analysis of the laws of change in terms of mathematical equations. Yet with the precision of their mathematical formulas they were also setting forth a primitive metaphysics of Becoming that was focused simply on the quantitative dimensions of physical reality. Thereby they implicitly dismissed the other qualitative dimensions both within and between individual entities that could be subjectively felt but not objectively measured for their intrinsic meaning and value. Paradoxically, then, only with new discoveries in various areas of science by the end of the nineteenth century were scientists forced to admit that they had lost all confidence in the viability of any deeper philosophical understanding of physical reality.

Stuart Kauffman, Terrence Deacon and many other contemporary philosophers of science have employed instead a basically mechanistic approach to physical reality while trying to explain the emergence of life from non-living components and the growth of self-consciousness from unconscious neural activity in the brain. That is, they initially agree that thermodynamic systems with low levels of organization and complexity will basically tend to merge with one another if they find themselves co-existing within a common field of activity. As a result, these lower-order systems will often end up automatically co-producing a higher-order morphodynamic system. Subsequently, two or more morphodynamic systems that likewise co-exist in a common field of activity will generally result in the emergence of a teleodynamic system with apparently lifelike properties or attributes. Yet from what transcendent source does the energy for all these transitions from lower-order systems to higher-order systems come if the ultimate components of lower-order systems are lifeless entities with no lasting influence on one another? Stuart Kauffman, for example, simply notes that life is a natural property of complex chemical systems, so that when the number of different kinds of molecules in a chemical soup passes a certain threshold, a self-sustaining network of reactions will suddenly appear (Kaufmann 1996, pp. 3–30). But knowing in advance that something will likely happen does not really explain how it happened. Kauffman later conceded that there must be an internal principle of self-organization in the universe that moves the cosmic process to He called this principle "creativity," progressively higher levels of size and complexity perhaps in imitation of Whitehead's notion of creativity as the principle of novelty within the cosmic process (Whitehead 1978, p. 21). But, unlike Whitehead, Kauffman did not inquire about the ontological status of creativity, where its energy comes from. Finally, in a new book Kauffman explains the emergence of life out of inanimate components in virtue of "Constraint Closure." By this term he means that the system literally sustains itself by constructing its own constraints.

> Such a system is a 'whole' and more than the sum of its parts… It is a Kantian whole with parts that exist for and by means of the whole. The heart, for example, pumps blood not simply to sustain itself but to guarantee the survival and continued prosperity of a human being. But later in this book Kauffman still attributes the origin of life to molecular diversity, i.e., a contingent combination of the right number and a sufficient diversity of molecules (Kaufmann 2019, p. x).

Terrence Deacon published in 2012 *Incomplete Nature: How Mind Emerged from Matter* in which he too analyzes the notion of constraint in the ongoing relations between dynamically interrelated systems. To his credit, he seems to have better recognized the philosophical implications of his theory in comparison with the matter-form composition of physical reality in Aristotelian metaphysics: "Being alive does not merely consist in being composed in a particular way. It consists in *changing* in a particular way (Deacon 2012, p. 175). But what is it that thus changes? Does one substitute a new and improved "substantial form" within Aristotelian metaphysics? Or does one compare and contrast entire systems of entities so as better to explain a given empirical situation? Yet systems as corporate entities with multiple parts or members do not exercise causal agency except in and through their constituent parts or members, That is, when two "orthograde" systems share a common field of activity and begin to merge into a third higher-order system, the two antecedent "orthograde" systems end up indirectly co- constructing the parameters for a new "contragrade" system simply by the way that their constituent parts or members interact with one another in adjusting to the demands of the other's mode of operation (Deacon 2012, pp. 332–339). No imposition of a new substantial form from an outside source is needed to make this happen. "Ententional properties like function and information will only be explained when we can demonstrate how they emerge from non-ententional [inanimate] precursors" (Deacon 2012, p. 237). Furthermore, phenomena at the quantum-level should not be used to explain mysterious and counter-intuitive events that cannot be explained at the level of ordinary experience in terms of cause-and-effect relations.

Jesper Hoffmeyer, however, proposes in line with the relatively new discipline of biosemantics that information is indeed traded between components of systems at the molecular and all higher-order levels of existence and activity within the natural world. Yet he too contends that atoms as the ultimate constituents of reality must be non-living entities (Hoffmeyer 2008, pp. 3–5, 31–37, 196–197). Accordingly, Kauffman, Deacon and Hoffmeyer each face the same logical dilemma, namely, how to explain the emergence of life from non-life. There is, however, good reason to think that the evolution of life is a far more complex issue than Charles Darwin's theory of natural selection would seem to demand. Evelyn Fox Keller, for example, in her book *The Century of the Gene* makes clear how scientific understanding of the nature and function of the gene has significantly changed. Genes evolve in response to one another within the genome and in response to changing environmental conditions (Fox Keller 2000, pp. 133–145). Still another natural scientist Celia Deane-Drummond makes clear that not just genes but entire organisms adapt to and modify their physical environment in much the same way that human beings create cultural "niches" for their continued prosperity and eventual survival (Deane-Drummond 2014, pp. 219–222). Likewise, Simon Conway-Morris, while investigating fossils that were found in the Burgess Schale region of British Colombia, came to the conclusion that roughly the same principles of internal self-organization seem to be present seem to be present in the evolutionary growth and development of widely different plant and animal species (Conway-Morris 2015, pp. 3–8, 297–300). Nature, in other words, seems to exhibit a tendency to "habit-taking", i.e., using the same method for growth and development until something better turns up. Conway Morris borrowed the phrase "habit-taking" from the evolutionary cosmology of Charles Sanders Peirce in which "mind" or subjectivity is somehow present at all levels of existence and activity within this world (Bracken 2010, pp. 1319–1331). Finally, Thomas Nagel takes issue with the dominant presupposition of most natural scientists that the ultimate constituents of physical reality are inanimate entities. The universe is rationally grounded not only through the universal quantitative laws of physics that underlie efficient causation but also through principles which imply that things happen because they are on a path that leads toward certain outcomes—notably, the existence of living, and ultimately of conscious, organisms (Nagel 2012, p. 67). Nature as a whole must somchow be alive, if its constituents interact with a sense of purpose and/or direction.

22.3 Classical Metaphysics in a New Format

Religiously oriented philosophers and theologians presumably should endorse Whitehead's proposal of an evolutionary cosmology that emphasizes God's intimate involvement in the cosmic process. Instead, many of them have chosen to update Aristotelian-Thomistic metaphysics by using metaphorical language more or less in line with a deliberately systems-oriented approach to the God-World relationship. But this strategy involves a major risk, namely, to overlook or even ignore profound differences in outlook between rival thought-systems. Thomas Aquinas, for example, carefully distinguished between the primary causality of God and the secondary causality of creatures in evaluating the providence of God in dealing with creation (Aquinas 1951, Q. 22 art. 1). It is indeed praiseworthy for God to exercise providence over creation (above all, over human beings with their God-given power of free choice). But God often uses creatures as fallible secondary causes to execute God's primary causality (Aquinas 1951, Q. 22 art. 2–3). In this way, God can predetermine that some events in creation happen simply by chance (Aquinas 1951, Q. art. 4). In the early twentieth century, A. N. Whitehead initially insisted that his philosophical cosmology be internally coherent and externally applicable to all available data. But he likewise stressed the provisional character of his entire system: "Words and phrases must be stretched toward a generality foreign to their ordinary usage; and however such elements of language be stabilized as technicalities, they remain metaphors mutely appealing for an imaginative leap" (Whitehead 1978, p. 4).

Both Aquinas and Whitehead, accordingly, recognized the risk involved in combining different world views into a comprehensive thought-system. To exemplify what I mean here, I review briefly the work of two major theologians of the past 50 years: the one Protestant and the other Roman Catholic, namely, Jürgen Moltmann and Karl Rahner. Moltmann, Emeritus Professor of Systematic Theology at the University of Tübingen in Germany, has written many books. But his first book to be published in English, *Theology of Hope*, was perhaps the most important because it critiqued both sides of an academic debate in Western Europe between Existentialists with their confident expectations of a this-worldly or utopian future for humanity, and by Essentialists mindful of God's eternal self-revelation to the world from beginning to end of Salvation History. For, on the one hand, Moltmann cites Ernst Bloch who forecast a prosperous but entirely this-worldly future for the human race: "Our knowledge as a knowledge of hope, has a transcendent and provisional character marked by promise and expectation, in virtue of which it recognizes the open horizon of the future of reality and thus preserves the finitude of human existence" (Moltmann 1967, p. 92). On the other hand, he also cites Karl Barth: "God cannot be proven, neither from the cosmos nor from the depths of human existence. He proves himself through himself. His revelation is the proof of God. No one reveals God but himself alone" (Moltmann 1967, p. 54). The God of the exodus and of the resurrection 'is' not eternal presence, but he promises his presence and nearness to him who follows the path on which he is sent into the future. YHWH, as the name of the God who first of all promises his presence and his kingdom and makes them prospects for the future, is a God 'with future as his essential nature', a God of promise and of leaving the present to face the future, a God whose freedom is the source of new things that are to come. Moltmann, accordingly, does not promise what the future of the God-world relationship will be, either for God or for the world of creation:

> This God is present where we wait upon his promises in hope and transformation. When we have a God who calls into being the things that are not yet, then the things that are not yet, that are future, also become 'thinkable' because they can be hoped for (Moltmann 1967, p. 30).

As a complement to *Theology of Hope,* Moltmann in a later book points to the importance of the suffering and death of Jesus on the cross for a deeper understanding of God's own identification with the human race in its feelings of Godforsakenness. In a book entitled *Night and* written by a Holocaust survivor, Elie Wiesel, Moltmann agrees with Wiesel that God is present in the suffering death of a young Jewish boy being hanged by the Nazis (Moltmann 1974, pp. 273–274). Moltmann also found inspiration in the writings of Abraham Heschel about God's heart-felt empathy for the suffering of the Jewish people:

> God in Auschwitz and Auschwitz in the crucified God –that is the basis for a real hope which both embraces and overcomes the world, and the ground for a love which is stronger than death and can sustain death… It is the ground for living and bearing guilt and Yet Moltmann also makes reference here to sorrow for the future of man in God (Moltmann 1974, p. 278).

Saint Paul's confident forecast of a new creation based on the risen Christ as depicted in his Epistles to the Ephesians and the Colossians. How much this is compatible with traditional Jewish expectations for the Coming of the Messiah in the power of the Spirit is, of course, debatable. Moltmann himself concedes that only a Trinitarian understanding of the Paschal Mystery will satisfy Christians: "Where for Israel immediacy with God is grounded in the presupposition of the covenant, for Christians it is Christ himself who the Fatherhood of God and the power of the Spirit" (Moltmann 1974, p. 275).

Rahner, a well- known Roman Catholic systematic theologian in the post-Vatican II Church, likewise tried to reconcile with one another two quite different approaches to reality: ontology and phenomenology. But did Rahner thereby effectively turn from analysis of Being (i.e., the objective God-World relationship using the categories of Aristotelian metaphysics) to analysis of Human Being (i.e., a phenomenological approach o self-awareness)? For example, with Immanuel Kant Rahner learned to distinguish between the *phenomena* of sense experience and the reality of things apart from the human mind (the *noumena).* But this new insight from Kant still confirmed his own belief in the ontological superiority of spirit over matter as found in the Thomistic doctrine of the *agent intellect* (i.e., the abstraction of intelligible forms over the data of sense experience (Aquinas 1951, I, Q, 83 art. 1). At the same time, he learned from another Scholastic thinker Pierre Rousselot that God is pure Spirit (McDermott 1997). Likewise, in reading an important work by Joseph

Maréchal, *Le Point de Depart de la Metaphysique,* Vol. 5, Rahner gained the insight that the dynamism of the human mind is toward full intelligibility and thus toward God as the perfect union of mind and spirit. But in his student years at the University of Freiburg in West Germany, Rahner was also heavily influenced by Martin Heidegger's notion of *Dasein* as the self-manifestation of Being-in-the-World (Heidegger 1962, p. 107). That is, Rahner identified *Dasein* with his own sense of God's abiding presence in the world of creation and the way that a Christian should respond.

In a book published some years before his death in 1984 (*Foundations of Christian Faith),* Rahner reflected on four basic assumptions of his own world view. With the first assumption, he asserts the necessity of a close link between the empirically based findings of philosophical anthropology and the faith-claims of Christian systematic theology. Here he was undoubtedly influenced by Martin Heidegger in *Being and Time,* as noted above. Next, he analyzed the connection between Man [sic]and Person: "Man experiences himself as subject and person insofar as he becomes çonscious of himself as the product of what is radically foreign to him" (Rahner 1978, p. 29). He then adds that the self-possession of a subject lies in "a conscious and free relationship to the totality of itself" (Aquinas 1951, I, Q. 76, art. 1). Is this consistent with Aquinas's understanding of the substance-accident approach to material reality? Rahner seems instead to be focused on an evolutionary account of human subjectivity rather than on an objective analysis of cause-effect relations within the God-World relationship.

Afterwards, in setting forth the topic of Man as Transcendent Being, Rahner notes that

> man is and remains a transcendent being, that is, he is that existent to whom the silent and uncontrollable infinity of reality is always present as mystery. This makes man totally open to this mystery and precisely in this way he becomes conscious of himself as person and as subject (Rahner 1978, p. 35).

Yet from a Thomistic perspective God is beyond human comprehension even when one senses therein the presence of God in the relative infinity of Being (Rahner 1978, p. 49). Here Rahner is clearly more focused on the subjective awareness of the presence of God for the individual than on the objective place of human beings within the classical God-World relationship set forth by Aquinas and his followers in the Scholastic tradition. Furthermore, the goal of the human quest for self-transcendence is not necessarily union with God in a heightened interpersonal relationship but success in this life in terms of money, power, prestige, etc. Likewise, this search for self-transcendence could also be what Jean Paul Sartre described as *mauve foi* in *Being and Nothingness*, a failed attempt to be other than what one actually is (Sartre 1992, pp. 86–119). Aquinas wrote the *Summa theologiae* in an age of faith. So, when he set forth his arguments for the existence of God in the *Summa,* he took it for granted that the Ultimate First Cause of things is what everyone calls God (Aquinas 1951, I, Q. 2, art. 3). But is that still the case? The cosmic process can be considered as Ultimate Reality in a non-religious sense.

Must one admit then that science and religion, reason and divine revelation, are sufficiently at odds with one another that a commonly shared world view is virtually impossible? In his book *Christianity and the Images of Science,* Granville Henry proposes instead three theses about the historical relation between religion and science. First, religion tends to accepts good science and usually finds a way to incorporate the latest scientific approach to reality into its interpretation of Sacred Scripture (Henry 1998, p. 21). Secondly, conflict between science and religion only occurs when scientists upon further research and reflection change their current understanding of physical reality in a way that at least seems to threaten how Scripture scholars still interpret those same texts in the old way (Henry 1998, p. 28). Thirdly, any new scientific discovery will necessarily involve new philosophical concepts and procedures that could be used by Scripture scholars if they take time to study the further implications for their own research and reflection (Henry 1998, p. 31).

As noted earlier in reviewing the work of Kauffman, Deacon and Hoffmeyer, systems-ordered thinking has become quite popular among these philosophers of science to explain how life and self-awareness have gradually emerged from the right combination of inanimate components (atoms) under the proper conditions. That is, instead of focusing on cause-effect relations between individual entities so as to explain evolutionary change, they studied cause-effect relations between organized groups of entities (systems) as the corporate agents of growth and evolutionary change. Is it possible that wholes and parts are reciprocally related so that (a) the parts influence the current governing structure of the whole, and (b) that current structure of the whole provides order and directionality to the parts in their successive interaction? Whether the "parts" are human beings living together in community, sentient animals that collectively organize to protect their young from predators, or a network of plants in a thriving vegetable garden, the organic unity of the system as a whole binds elementary systems into intermediate systems that are in turn hierarchically ordered vis-à-vis one another so as to constitute an ever-growing totality of systems within systems.

But is there an underlying metaphysical principle at work here to account for the way that some (but not all) systems self-organize (progressively develop in size and complexity)? Could the difference between systems be due to the way in which parts and wholes are related to one another so as to constitute either a quantitative or a qualitative totality? That is, if a system is always composed of uniform parts or members, then the entity is a quantitative whole, nothing more than the numerical sum of those parts or members. But, if the whole is composed of different kinds of parts or members, then the most complex part or member of the system tends to become the principle of order and mutual intelligibility for all the other parts or members in their relations with one another. Admittedly, this qualitative relation is also reflected in the priority of form over matter in classical metaphysics. But there is a key difference between the two types of qualitative wholes. Classical metaphysics is based on top-down causality coming from an outside energy-source (God or a deft human craftsman), whereas process- or systems-oriented metaphysics is based instead on bottom-up causality as a result of the ongoing interaction of its parts or members with one another.

Several new books on the so-called non-dual relation between matter and spirit, the natural and the supernatural, have appeared in recent years. For example, in his Gifford Lectures for 2019–2020 Michael Welker (University of Heidelberg, Germany) claims that the mental gap between the Infinite and the finite is so great that an authentic experience of the Divine Spirit can only be inferred from linking it to multiple interrelated instances of the human spirit at work in human life (Welker 2021, pp. 24–29). Welker, of course, is here reversing the traditional understanding of the principle of analogy in classical metaphysics. That is, the abstract notion of divine goodness ought to be rethought in terms of one's understanding and appreciation of human goodness at work in this world, not vice-versa. One learns to accept a "multi-modal" or non-dual understanding of the relation between the divine and the human in thus learning how to make solid judgments about the workings of justice, freedom, truth and peace in civil society today (Welker 2021, pp. 29–35). Along the same lines, David McDuffie authored a book entitled *Nature's Sacrament: The Epic of Evolution and a Theology of Sacramental Ecology:* "It is through the acceptance of the broader sacramentality of a sacramental universe where the Christian concept of sacramental grace is seamlessly brought into a relationship with the ecological value inspired through a proper interpretation of the evolutionary epic" (McDuffie 2021, p. 28). McDuffie's argument then is that a broader understanding of evolutionary sacramentality allows Christians to immerse themselves deeper into the world with a greater sense of awe. Finally, Christian Barrigar has written a book entitled *Freedom All the Way Up* in which he claims that God created the universe to provide the proper existential conditions for the gradual emergence of *agape*-capable beings who are engaged in *agape*-love relations not only with God but with one another. Yet

> the route of *agapic* freedom must be consciously chosen. Even though we have evolved neural dispositions toward cooperation and empathy, along with rational. Self-reflective and volitional capacities, all of which make our *agape*-capabilities of self-giving and self-emptying possible, nonetheless our dominant evolutionary dispositions are toward selfism (Barrigar 2017, p. 153).

Not only God's saving grace, but also persistent work by a human being on a change of habits, is needed to reform one's basic character or way of life.

22.4 The Urgent Need for an Acceptable World View

Given Toulmin's sense of the urgency in coming up with a new philosophical cosmology that would be acceptable to scientists, philosophers and theologians, I turn finally to the example of a remarkable religious leader who has grasped the gravity of the current environmental crisis and tried to address it from his own religious perspective. Pope Francis, current head of the Roman Catholic Church, has written an encyclical letter on the environmental crisis and expressed his own views on what the Christian response to it should be (Pope Francis 2015a). He makes frequent reference to the various systems (economic, political, social and racial/cultural) that exercise great influence both positively and negatively on the behavior of human beings in today's world. At the same time, Francis did not pretend to speak as a professional economist or environmentalist. The great bulk of the references in his encyclical, for example, are to previous Church pronouncements issued by his predecessors as Pope or by episcopal conferences in various parts of the world. So he could have been widely criticized by the scientific community as a rank amateur in issues beyond his own competence as a spiritual leader. Yet paradoxically the encyclical letter was well received by most environmental scientists even though some of them expressed disappointment that Francis did not more explicitly deal with the controversial issue of population control (Pope Francis 2015b).

As the well-known environmentalist Bill McKibben noted in a recent commentary on the encyclical,

> the empirical data about climate change make it clear that the moment is ripe for this encyclical. The long line of brown-robed gurus, of whom Francis is the latest, now marches next to scientists in lab coats; instead of scriptures, the physicists and chemists clutch the latest printout from their computer models, but the two ways of knowing seem to be converging on the same point (McKibben 2015, p. 8).

McKibben thereby implicitly concedes that progress in science and technology cannot solve all environmental issues. "Pope Francis, in a moment of great crisis, speaks instead to who we could be. As the data suggest, this may be the only option we have left" (McKibben 2015, p. 9). For, as Alister McGrath (2016, p. 8) comments,

> [w]e have to use a variety of research methods to do justice to the important questions of life […] Insisting that we use only scientific methods, forms and categories confines us to a narrow world that excludes meaning and value, not because they are *absent* but because this research method *prevents them from being seen.*

Contemporary human beings need a bigger picture of reality "that does more than simply create space for science and theology but allows the nature, limits, and benefits of their interaction to be grasped" (McGrath 2016, p. 3).

References

Aquinas T (1951) Summa theologiae
Barrigar CJ (2017) Freedom all the way up: god and the meaning of life in a scientific age. Friesen Press, Victoria
Bracken JA (2010) Feeling our way forward. Theol Sci 8(3):1319–1331
Conway-Morris S (2015) The runes of evolution: how the universe becomes self-aware. Templeton Foundation, West Conshohocken
Deacon TW (2012) Incomplete nature: how mind emerged from nature. WW Norton, New York
Deane-Drummond C (2014) The wisdom of the liminal. Eerdmanns, Grand Rapids
Fox Keller E (2000) The century of the gene. Harvard University Press, Cambridge
Heidegger M (1962) Being and time. Harper & Row, New York

Henry GC (1998) Christianity and the images of science. Smyth & Helwys, Macon
Hoffmeyer J (2008) Biosemiotics: an examination into the signs of life and the life of signs. Scranton University Press, Scranton
Jones WT (1969) A history of western philosophy. Harcourt, Brace & World, New York
Kaufmann S (1996) At home in the universe. The search for the laws of self-organization and complexity. Oxford University Press, New York
Kaufmann S (2019) A world beyond physics: the emergence and evolution of life. Oxford University Press, New York
Lyotard JF (1984) The postmodern condition: a report on knowledge. University of Minnesota Press, Minneapolis
McDermott JM (1997) De Lubac and Rousselot. Gregorianum 78:735–759
McDuffie DC (2021) Nature's sacrament: the epic of evolution and a theology of sacramental ecology. Christian Alternative Books, Winchester
McGrath A (2016) Enriching our vision of reality: theology and the natural sciences in dialogue. Templeton Press, West Conshohocken
McKibben B (2015) Introduction. In: Cobb JB Jr, Castuera I (eds) For our common home: process-relational responses to Laudato Si. Process Century, Anoka
Moltmann J (1967) Theology of hope. Harper & Row, New York
Moltmann J (1974) The crucified god: the cross of Christ as the foundation and criticism of Christian Theology. Harper & Row, New York
Nagel T (2012) Mind and cosmos. Why the materialist neo-darwinian conception of nature is almost certainly false. Oxford University Press, New York
Pope Francis (2015a) On care for our common home (Laudato Si'). United States conference of Catholic Bishops, Washington, DC
Pope Francis (2015b) Hope from the pope. Nature 522:391
Rahner K (1978) Foundations of Christian faith: an introduction to the idea of Christianity. Seabury/Crossroads, New York
Sartre JP (1992) Being and nothingness: an essay o phenomenological ontology. Washington Square, New York
Toulmin S (2022) The return to cosmology: postmodern science and the theology of nature. University of California Press, Berkeley
Welker M (2021) In God's image: an anthropology of the spirit. Eerdmans, Grand Rapids
Whitehead AN (1978) Process and reality: an essay in cosmology. Free Press, New York

Part III
From Pantheism to Ethics and Politics

Chapter 23
Pantheism: Destruction of Boundaries?

Gabriel Vidal

Abstract Pantheism has often been associated with the concept of blurring or dissolving boundaries. The prevailing notion suggests that pantheism removes the barrier that separates God and the world, consequently eradicating all other boundaries, distinctions, and limits. These include distinctions between animals and humans, nature and artificial constructs, and matter and spirit, among others. This understanding of pantheism has inspired movements like ecofeminism, which aims to challenge the boundaries responsible for oppressive binaries that underpin patriarchal systems. However, there are critics who view this blurring of boundaries as a threat to natural distinctions and boundaries. Surprisingly, this association between pantheism and the erasure of boundaries has not been explicitly explored. On one hand, there has been a lack of in-depth exploration regarding the meaning of limits or boundaries within this context. On the other hand, the question remains unanswered as to whether the central idea of pantheism, namely the identification of God and the world, inherently implies the dissolution of all other distinctions. In this paper, I aim to demonstrate that this implication is not necessarily apparent in one of the primary sources of pantheism, particularly the philosophy of Spinoza. My hypothesis is that within Spinoza's philosophy, especially in his treatment of physics, there exists a recognition of the existence of limits, boundaries, and determinations. This recognition indicates that pantheism is not necessarily committed, at the conceptual level, to the eradication of boundaries, but rather it could be compatible with their existence and acknowledgement.

Keywords Pantheism · Ecofeminism · Spinoza · Boundaries · Destruction

---

G. Vidal (✉)
Philosophy Department, Pontificia Universidad Católica de Chile, Santiago de Chile, Chile
e-mail: gtvidal@uc.cl

L. Valera (ed.), *Pantheism and Ecology*, Ecology and Ethics 6,
https://doi.org/10.1007/978-3-031-40040-7_23

23.1 Love, Limits, and Destruction

The tenet of pantheism, namely, that God and the world coincide or, even, are identical, was controversial, and during Modernity it even had a pejorative connotation (Moran 1990). The major heresy of pantheism, which even caused Spinoza's excommunication, is its denial of the absolute transcendence of God, typically attributed to classical theism (Levine 1994). The heresy of this notion is connected to its demeaning God, absolutely perfect and unscathed, to the mundane level, susceptible to imperfection and corruption. Briefly, the entire spiritual nature of God is demeaned to mere physical matter.

Notwithstanding that, it is important to note that the religious motivations of pantheism are not limited to the purpose of trivializing God or placing him in a worldly level. In this sense, the interpretation of the remarkable Dutch theologian, Abraham Kuyper, is very interesting. At the end of the nineteenth century, he stated that, in fact, one of the main religious motivations of pantheism is love. It is deep love of God that inspires our desire to come near him and build a bridge toward that unreachable transcendence:

> This is shown by religious pantheism, which, afraid of a God "afar off" has no peace even with God "at hand" but in the prayer-mystery here seeks to penetrate the being of God, and, in the hereafter, yearns after identification with the divine Being, until at length every boundary between God and the soul is lost (Kuyper 1893, p. 12).

That love might be wrongly based, as suggested by the passage, insofar as it would prove impossible to eliminate such transcendence. We cannot go into this matter here, but it is interesting to point out that this very same spirit may have inspired a good part of environmental ethics, so much so that it has become a popular paradigm in that discipline (Wood 1985). The notion that pantheism provides a gesture that re-sacralises nature by considering everyone and everything part of God. Insofar as everything partakes this divine flame it proves possible to ground its intrinsic value and then protect it, at the normative level (Valera and Vidal 2022). In this sense, this desire to unite with God may be understood not as desacralization, but as the action of turning the world into something divine: every entity would be sacred.

But, according to Kuyper, it is not only this that derives from holding pantheism. Apparently, when pantheism eliminates the boundary between God and the world, it cancels all distinctions, differences, and limits of reality.

> This is most clearly demonstrated by philosophical pantheism, which systematically fuses every thesis and antithesis into a synthesis, and, by the tempting notion of identity, explains everything which seems dissimilar as similar and, in the end, as being of like essence. [...] And here, also, both with spiritual and natural phenomena, are denied all real differences of kind, together with independence of origin, and every deeper distinction of being, [...] every line which marks a boundary is wiped out, and every boundary post which divides the jurisdiction is leveled to the ground (1893, pp. 12–14).

In Kuyper's view, this implies the corrosion and degrading of reality; therefore, despite being originated by a gesture of love, pantheism is negatively viewed in favour of classic theism:

> Faith in the living God stands or falls with the maintenance or removal of boundaries. God created the boundaries. Lie himself is the chief boundary for all his creatures, and the effacement of boundaries is virtually identical with the obliteration of the idea of God (1893, p. 16).

In the end, Kuyper blames pantheism for being a concept that destroys limits and values it negatively. The temptation to come near God ends as Icarus trying to reach the sun: in tragedy.

23.2 Ecofeminism and Pantheism

Precisely against this assessment, it is odd to see that this last move -destroying limits- has also been used by environmental ethics. Specifically, by one of its recent branches that carries out this proceeding of weakening or even erasing boundaries to deal with some normative issues. This is Ecofeminism, a perspective that became known in the mid 70s:

> Central to ecofeminism is the belief that the oppression of women and nature are interconnected and are driven by patriarchal power relations that intersect with other inequalities, such as global power imbalances between nations and classism, racism, heterosexism, and ageism. For ecofeminists, the normative masculine attitudes and activities that shape women's oppression – aggression, conquest, possession, and control – also contribute to the destruction of the environment (Wachholz 2011, p. 289).

Although several sub-branches have come out during its development, we may find a common and frequent issue around the analysis of the way some patriarchal attitudes would seep into philosophy and the canonical view of the world. It is the notion that the dominant patriarchal perspective applies an inflexible categorization of reality through which it generates several binaries such as man/woman, culture/nature, God/world, foreigner/citizen, etc. (Estévez-Saá and Lorenzo-Modia 2018). Not only do these dualisms introduce an arbitrary split between each part of the opposing poles, but they would also be hierarchically organised in such a way that they establish the superiority of one over the other (Warren and Cheney 1991). In the end, this would be the intellectual grounding of several actions of dominion and oppression over minority groups and of man's destruction of nature (Besthorn and McMillen 2002). Consequently, the patriarchal view considers man and woman as separate and the masculine side as superior, which results in the dominion of the latter over the former. Animals as different and below human beings; the environment or nature as separate and at man's disposal.

In fact, the conceptual tool to carry out this split-hierarchization between the world's entities is the boundary. The patriarchal viewpoint draws arbitrary and oppressive boundaries where, in fact, there is an indissoluble unity (Morton 2010). That is why the ecofeminist project envisages amending this procedure which has historically resulted in oppression of individuals and nature. The key would lie precisely in recovering something considered as informed by the feminine experience that has been silenced:

> The boundary-less, free-floating, nondiscriminating sense of oneness that females experience could more accurately be called *le petit satori* (the little glimpse of enlightenment). In a culture that honored, rather than denigrated such "body truth," the holistic realities would be guiding principles of ethics and structure (Spretnak 1989, p. 129).

In this sense, the ecofeminist project becomes an ethical project that crosses boundaries in general: sexual, gender, racial, between nations, between men and women, nature and culture, re-establishing an original unity which, simultaneously, does away with oppressions.

As mentioned above, the idea that God and the world coincide allows for the re-sacralisation of Nature and seeing each entity as expressions of an only divine substance. This also restores the original unity which, according to ecofeminism, has been split by the strict binaries established by the patriarchal worldview. This has been voiced by several ecofeminist theologians, particularly by Sallie McFague (1993), that uses the metaphor of the multiplicity of entities that make the world as the body of God:

> Our proper place as inspirited bodies within the larger body, within the scheme of things, and a way of seeing both the immanence and the transcendence of God – God as the inspirited body of the whole universe (McFague in Van Dyk 1994, p. 178).

Therefore, several liberation theologies have found a fruitful source for their normative interests in pantheism, namely, the notion that God and Nature coincide. By acknowledging God in all the aspects of the world we perceive them as objects of reverence and care and no longer of oppression-dominion (Birch et al. 2007). This is the normative force that ecological and feminist theologies seem to perceive in the notion of pantheism.

It is obvious that this idea fully coincides with radical holisms, such as animism, insofar as the world is not a mere aggregate of things, but a multiplicity that makes a body and initiates a spirit. And, as acknowledged by radical holisms and the ecofeminist proposal, its assimilation of pantheism is paired by a relativisation or even an erasure of boundaries.

23.3 Transgression of Boundaries at the Ontological and Cosmological Level

Mary-Jane Rubenstein (2018) acknowledges it in her relatively recent writings. She develops an ecofeminist theology which attributes to pantheism and Spinoza a much more radical relevance than the divinisation of nature. Rubenstein remarks that Spinoza himself was persecuted because his doctrine was seen as a pantheism which was considered a monstruous and heretic doctrine. Its monstruous and heretical essence was not merely due to the fact that it seemed to contradict the traditionally established transcendence of God, but to the fact that it carries transgression up to an ontologic-cosmic level by weakening or cancelling the boundaries between God and the world. In fact, this combination implies bringing back binaries that imply –at least for classical theist tradition– falling into obvious contradictions: God is one while the world is multiple; God is perfect and the world is imperfect; God is immaculate, while the world is plagued by evil, among other couples of contradictions (Levine 1994).

According to Rubenstein, however, rather than with the conceptual difficulty of the need to reconcile these contradictory traits of God and the world, Spinoza's transgression is more closely connected to the fear of patriarchal Western tradition of radical limits-transgressions and perverse categorial combinations. The point is that these categorial combinations question patriarchal hierarchies and binaries and would therefore weaken the tyrannical power held against women and nature. In other words, substantial monism where everything multiple is dissolved into one sole substance or divine unity perishes in the face of the multiplicity of the world, directly threatening patriarchal hegemony since it questions the conceptual categories on which it is grounded:

> If pantheism were seriously to be entertained, the whole western symbolic, constituted as it is by the binary polarities which run through it like a fault – line, would thereby be brought into question. Pantheism rejects the split between spirit and matter, light and darkness, and the rest; it thereby also rejects the hierarchies based on these splits (Jantzen 1998, p. 267).

In other words, every split between things would fade away; consequently, crossing boundaries reaches a cosmologic and ontological level, a total subversion of the conceptualisation of reality by the West, according to Rubenstein.

Several of the reactions both after and during Spinoza's time seem to prove Rubenstein's point: blurring frontiers seen by different authors end either in a chaotic multiplicity or in an undifferentiated unity. On the one hand, Hegel, for example, sees in Spinoza the return of Parmenides' doctrine since multiplicity is dissolved in the perfect eternity of the one substance; in other words, the world is eliminated by the prevalence of God. That leads the German philosopher to state that Spinoza's philosophy is an a-cosmism intoxicated by the idea of God (Melamed 2010). On the other hand, other reactions hold that it means God falls into the becoming and mutability of the world. Therefore, God is left aside in view of the priority of the cosmos. As opposed to that, in this interpretation Spinoza's pantheism is atheism intoxicated by the cosmos (Colie 1959). According to Rubenstein, indecision between both interpretations does not suggest a possible ambiguity in Spinoza's philosophy. Contrarywise, it witnesses to the strength of his philosophy to de-stabilise the traditional categories of philosophy through a consistent philosophical system.

By crossing the most fundamental barrier, between God and the world, Spinoza goes to the core of Western patriarchal thought, cancelling all other traditional and oppressive divisions and binaries. The opposition and resistance to his thinking, which actually caused his ex-communion, are proof of what the Western world acknowledges in Spinoza. Criticism and concerns against pantheism voiced by readers of Spinoza such as !all dividing lines are squashed, all differences vanish, all life, all thought and all reason are beaten, piled up and crushed in a monstruous mass… a dreadful chaos! (Rubenstein 2018, p. 16) rather than objections with a well-grounded metaphysical basis, they are merely a defensive reaction of the Western building, according to Rubenstein.

A description of abnormality by Michel Foucault (2004, p. 63), explains why mixtures and border-crossings allegedly carried out by pantheism prove monstruous for the philosophical tradition:

> The monster is essentially a mixture. It is a mixture of two realms, the animal and the human... of two species... of two individuals ... of two sexes ... of life and death... Finally, it is a mixture of forms... the transgression of natural limits, the transgression of classifications, of the table, and of the law as table: this is actually what is involved in monstrosity.

Rubenstein does not perceive intrinsic malice in this monstrosity, rather, she sees the liberating force of the representation of the god Pan, precisely a hybrid of ram and human (2018, p. 29), to restore a division introduced by patriarchal thinking:

> The ontological distinction between God and creation does not merely separate the two terms; rather, it establishes the absolute supremacy of the former over the latter. In turn, this logic of mastery secures the rule of everything associated with this God over everything associated with the material world. Again, then, spirit, masculinity, reason, light, and humanity become unconditionally privileged over matter, femininity, passion, darkness, and animal-vegetal-minerality (Rubenstein 2018, p. 10).

Therefore, this presumed monstrosity actually disarms the presumption of hierarchy and split that has resulted in various oppressions in the Western world: colonialism, racism, anthropocentrism, 'machismo' and nationalism. In this sense, pantheism represents the perfect conceptual tool for the ecofeminist project and, actually, the full implementation of its boundaries-disruption project.

So, here we come across a peculiar fact. Kuyper seems to represent that supposedly backward thinking that aims to preserve the limitations that Rubenstein tries to cancel. Despite which, Kuyper and Rubenstein agree on the more profound ontological issue: pantheism is, in fact, a demolisher of boundaries. They only disagree in their normative assessment of such corrosion of boundaries, but the key ontological issue is not controversial. What would happen if we should question the common tenet?

23.4 Boundaries as Determinations: Constrictive or Constitutive?

Up to now we have seen how and why ecofeminism and, especially, Rubenstein's more radical version eliminates and relativises the boundaries that would produce oppressive binaries and divisions and how they get their inspiration from pantheism. Kuyper also admits this corrosion of boundaries, but he doesn't seem to arrive to a negative evaluation of boundaries like Rubenstein's. Nevertheless, there has not been a detailed account of the meaning attributed to limits or boundaries. We may not be describing geometric or physical limits; they are understood differently; namely, as determinations. Determinations in the sense that, in this case, a limit determines what an entity is and is not. When drawing the boundary, for example, it determines what human nature is, what a nation is, what a living being is and, on the other hand, it also excludes what the entity is not; for example, human being implies being a non-plant.

It seems to be the case that ecofeminism understands these determinations as mere restrictions which befall arbitrarily upon a given entity. However, it might also be the case that such determinations, or some of them, are constitutive. In other words, they draw limits that actually belong to that entity and do not befall arbitrarily upon it. In fact, they cause the entity to be the entity it is, they grant it its identity (Marcos 2001). Should this be the case, would crossing boundaries not be liberating? Rather, would it infringe the integrity of the entity? In order to answer these questions, it is useful to go further into Spinoza's philosophy, which inspired ecofeminism. It might be the case that some of the concerns of the authors who oppose pantheism because they fear the dissolution of things are valid, since they are thinking of limits as constitutive. In that case, the motivation of such reaction is not merely adhering to and defending a patriarchal system. I don't expect to provide a final answer to the matter, but I do wish to show how limits may be interpreted as constitutive and not only as constrictive.

Firstly, it is relevant to note that although traditionally Spinoza has been labelled as a pantheist, based on his formula *Deus sive Natura*, the co-belonging or identification between God and Nature is not so obvious. Although Spinoza acknowledges that there is an immanent connection between God and the world, he constantly stresses and acknowledges that they are distinct and, besides, that God enjoys priority and prevalence, which has come to be known as panentheism (Valera and Vidal 2022). As opposed to pantheism, in panentheism it does not seem to be the case that all differences, borders and limits are diluted, although it is still possible to hold the metaphor that inspires other feminist theologies, namely, the world as the body of God.

Besides, Spinoza himself acknowledges the value of limits as determinations when he acknowledges -a statement that has also been one of the most influential remarks in the history of philosophy – that every determination is negation (Melamed 2012). In other words, it seems that Spinoza believes there are certain borders that, on one hand, constitute me, and on the other, deny and exclude what I am not and limit me. Nevertheless, these dialectics would be necessary to preserve the integrity of an entity.

Notwithstanding that, if we probe further into Spinoza's metaphysics, we can notice that, effectively, he holds a relatively fluid "physics". The individuation criteria that Spinoza describes in *Ethics* are simply a proportion of motion and rest (EIIa1, a2, 11) namely, a sort of organization of the parts in one entity. Not only that, but Spinoza also explains that such relations of movement and rest, under quite unrestricted conditions may be communicated to and shared with other individuals, which results in fusions and combinations that, once again, seem to recover the limits-transgression typically attributed to his doctrine:

> Lemma 4. If from a body, or an individual which is composed of several bodies, certain bodies be separated, and if at the same time their place is taken by an equal number of parts from another body of the same nature, the individual will retain its nature as before, without any change of form.
>
> […] Lemma 5. If the parts composing an individual [become] greater or smaller, but in such a proportion that they all preserve the same relative rate of motion and rest as before, the individual will retain its nature as before without any change of form.
>
> […] Lemma 6. If [some of] the bodies composing an individual are compelled to change the direction of their motion, but so that they can continue this motion, and communicate it to each other in the same ratio as before, [then] this individual will retain its nature without any change of form (EII L4–6).

These comments have allowed for Spinoza to be considered a proto-precursor of a processual ontology (Di Poppa 2010).

However, all this excessive fluidity seems compensated, while at the same time recovering its compromise with determinations, when Spinoza describes his doctrine of conatus:

> The effort[1] by which everything strives to persevere in existing, is nothing but the actual essence of that thing.
>
> Dem. From the given essence of anything certain consequences necessarily follow, and things can be nothing else than what necessarily follows from their determinate nature. Therefore the power or effort of anything by which, either alone or together with other things, it does or strives to do something, i.e. the power of effort by which it strives to persevere in existing, is nothing else than the given or actual essence of that thing (EIII P7).

Thus, conatus must be understood as the effort a thing makes to maintain or persevere in its being, and that is nothing else than the very essence of the thing. This can clearly be understood as the process through which an entity self-imposes limits and determinations that allow it to preserve its integrity. In fact, conatus retrieves the notion of determination as negation, insofar as it also actively excludes what the entity is not:

> Conatus is essentially the striving of an individual mode to resist the potentially dissipating influences of external impingements which may dangerously alter the harmony of motion characteristic of that individual, insofar as the individual as a whole is concerned (Gabhart 1999, p. 615).

Therefore, all this may suggest that Spinoza's pantheism (or panentheism) may not be the best inspiration for the ecofeminist enterprise of crossing limits. Notwithstanding that, should ecofeminism wish to go on finding a source of inspiration in Spinoza, it may turn into an invitation to welcome the suggestion that not all determinations are mainly restrictive, some are constitutive and, therefore, the purpose of crossing borders may not simply brush away all divisions and differentiations without distinctions.

---

[1] Or *conatus*.

23.5 Artificial, Natural Boundaries and Destruction

At this point it may be useful to dwelve into certain distinctions, namely, between artificial and natural limits. Artificial boundaries are those that come about because of human conventions, for example, boundaries between countries. It is obvious that these are arbitrary limits and they come befall into an entity from outside, as a restriction. That applies also to many of the divisions and stereotypes denounced by ecofeminism: that may be the case, for example, of racial, class and other stereotypes. But there are some limits that do not qualify simply as due to conventions: they happen where there are true material discontinuities. These limits do not come from outside, they belong to the entity, they hold it together and they differentiate it from all others because they are constitutive (Smith and Varzi 2000).

Having said that, it seems ecofeminism does not pay explicit attention to this distinction when it comes to work on crossing limits. We may illustrate this with one example in particular. One of the more recent compilations on ecofeminism (Adams and Gruen 2021), includes a chapter that deals specifically with the transgression of limits demonstrated by nature itself in the case of an organism that changes sex from female to male:

> In 2017, the premiere episode of BBC's documentary series Blue Planet II opens with the sexual transformation of an Asian sheepshead wrasse (Semicossyphus reticulatus), colloquially known as the kobudai in Japan. [...] The edited footage, roughly five minutes in length, begins with a playful tone but takes a sharp eerie turn as Attenborough describes how "diminutive" kobudai 10 years old and older stop mating with kobudai of different sexes, and upon reaching a "critical body size," they undergo a "dramatic transformation" of hormonal and enzymatic change (Kirts 2021, pp. 373–374).

The scene of the transformation is played in a more sombre setting with dissonant and tense music, as if describing a dark and perverse process. This leads the author to reflect:

> The scene is frustrating to watch. The filmmakers highlight a biological process that is standard in kobudai communities yet rarely seen on mainstream television, but rather than portray sex mutability as fascinating and common- place, this episode falls into a pejorative cisgender trope, depicting the kobudai's ability to change sex as terrifying. [...] That there is no room for these species in heteronormative binary sex (and corresponding gender) categories is not a new problem (Kirts 2021, p. 374).

All these reflections lead the author, precisely in line with other authors of queer criticism of scientific discourse (Alaimo 2010) to arrive to the conclusion that this kind of queer organisms are natural entities that *transgress limits*. The conclusion, however, seems ambiguous in view of the distinctions we have analysed because we might ask: transgressors of which limits?

It is obvious that carrying out a biological process which is natural in that organism does not transgress any natural limits. The entity is faithful to its constitutive boundaries and merely follows its nature. From this point of view there is no transgression. Precisely, the limits apparently questioned by the entity are those denounced by the author as 'heteronormative binary categories', but those are the ways employed by the makers of the documentary to conceptualise the sexual identity of the organism. They are not constitutive limits of the *kobudai* itself. The *kobudai* subverts the notion that organisms generally remain strictly as male or female throughout their development. But that notion is not a constitutive limit of the *kobudai*. It is simply the prejudice of the authors of the documentary and, consequently, corresponds to a conceptual and subjective limitation which reaches *kobudai* from outside. However, the description mentions limits-transgressions in general, with no distinction between natural constitutive limits and artificial ones that are, in fact, built as constrictive stereotypes.

It is probably positive to question conceptual limits that cause stereotypical and prejudiced ways to describe certain natural processes, but if we do not do this differentiation, we sweep away constitutive limits too. And discarding these limits it no longer innocuous, as I will show in another example.

There is quite a well-known story, amplified by urban legends: the case of Hisachi Ouchi, a worker in a nuclear-power plant in Japan. Due to negligent practices by the company, Hisachi found himself involved in a fatal work-accident. When he made a mistake while pouring a mixture of radioactive chemicals, there was a critical liberation of radiation. Hisachi fainted during some time, but was apparently unharmed. He was driven to a hospital with no symptoms of the accident, with the exception of a light tan in his skin which was not there before. It looked like he went out in the sun without sun protection. But when he was taken to a hospital specialised in radiation and underwent a cell-study, doctors noticed that the radiation was destroying the genetic material of each one of his cells, keeping the rest of the structures apparently unscathed. Because of the lack of genetic material, Hisachi's body was unable to synthetise new proteins or carry out any kind of cell proliferation or division. The apparently unscathed body began to show ongoing and slow signs of degradation, starting by his skin. Eventually, Hisachi's body began to falter up to the point that he had to be connected to mechanical ventilation and his skin had completely vanished. His muscles were visible. Litres of fluids were constantly administered into his body since, with no skin, they outpoured. Nurses wrapped his body completely with new gauze everyday. He underwent several heart arrests and was always revived. His family kept a naïve hope that he would recover. After 83 days in agony and almost superhuman medical efforts, Hisachi died (Nippon Hoso Kyokai 2015).

Hisachi's case stands in radical contrast to the case of the *kobudai*. While the transformation of the fish does not imply transgression of any natural limits, the case of the Japanese worker may be described as a dissipation of constitutive limits. Hisachi loses his capacity to live as a self- contained entity. It is interesting to note that precisely the first barrier to disappear is his skin, which is colloquially seen as the boundary that separates us from our environment. As opposed to what some eco-feminisms seem to suggest, the blurring of these natural limits does not end in a harmonious reunification with the environment. It ends in the destruction of the entity. This can also be expressed in Spinoza's language: it is as if Hisachi had lost his *conatus* and, consequently, the possibility of remaining alive and persevering in his being. His unique proportion of movement and rest smashes down.

The example shows that transgression of all limits is not necessarily positive, as Kuyper fears. Crossing limits may result in correcting prejudices and wrong conceptualisations or, for example, in dissipating oppressive national boundaries, or discriminatory racial prejudices or others. This may obviously be very positive and revindicate oppressed groups. But blurring natural boundaries is a very different matter. Under this light, combinations and chimeras described by Rubenstein seem to recover their monstruous character. In fact, many of them may endanger the constitutive integrity of particular beings. If pantheism blurs everything in this manner, then it truly seems terrifying. But we have already seen that Spinoza himself, whom this idea is attributed to, does not think along this line. Maybe fear of these transgressions is not only connected to resistance by Western patriarchal constructions to be questioned, but also to the intuition that blurring all limits endangers the wholeness of several entities. Entities whose care and protection is precisely the mission of environmentalism.

This does not mean, however, that boundary questioning should be abandoned; it implies, rather, that this task must also differentiate precisely those limits that mark binaries and arbitrary divisions, from constitutive ones. In fact, there are recent ecofeminist thinkers that seem to acknowledge the constitutive character of some limits and, therefore, re-establish the connection between individuals and the environment through acknowledging semi-permeable limits (Alaimo 2016), which determine the traits of an entity, but without closing it to relationships and interchange of materials, energy, and information. Annemarie Mol (2008, p. 30), for example, when describing the process of food intake, stresses that "neither narrowly closed, nor completely open, a diner has semi-permeable boundaries."

In ecofeminism, specially in Mary-Jane Rubenstein's pan-theologies, the issue of limits is shown as particularly controversial in view of the discussions we have gone through. This allows us to point at something that, although the literature on environmental ethics seems to insinuate, is not thematically discussed. The tendency towards holisms, common in current environmental ethics, typically ends in ontologies that reject or relativise boundaries, but refrain from studying deeply what those limits are or the negative consequences of rejecting them. On the other hand, there's a deficit of environmental ethics based on the acknowledgement and promotion of the existence and usefulness of limits. In that sense, there seems to be an imbalance among the stances in the debate. That is why it seems important to listen to ontologies that argue in favour of limits, particularly if we acknowledge the existence of constitutive limits. In fact, it is possible, as insinuated in this chapter, to develop an environmental ethics that grounds the protection of the environment and living beings on the acknowledgement and protection of its constitutive limits.

23.6 Conclusions

The criticism I have made seems to concede Kuyper's initial point, namely, that it is dangerous to sweep away all boundaries, especially since many of them are constitutive. However, Kuyper also errs in negatively assessing Pantheism as a destroyer of boundaries, since it may be the case that, in fact, pantheism allows one to respect and care for the boundaries of entities if understood otherwise.

The concept of limit or boundary has a pre-eminently metaphorical use, it refers to other matters whose original meaning is more precise and which are different from each other, even if they are related. In some of the mentions we have made here there is already more than one meaning. In some cases limit have been understood as limitations, as something that restricts and hinders the full expression of an entity; at other times it has been understood as a determination or as a property, so that its function is rather constitutive. In other instances it has been used as a topological limit, that is, the limit marks where the extension of an entity begins and ends and, therefore, confronts other topological issues such as fusions or combinations. And probably, there are other possible meanings for which to use limits as metaphors. This proves that the concept of limit is polysemous and it is necessary to be clear about how we are using the metaphor; otherwise, categorically different and sometimes even unrelated issues are confused. It seems to me that this is the case when we speak, on the one hand, of a topological or even ontological limit that separates things, and that conceptual limit that marks the distinction between concepts, for example, god/world, man/woman, and on the other hand, the limit as oppression. In this case, limit means different things according to the context of statements.

This makes us question even more that this first removal of a limit, implies the removal of all the rest of limits. The implication simply does not exist, because there was no removal in the first place, only a correction to an ontology. What happens to the rest of the limits of things, or whatever we mean by their limits, is not something that depends conceptually on the boundary between God and world necessarily. That my ontology is now pantheistic does not in any way imply that any boundary has been removed. It only means that the boundaries that already existed and separated things exist within the context of a single all-encompassing entity, and no longer two, which were God and world. The removal of a boundary at one categorical level does not necessarily imply its removal at another, even if they are related or one contains the other. For example, if we wanted to merge the categories of animal and plant, removing their boundary, it does not magically follow that now a larch and a dog are the same thing or that they are no longer separate, it simply happens that they are now subsumed under one and the same category. The same thing happens in the case of pantheism: it does not eliminate the differences between things, it simply subsumes that multiplicity into a single category, which is the fusion between God and the world, instead of the two separate ones. As we saw, Spinoza's ontology itself, considered a pantheism par excellence, does not sweep away boundaries. Even if we were to affirm such a thing, Spinoza's pantheism is not the only possible version of pantheism.

This does not mean that it cannot be argued that it is indeed the case that pantheism is boundary destroying, but it is not a given, such an implication must be made explicit and substantiated, whereas Kuyper and Rubenstein simply assume it. In this case what is assumed is that pantheism is what, in contemporary terminology, is called existence monism, in which there is only one thing and nothing but one thing, as Parmenides affirmed (Schaffer 2007). It is evident that there are no separations or distinctions, only an undifferentiated continuity. But it is necessary to note that, as Jonathan Schaffer identifies, pantheism can also be interpreted as a monism of priority: fundamentally there is only one fact, but it is composite, that is, it has parts (2010). In this cosmology there is multiplicity of things, which form a single composite entity and which possesses the highest ontological status, namely the whole. The whole takes precedence over the parts, but the parts still exist, i.e., distinct and separate things, demarcated by boundaries. In such a pantheism, everything would be subsumed under God-world, but there is no elimination of boundaries. Indeed, as Schaffer suggests, this would be a reasonable way to interpret Spinoza. And the normative power for environmental ethics does not end with this, on the contrary: it makes everyone and everything part of a single, all-encompassing, divine reality and calls us to care for the intrinsic value of that cosmic citizenship. And to protect each and every one it is necessary for there to be many. And for there to be more than one thing requires that they be separate and for there to be variety requires that they be distinct; that is, boundaries are required.

Therefore, that pantheism is boundary-destroying is a conceptually hasty implication. Only one interpretation of pantheism seems to presuppose a world without distinctions, that of existential monism. Pantheism does not necessarily imply the destruction of boundaries, although its historical usage has been such, since it can also be interpreted as priority monism.

References

Adams CJ, Gruen L (2021) Ecofeminism: feminist intersections with other animals and the earth. Bloomsbury Publishing, London

Alaimo S (2010) Eluding capture: the science, culture, and pleasure of "queer" animals. In: Erickson B, Sandilands C (eds) Queer ecologies: sex, nature, politics, desire. Indiana University Press, Bloomington, pp 51–72

Alaimo S (2016) Exposed: environmental politics and pleasures in Posthuman times. University of Minnesota Press, Minnesota

Besthorn FH, McMillen DP (2002) The oppression of women and nature: ecofeminism as a framework for an expanded ecological social work. Fam Soc 83(3):221–232

Birch C, Eakin W, McDaniel JB (2007) Liberating life: contemporary approaches to ecological theology. Wipf and Stock Publishers, Eugene

Colie RL (1959) Spinoza and the early English deists. J Hist Ideas 20(1):23–46

Di Poppa F (2010) Spinoza and process ontology. South J Philos 48(3):272–294

Estévez-Saá M, Lorenzo-Modia MJ (2018) The ethics and aesthetics of eco-caring: contemporary debates on ecofeminism(s). Women Stud 47(2):123–146

Foucault M (2004) Lectures at the Collège de France, 1974–1975. Picador, London

Gabhart M (1999) Spinoza on self-preservation and self-destruction. J Hist Philos 3(4):613–629
Jantzen G (1998) Becoming divine: towards a feminist philosophy of religion. Manchester University Press, Manchester
Kirts L (2021) Upsetting boundaries: trans queer interspecies Ecofeminisms. In: Adams CJ, Gruen L (eds) Ecofeminism: feminist intersections with other animals and the earth. Bloomsbury, London, pp 374–428
Kuyper A (1893) Pantheism's destruction of boundaries. Methodist Review, New York
Levine MP (1994) Pantheism, theism and the problem of evil. Int J Philos Relig 35(3):129–151
Marcos A (2001) Ética ambiental. Universidad de Valladolid, Valladolid
McFague S (1993) The body of god: an ecological theology. Fortress Press, Michigan
Melamed YY (2010) Acosmism or weak individuals?: Hegel, Spinoza, and the reality of the finite. J Hist Philos 48(1):77–92
Melamed YY (2012) "Omnis Determinatio Est Negatio" determination, negation and self-negation in Spinoza, Kant, and Hegel. In: Forster E, Melamed YY (eds) Spinoza and German idealism. Cambridge University Press, Cambridge
Mol A (2008) I eat an apple. On theorizing subjectivities. Subjectivity 22(1):28–37
Moran D (1990) Pantheism from John Scottus Eriugena to Nicholas of Cusa. Am Cathol Philos Q 64(1):131–152
Morton T (2010) Queer ecology. PMLA/Publ Mod Lang Assoc Am 125(2):273–282
Nippon Hoso Kyokai (2015) A slow death: 83 days of radiation sickness. Kodansha USA, New York
Rubenstein MJ (2018) Pantheologies: gods, worlds, monsters. Columbia University Press, New York
Schaffer J (2007) From nihilism to monism. Australas J Philos 85(2):175–191
Schaffer J (2010) Monism: the priority of the whole. Philos Rev 119(1):31–76
Smith B, Varzi A (2000) Fiat and bona fide boundaries. Philos Phenomen Res 60(2):401–420
Spretnak C (1989) Toward an ecofeminist spirituality. In: Plant J (ed) Healing the wounds: the promise of ecofeminism. New Society Publishers, Gabriola, pp 127–132
Valera L, Vidal G (2022) Pantheism, panentheism, and ecosophy: getting back to Spinoza? Zygon 57(3):545–563
Van Dyk L (1994) The body of god: an ecological theology. Theol Today 51(1):176
Wachholz S (2011) Ecofeminism. In: Chatterjee DK (ed) Encyclopedia of global justice. Springer, Dordrecht, pp 289–290
Warren KJ, Cheney J (1991) Ecological feminism and ecosystem ecology. Hypatia 6(1):179–197
Wood HW (1985) Modern pantheism as an approach to environmental ethics. Environ Eth 7(2):151–163

Chapter 24
Intrinsic Values, Pantheism, and Ecology: Where Does Value Come From?

Luca Valera

Abstract Pantheism (or panentheism) is at the basis of ecological thinking and presumes some sort of ethics focused on the conservation of the environment. Besides, since the main interest of environmental philosophy in general is the ethical issue (not the ontological or cosmological one), we could also state that environmental ontology derives from an ethical interest in the preservation of nature. In this paper I try to walk the path that leads from interest or respect for nature (i.e., the ethical issue), to the very idea of nature underlying it (i.e., the cosmological question). In this sense, I mainly focus on the intrinsic ecological value which constitutes one of the critical issues in most approaches and paradigms of environmental ethics, showing some possible aporias of this concept. Finally, I show how declarations (or tenets) about the intrinsic value of nature –or, more specifically, of species– should be based on an adequate cosmology, in order to be consistent. The pantheist hypothesis is precisely one of them.

Keywords Intrinsic values · Pantheism · Arne Næss · Cosmology · Environmental ethics

24.1 Introduction: Preliminary Ideas on Ecology and Pantheism

Ecology has a special connection to pantheism. This is probably due to Næss's reflections on philosophy and ecology. He based his "ecosophy" on Spinoza's philosophical speculation and, mainly, on the notion of *Deus sive Natura*. He writes,

L. Valera (✉)
Department of Philosophy, Universidad de Valladolid, Valladolid, Spain

Center for Bioethics, Pontificia Universidad Católica de Chile, Santiago de Chile, Chile

Cape Horn International Center –CHIC, Puerto Williams, Chile
e-mail: luca.valera@uva.es; luvalera@uc.cl

L. Valera (ed.), *Pantheism and Ecology*, Ecology and Ethics 6,
https://doi.org/10.1007/978-3-031-40040-7_24

interpreting Spinoza: "The characterization of Spinoza as a 'pantheist' is often interpreted in the direction that he identifies God with the universe or the world. Therefore, a better term for his view is panentheist ('God in all')" (Næss 1983, p. 684). Furthermore, such connection may be probably traced back to the origins of ecology itself, since the evolutionary philosopher and biologist, the father of ecology Ernst Haeckel, (Haeckel 1866), argues that his ecological worldview was developed from both Darwin's evolutionary theories and Spinoza's monism, which he interprets as a "pantheist" (Haeckel 1894, p. 4, 80). Along the same line, it is worth keeping in mind that one of the North American "fathers" of ecological conservationism (Wilson 1988), John Muir, was defined as a "pantheist prophet" (Wood 1988)[1] and that his worldview influenced Potter's bioethics (Muzur et al. 2016), which is based on Teilhard de Chardin, another theologian whose thinking coincided with Muir's one.

To sum up: there is a relevant tradition that connects the ecologic worldview with pantheism,[2] so much so that some have defined pantheism as the "hidden religion of America" (Everson 1976), particularly linked with environmental conservationism (Fox 1981). Forrest's (2010, p. 470) statement clearly expresses this connection: "When we humans destroy the natural world we are wounding God": if God coincides with the cosmos –if God *is* the cosmos, *Deus sive Natura*[3]– any of our interventions in the natural world is a modification of God. Thus, from the ethical perspective (i.e., how should I treat nature?), we can reach the cosmological or theological question (i.e., what is the relationship between the part and the whole – or, maybe, between the finite modes and infinite God?). In this regard, the pantheism at the basis of ecological thinking presumes some sort of ethics focused on environmental conservation. Besides, since the main interest of environmental philosophy in general is the ethical issue (not the ontological or the cosmological one), we could even state that environmental ontology derives from an ethical interest in natural preservation. Indeed, Mander (2022) argues: "Since nature is taken as intrinsically valuable, and because relating appropriately to nature presupposes its preservation and protection; nature in general and environmental issues in particular, are important to the pantheist."

---

[1] On the issue of Muir's "eco-theological" speculation, Framer (2008, p. 859) writes: "Muir developed a concept of 'God' synonymous with beauty and harmony-universal principles of nature. In a clever turn, Worster employs Linnaean taxonomy to describe Muir's belief system: Pantheism muirii var. sierra. In today's world when science so often gets dragged into bipolar debates between theists and atheists, Muir offers a historical example of a third way. He felt equally comfortable with the language of science and the language of religion. For him, holism was a spiritual as well as a scientific pursuit."

[2] In this paper, I will not differentiate between pantheism and panentheism. Regarding the issue I will face here, it is enough to analyze the monist vision underlying pantheism that insists on the intrinsic unity between God and/or Nature (*Deus sive Natura*). For a more detailed account of that distinction, please have a look at Valera and Vidal (2022).

[3] The connection of Spinoza's worldview with the philosophy of ecology is quite clear and well known –probably also due to the philosophical speculation of Arne Næss (1977). For further detailed analysis on this point, please see Valera and Vidal (2022).

Therefore, the cosmological issue appears to be of secondary importance as regards ethics.[4] Although the grounding of values and duties towards the natural world are included in the general worldview, in most environmental ethics reflections such vision is not expressed in the connections and links between the various living beings, or in the interconnections of dependence among them (Valera et al. 2021). In the following, therefore, I will try to walk the path that leads from interest or respect for nature (i.e., the ethical issue), to the very idea of nature underlying it (i.e., the cosmological question). In this sense, I will mainly focus on the intrinsic ecological value, which constitutes one of the critical issues in most environmental ethics paradigms. Finally, I will show how declarations (or tenets) about the intrinsic value of nature –or, more specifically, of species– should be based on an adequate cosmology, in order to be consistent. The pantheist hypothesis is precisely one of them.[5]

24.2 Respect for Nature: Where Does Value Arise From?

Why *must* we respect nature? And, above all, must we respect *every* form of natural life? Obviously, if we look at the problem from afar and hastily, the second question immediately poses an aporia –or a practical problem: we cannot respect *all* the forms of life equally, because that would turn every action practically impossible. The life (of some) requires the death (of others), and that is a fact. Nutrition or, more generally, the fight for survival is a good example of that need to "put an end to another being's life" in order to keep one's own life. In that sense, the forms of life are implicitly selfish or "speciesist"[6]: in general, they value their own existence (or that of their own species) more that the existence of other living beings. This is one of the most obvious problems of some line of ecologist thought –well represented by the ideas of Arne Næss and John Muir,[7] for example –: an exaggerated optimism, both as regards human nature as that of other living beings. I have intended to clarify this point in another context: "Næss has not considered this latest aspect to its further implications, and the idea of an essentially positive relationship seems to be defined, therefore, more as an initial hypothesis than as a result of a philosophical

[4] Along this line, Arne Næss may be an exception, since he is interested in developing a consistent and systematic worldview, where ethics should be interpreted as a corollary to that system of thought (Valera 2018).

[5] Another possibility of grounding the existence of intrinsic values, within the framework of a Christian cosmovision, has been explored by Scheid (2016).

[6] I am obviously referring to Peter Singer (1975) here.

[7] On this point, Worster (2005, p. 13) argues: "By criticizing those who failed to treat the world as a holy place, however, Muir did not become a glowering pessimist about his fellow humans. On the contrary, he regarded every individual as potentially his equal, capable of sharing the same innate feeling for nature that he felt. In that benevolent optimism he included all women, children, and men."

argumentation. It is evident, however, that this is a crucial issue in order to understand Næss's environmental ethics, since the following proposals –for example, the idea that the realization of the other promotes self-realization– derive entirely from that assumption. In this regard, the philosophy of perception developed by Næss is too optimistic and lacks in sufficient argumentative elaboration. This leads to an essentially optimistic anthropology –human nature is essentially positive, even though human behaviors may accidentally be destructive. These claims should have been further explained or justified in his works. This is a particularly important point, since much of Næss's ontological and ethical considerations depends on his theory of perception and anthropology –for instance, the process of self-realization is logically dependent on the process of identification" (Valera 2018, p. 668). If it is true that humans are not necessarily the planet's cancer –to recall a striking image by the father of bioethics, Potter (1971)– we cannot state either that they are originally "good" beings, that they merely need to return to a pristine state or original life, in greater contact with nature (i.e., the myth of the noble savage).

A middle way between the cancer and the myth of the noble savage is the idea that posits that human beings may develop some capacities, attitudes, or virtues –a "second" nature, which are added to their "original nature"– that would allow them to respect nature and moderate their egotistic, utilitarian, or self-centred behaviours. On this basis, it is easy to understand the importance, centrality, and relevance of "Environmental Virtue Ethics" (Sandler 2004, 2007; Cafaro and Sandler 2005) among the various current environmental ethics paradigms (Valera et al. 2021). Consequently, the respect for nature would be the consequence of an ecological education –or of a new and more mature worldview, as Næss (1984) argues– that would positively transform our attitude toward the environment. Briefly, it would mean modifying human actions, even before changing the environment.

Within this context, we should ask, yet again: why should we modify our actions to fulfil ends that are extrinsic to human nature itself (e.g., the preservation of nature)? Or, more briefly, why should we respect nature? In fact, the issue of virtues is necessarily connected to the issue of values, in a post-modern context, while in the ancient world the ethical centre was the good or end to achieve. Therefore, our attention moves from human actions (and their structure) to the values we may achieve or promote (and their structure). Consequently, the focus lies on the object of the action, not on the action itself. The question now turns into: is nature worthy of being respected *per se*? Or: is nature only valuable when there is someone who values it?

24.2.1 Intrinsic Values and Aporias

To answer to the previous question there are apparently two main possibilities: intrinsic values and instrumental values. The difference between them is clearly explained by Sandler (2012, p. 4):

> Intrinsic value is the value that an entity has in itself, for what it is, or as an end [...]. The contrasting type of value is instrumental value. Instrumental value is the value that something has as a means to a desired or valued end. Instrumental value is always derivative on the value of something else, and it is always conditional.

Although this difference appears to be clear –i.e., the presence or absence of a subject that evaluates a natural object as desirable– the fact is that there are at least some doubts on this topic. Indeed, different experts in environmental ethics have pointed out the issue of the "necessary valuer" and, thus, the anthropogenic origin of value[8]:

> Moralists among environmental ethicists have erred in looking for a value in living things that is independent of human valuing. They have therefore forgotten a most elementary point about valuing anything. Valuing always occurs from the viewpoint of a conscious valuer (Norton 1991, p. 251).

This issue brings us back again on the term "intrinsic," associated to the issue of value. In fact, if we consider what we have just said, the term "intrinsic" would be "misleading."

Bearing that in mind, Holmes Rolston III (2003, p. 144) proposes to replace it by its opposite, "extrinsic," since the "ex" would more precisely point at the "anthropogenic" origin of value itself:

> What is meant is better specified by the term *extrinsic*, the ex indicating the external, anthropogenic ignition of the value, which is not *in*, *intrinsic*, internal to the no sentient organism, even though this value, once generated, is apparently conferred on the organism. In the *H-n* encounter, value is conferred by *H* on *n*, and that is really an extrinsic value for *n*, since it comes to *n* from *H*, and likewise it is an extrinsic value for *H*, since it is conferred from *H* to *n*. Neither *H* nor *n*, standing alone, have such value. We humans carry the lamp that lights up value, although we require the fuel that nature provides. [...] Humans are the measurers, the valuers of things, even when we measure what they are in themselves.

These reflections seem to indicate that value exists since there is a previous relationship between a human being and another being. Therefore, when we talk about "values," we always indicate that it is something "for a perceiver:" something has value *for someone*. Consequently, the value is a relational property –i.e., a property between the subject and the object– and one of the parts of this relationship is necessarily "someone," namely, an evaluating subject.[9] Indeed, stating that something has "intrinsic value" (or "extrinsic") is not equivalent to say that it has value even in the absence of an evaluating subject; rather, it means that something has value for that

---

[8] Taylor (2011, p. 74), one of the first and most relevant holders of the notion of an intrinsic (or, rather, inherent) worth says: "Whatever the basis of the inherent value placed on something may be, whether aesthetic, historical, cultural, or a matter of personal sentiment, wonder, or admiration, *the inherent value of anything is relative to and dependent upon someone's valuing it*."

[9] It is worth noticing that this statement is fully compatible with Arne Næss's (2010) philosophy. He highlights the importance of Sextus Empiricus' thought to counter an "objectivistic" view of reality, which permanently intends to achieve the "*Ding an Sich*." As is the case of all things, for Næss value is inserted within a Gestaltic view, or in a network of interrelations between different living beings, interpretations, perceptions, emotions, evaluations, etc.

(or another) subject, not only as a means, but as an end in itself. We cannot certainly forget the relational perspective, which is impossible to eliminate (Sandler 2013, p. 356): the value always comes from a relationship, and it only acquires sense within that context. If we eliminate one of the basic elements of the value –i.e., its necessarily "subjective" configuration– its sense is completely lost, turning the very existence of intrinsic value into a *petitio principii.*[10]

24.2.2 Beyond Aporias: An Attempt at Explanation

Once we have discussed the issue of the necessary link between value and evaluator, it is time to ask if it is true that environmental ethics was built on an obvious logical error –as the one I have just mentioned– or if there is another way to justify the existence of those intrinsic values. The large number of interpretations and supporters of intrinsic values, in fact, lead us to think that there are good reasons behind an apparently obvious aporia.

A good point to understand the good reasons of intrinsic value should be found in Callicott (1989, p. 151): he argues that species "may not be valuable *in* themselves but they may certainly be valuable *for* themselves." In this sense, following Callicott, "value is, to be sure, humanly conferred, but not necessarily homocentric." The theoretical justification of intrinsic values would lie in the difference between "*in*" and "*for*." Although it is true that the evaluation always depends on us, the humans, it may also be true that some realities express an evident value. The point, then, is to show that evidence. To do that, I find interesting to refer to one of the "fathers" of environmental ethics, Paul Taylor (2011, pp. 71–80). He developed an ethical theory based on the *inherent worth* of all living beings, opposing to an ethics based on merits (virtues). In his opinion, species are valuable *per se*, beyond their instrumental value –i.e., their value *for* "someone"– or their inclusion in an "economy" of nature. The reasons of that statement should be found in the difference between inherent and intrinsic value, forwarded by Taylor (2011, p. 73). We can note the details of such a distinction:

> Intrinsic value. When humans or other conscious beings place positive value on an event or condition in their lives which they directly experience to be enjoyable in and of itself, and when they value the experience (consider it to be good) because of its enjoyableness, the value they place on it is intrinsic. The experience is judged to be intrinsically good. Intrinsic value is likewise placed on goals that conscious beings seek to bring about as ends in themselves, and also on interests they pursue as intrinsically worthwhile. Experiences, ends, and interests may be valued not only intrinsically but also as means to further ends. They would in that case have both intrinsic and instrumental value for the beings concerned. [...] But as long as any activity is carried on for its own sake or as an end in itself, whether or not it is also valued for its consequences, it has intrinsic value for those who find satisfaction in it.

[10] Some of the aporias and problems concerning intrinsic values –within the framework of biology of conservation– are well summarised in Hyde's (2018) paper.

On the other hand, when we talk about inherent worth, we refer to

> the value we place on an object or a place (such as a work of art, a historical building, a battlefield, a "wonder of nature," or an archaeological site) that we believe should be preserved, not because of its usefulness or its commercial value, but simply because it has beauty, or historical importance, or cultural significance. When an object or place is valued in this way, it is considered wrong to destroy or damage it, or to allow it to deteriorate through neglect. [...] It is something precious to people, something they hold dear. Quite aside from whatever practical usefulness or commercial value it might have, it is held in high esteem and considered important because it is the kind of thing it is.

To sum up: the very difference between both lies in the focus of our attention: if we consider the object in connection with possible enjoyable human experiences (i.e., if it causes pleasure as an end in itself, not as a means toward another end), value will be intrinsic, while, if we consider the object, beyond any human end or pleasure, it will be inherent. Consequently, especially in the case of intrinsic value, the difference with instrumental values is very elusive: in the case of intrinsic values, what matters is the impact of that reality on human life –or on a conscious being, as Taylor states– at the level of pleasurable experiences, when it is enjoyed as an end in itself and not as a means in connection to another end. In the case of inherent worth, the importance of that reality is stressed, leaving aside the possibility of its being part of a context of human perceptions –in other words, the need of its existence is emphasized. As Stephano (2017, p. 321) writes, "inherent worth [...] refers to a being's status as a moral subject on the basis of having a good of its own." Two examples of such distinction may be: *wilderness* (intrinsic worth),[11] which is pleasurable –though not immediately useful for a conscious being; and *biodiversity* (inherent worth), which is something that *must* exist,[12] beyond any evaluation about its need for human life or the life of a conscious being– since it doesn't provide any pleasant experience. In the latter case, then, qualifying an object as "valuable" is not seen as the result of an *assessment*, based on an evaluation of its properties –as in the case of "realist" or "finalist" ethics– or the connection between the object itself and the evaluator –as in the case of utilitarianism– but it is an *a priori* postulate, similar to Soulé's stance (1985, p. 731): "Species have value in themselves, a value never conferred nor revocable, but springing from a species' long evolutionary heritage and potential or even from the mere fact of its existence." Being a postulate, however, it would appear to be closed to any enquiry about the deep reasons of its content. So, as Sandler (2007, 2012) argues, "it is not easy to explain why

[11] On the topic of the insistence on the intrinsic worth of wilderness, Scruton (2012, p. 197) writes: "From the beginning of the environmental movement in America the wilderness has been regarded as something valuable in itself, and not by virtue of any use to which it might be put. It is precisely as wilderness –i.e. as a place removed from human uses– that the wilderness has a claim on us."

[12] In fact, Soulé (1985, p. 731) argues that the sentence "ecological complexity is good" is one of the central postulates –and the word postulate has an eminent value here– of conservation biology. Another one is: "Biotic diversity has intrinsic value, irrespective of its instrumental or utilitarian value. This normative postulate is the most fundamental. In emphasizing the inherent value of nonhuman life, it distinguishes the dualistic, exploitive world view from a more unitary perspective" (Soulé 1985, p. 731).

natural-historical properties of species and systems are objectively value adding." Yet again, we would have a *petitio principii* or the statement of an indisputable postulate (Soulé 1985, p. 732).

24.3 Intrinsic Values: A Consistent Pantheist Groundwork

Nevertheless –and we go back to the initial hypothesis of this paper– there is another option to find a consistent groundwork for intrinsic values, without referring to a postulate– which would also be a legitimate option, obviously. Indeed, Soulé (1985, p. 732) admits that the "endless scholarly debate will probably take place about the religious, ethical, and scientific sources of this postulate and about its implications for policy and management." That debate has actually developed over the past recent years. One of the hypotheses that has become more relevant on account of its consistency, within this context, is pantheism (Valera and Vidal 2022). The hypothesis was held by Muir: "Every species [...] demands respect, and every creature has a right to life, liberty, and the pursuit of happiness. Every form of life, like every group of people, is equal in the eyes of the Creator; indeed, all species are in some sense 'people'" (Worster 2005, p. 11). From this point, we can conclude that the dignity –or value– of each living being is based on its participation to the divine substance. Respect is owed to each living being precisely because it is part of God, sharing God's essence in the same degree as the other living beings. Consequently, the measure of value –or that of dignity– is precisely and exclusively the ontological participation to the Whole –to the unity of the Whole.[13] Such unity is, thus, one of the main principles of the ecologic cosmovision: it is worth remembering that one of the basic postulates of ecology is "Species are interdependent" (Soulé 1985, p. 729) or "Everything is connected to everything else" (Commoner 1971). In this regard Levine (1994, p. 132), explicitly connecting ecology with pantheism, states:

> The belief in a divine Unity, and some kind of identification with that Unity, is seen as the basis for an ethical framework (and "way of life") that extends beyond the human to non-human and non-living things. The divine Unity is, after all, "all-inclusive." It is, I shall argue, not accidental that pantheism is often taken to be inherently sympathetic to ecological concerns.

Therefore, the most essential fact concerning pantheism is the intimate unity between God (or Nature as an organic totality) and natural entities and that is why "the pantheist's ethic, her environmental ethic and her ethics more generally, will be metaphysically based in terms of the divine Unity" (Levine 1994, p. 132). From here, the idea of intrinsic value may be deduced as essentially linked to that unity:

[13] Grula (2008, p. 160) admits that unity is the basic principle held by the pantheist cosmovision: "The definition of pantheism I use here is: the doctrine that God is not a personality or transcendent supernatural being but that all laws, forces, manifestations, and so forth of the self-existing natural universe constitute an all-inclusive divine Unity."

> Intrinsic value is, of course, value that is non-derivative. But, what determines the intrinsic goodness in a person's life will, for the pantheist, rely on that person's relationship to the Unity. A person's "good" is partially constituted by the divine Unity of which everything is a part. In pantheistic terms it makes little sense to speak of the intrinsic value of a human life as measured against a standard independent of how that life affects others, since for the pantheist all such value, even so-called "intrinsic value," is partly derivative. The standard of intrinsic value and perfection cannot be determined without reference to the divine Unity. The essential nature and well-being of a person, or anything else, cannot be analysed apart from its context in relation to the Unity and everything it includes (Mander 2022).

Not only, then, the idea of intrinsic value cannot be determined without reference to divine Unity but, precisely, pantheism is the most consistent way to determine it.

Nevertheless, those ideas do not yet justify the notion that to all living beings should be granted the same respect. Although they admit that all living beings *are worthy of respect,* insofar as they are part of God, they do not justify the notion that all are *equally* worthy of respect. In fact, Taylor (2011, p. 154) states:

> It seems to me that the negative statement, "It is not the case that humans are superior to other forms of life," does not follow by logical necessity from the three premises: "Humans are members of the Earth's Community of life;" "All living things are related to one another in an order of interdependence;" "Each organism is a teleological center of life."

Although Taylor's critique seems to be more than appropriate[14] –denying the superiority of human beings over the other living beings does not entail the equality of all living beings– notwithstanding that, his proposal does not appear to be very consistent: "In the absence of any good reasons for accepting it, the idea that humans are superior to all other living things would appear simply as an irrational bias in our own favor. And the same way of thinking would apply to any claim that one species is inherently superior to another. Such a claim would appear completely arbitrary and unreasonable from the viewpoint of the first three elements of the biocentric outlook" (Taylor 2011, p. 155). Based on that, Taylor (2011, p. 155) infers the principle of "species-impartiality":

> This is the principle that every species counts as having the same value in the sense that, regardless of what species a living thing belongs to, it is deemed to be prima facie deserving of equal concern and consideration on the part of moral agents. Its good is judged to be worthy of being preserved and protected as an end in itself and for the sake of the entity whose good it is. Subscribing to the principle of species-impartiality, we now see, means regarding every entity that has a good of its own as possessing inherent worth –the *same* inherent worth, since none is superior to another.

Obviously, the notion of the intrinsic value of all living beings also implies some non-irrelevant ethical consequences. But the point is to find out if they are duly justified. For example, Næss (1989, p. 184), argues: "A principal point in

[14] It is the same objection offered by Næss (1989, p. 167) about the notion of "relative intrinsic value," which would be "rankable," held from different perspectives and several arguments (i.e., the presence of a soul in the living being; the presence of reason, conscience, etc.): "None of these standpoints, so far as I can see, have been substantially justified. They may appear to be reasonable at first glance, but they fade after reflection and confrontation with the basic intuitions of the unity of life and the right to live and blossom."

'egalitarianism in the biosphere' is thereby won: every living being is equal to all others *to the extent* that it has intrinsic value." Such assertion makes some sense, for example, in a pantheist –or panentheist– viewpoint, as the one offered by Næss (1983, p. 684). If, at the ontological level, all living beings, as finite modes of the only divine substance, participate in the same way[15] –i.e., without differences– of God's substance, there will be no need to differentiate their value. In fact, from a pantheistic point of view, the value of those beings is defined mainly by their unity with substance –i.e., the only really existent and subsistent being: "Since Unity is predicated upon some evaluative consideration (e.g., the divine Unity being constituted on the basis of 'goodness'), value is a focal point for the pantheist and a principal concern" (Levine 1994, p. 133). On that basis, then, it is possible to infer the equal value of living beings: ultimately, the goodness of finite modes coincides with God's goodness –being part of God, they also acquire God's characteristics, albeit incompletely.

Notwithstanding that, not all experts in environmental ethics acknowledge this need of a cosmological grounding –such as that of pantheism– consistent with the idea of the intrinsic value of all living beings. Rather, they choose to take it as a non-renounceable postulate. The problem is that such postulate conforms the groundwork of their environmental ethics: in this regard, the criticism of such paradigms is quite obvious.

Acknowledgments This research has been supported by Project n° 1210081 (ANID/Fondecyt Regular); and Project ANID/BASAL FB210018.

References

Cafaro P, Sandler R (2005) Environmental virtue ethics. Rowman & Littlefield, Lanham
Callicott JB (1989) In defense of the land ethic: essays in environmental philosophy. State University Press of New York Press, Albany
Commoner B (1971) The closing circle. Nature, man, and technology. Bantam, New York
Everson W (1976) Archetype west: the Pacific coast as a literary region. Oyez Press, Berkeley
Farmer J (2008) The great American pantheist. Science 322(5903):859–860
Forrest P (2010) Spinozistic pantheism, the environment and Christianity. Sophia 49(4):463–473
Fox S (1981) John Muir and his legacy: the American conservation movement. Little Brown, Boston
Grula JW (2008) Pantheism reconstructed: Ecotheology as a successor to the Judeo-Christian, enlightenment, and postmodernist paradigms. Zygon: J Relig Sci 43(1):159–180
Haeckel E (1866) Generelle Morphologie der Organismen. Allgemeine Grundzige der organischen Formen-Wissenschaft, mechanisch begriindet durch die von Charles Darwin reformirte Descendenz-Theorie, 2 vols. Reimer, Berlin
Haeckel E (1894) Monism as connecting religion and science: the confession of faith of a man of science. Adam and Charles Black, London
Hyde D (2018) Is there a need for intrinsic values in conservation biology? Australas J Log 15(2):498–512

[15] I tried to describe what this participation means, in the paper: Valera and Vidal (2022).

Levine MP (1994) Pantheism, ethics and ecology. Environ Values 3(2):121–138
Mander W (2022) Pantheism. In: The Stanford Encyclopedia of Philosophy. Available at: https://plato.stanford.edu/archives/spr2022/entries/pantheism/
Muzur A, Rinčić I, Sodeke S (2016) The real Wisconsin idea: the seven pillars of van Rensselaer Potter's bioethics. J Agric Environ Ethics 29:587–596
Næss A (1977) Spinoza and ecology. Philosophia 1:45–54
Næss A (1983) Einstein, Spinoza, and God. In: van der Merwe A (ed) Old and new questions in physics, cosmology, philosophy, and theoretical biology. Essays in Honor of Wolfgang Yourgrau. Springer, Dordrecht, pp 683–687
Næss A (1984) The arrogance of Antihumanism? Ecosophy 6:8–9
Næss A (1989) Ecology, community and lifestyle. Outline of an ecosophy. Cambridge University Press, Cambridge
Næss A (2010) The world of concrete contents. In: Drengson A, Devall B (eds) The ecology of wisdom. Writings by Arne Næss. Counterpoint, Berkeley, pp 70–80
Norton BG (1991) Toward unity among environmentalists. Oxford University Press, New York
Potter VR (1971) Bioethics: bridge to the future. Prentice-Hall, Englewood Cliffs
Rolston H III (2003) Value in nature and the nature of value. In: Light A, Rolston H III (eds) Environmental ethics. An Anthology. Blackwell, Malden, pp 143–153
Sandler R (2004) Towards an adequate environmental virtue ethic. Environ Values 13(4):477–495
Sandler R (2007) Character and environment: a virtue-oriented approach to environmental ethics. Columbia University Press, New York
Sandler R (2012) Intrinsic value, ecology, and conservation. Nature 3(10):4
Sandler R (2013) The ethics of reviving long extinct species. Conserv Biol 28(2):354–360
Scheid DP (2016) The cosmic common good. Religious grounds for ecological ethics. Oxford University Press, New York
Scruton R (2012) How to think seriously about the planet. The case for an environmental conservatism. Oxford University Press, New York
Singer P (1975) Animal liberation: a new ethics for our treatment of animals. HarperCollins, New York
Soulé ME (1985) What is conservation biology? Bioscience 35(11):727–734
Stephano O (2017) Spinoza, ecology, and immanent ethics: beside moral considerability. Environ Philos 14(2):317–338
Taylor PW (2011) Respect for nature. A theory of environmental ethics (25th anniversary edition). Princeton University Press, Princeton/Oxford
Valera L (2018) Home, ecological self and self-realization: understanding asymmetrical relationships through Arne Næss's ecosophy. J Agric Environ Ethics 31:661–675
Valera L, Vidal G (2022) Pantheism, panentheism, and ecosophy: getting back to Spinoza? Zygon: J Relig Sci 57(3):545–563
Valera L, Leal Y, Vidal G (2021) Beyond application. Case Environ Eth Tópicos (México) 60:437–459
Wilson J (1988) The father of conservation. The Highlander – Mag Scott Herit 26(2):1–11
Wood H (1988) Pantheist prophets: John Muir 1838–1914. Pantheist Vis 9(2)
Worster D (2005) John Muir and the modern passion for nature. Environ Hist 10(1):8–19

Chapter 25
Humans Are *Humus*: An Analysis of Boff's Panentheistic Ecotheology in the Framework of the Biocultural Ethic

Ricardo Rozzi

Abstract I have proposed a biocultural ethic grounded on the conservation of the vital links among (i) the habits of caring and conservation of land, freshwater, and marine ecosystems, (ii) the daily interactions with diverse human and other-than-human co-inhabitants, and (iii) their shared habitats. This is the 3Hs triad (habits, co-inhabitants, and habitats) of the biocultural ethic. The conservation of these habitats is the condition of possibility for the wellbeing of the communities of co-inhabitants, and the continuity of complex interrelationships between biological and cultural diversity. However, today a vast majority of human and other-than-human co-inhabitants suffer from oppression derived from territorial displacements, degradation of native habitats, climate change and other processes that are characterizing the current era as a Necrocene (i.e., an era of marked by death). Brazilian liberation theologian Leonardo Boff has stood up in defense of communities of co-inhabitants conserving and protecting access to their ancestral lands, affirming that "not only the poor cry; also the lands cry, the waters cry, nature cries". Hence, both the biocultural ethic and the ecological approach of Boff's theology of liberation contribute to counter the oppression of the current Necrocene (an age of death and extreme extermination) and foster, instead, a biocene (an age of life, encompassing its biological and cultural diversity). I identify two core concepts and values of my proposed biocultural ethic with Boff's perspective ecotheology of liberation. First, Boff affirms that social justice is a necessary condition for accomplishing ecological justice; one involves the other. Boff's statement converges with my demand for the conservation of native habitats and access to them by local communities of human and other-than-human co-inhabitants as a biocultural ethical imperative. Second, the concept of co-inhabitation implies that human beings live ethically when they decide to stop placing themselves above all others, and decide instead to stand together with others. Boff's concept of a *dignitas terrae* adds a demand for a mate-

---

R. Rozzi (✉)
Cape Horn International Center (CHIC), Universidad de Magallanes, Puerto Williams, Chile

Sub-Antarctic Biocultural Conservation Program, Department of Philosophy and Religion and Department of Biological Sciences, University of North Texas, Denton, TX, USA
e-mail: Ricardo.Rozzi@unt.edu

L. Valera (ed.), *Pantheism and Ecology*, Ecology and Ethics 6,
https://doi.org/10.1007/978-3-031-40040-7_25

rial and spiritual reconnection with both the exterior and the interior nature of each human being and society with the Earth as a whole. His eco-theological path is based on a Christian-Franciscan panentheism rooted in care for and tenderness in encounters with other beings. Boff's path converges with my path oriented by a biocultural ethic that does not value and treat other-than-human beings as mere natural resources, but as *co-inhabitants* with whom we share the habitats and have inherent moral value.

Keywords Leonardo Boff · Ecotheology · Biocultural ethics · Co-inhabitants · Liberation theology

25.1 Humans Are Humus

The word *human* comes from the Latin *humus* (the most fertile layer of soil). Recovering the historic memory or the deep meaning of our name as a species should stimulate our consciousness that we humans are inextricably connected to earth. This understanding has had both a material and symbolic-linguistic meaning since the origins of Western civilization.

The material dimension is corroborated today by biogeochemical sciences that demonstrate that the molecular constitution of human bodies has a chemical composition similar to humus or organic matter in the soil (Schlesinger and Bernhardt 2013). In the human body just four chemical elements correspond to 99% of the total atoms: hydrogen (63%), oxygen (26%), carbon (9%), and nitrogen (1%) (da Silva and Williams 2001). The very large percentage of oxygen and hydrogen atoms arise from the highwater content −45% to 75%– in the human body (Jéquier and Constant 2010; Ohashi et al. 2018). Water is the matrix of all living systems (Szent-Gyorgyi 1957). The biosphere means "life on Earth," and is formed by all living beings with whom humans coinhabit the planet. Yet the biosphere is an integrated system that is coupled in complex ways to other major Earth systems: the geosphere, hydrosphere, and atmosphere. The chemical elements and compounds essential to life circulate through these systems. Water, the hydrosphere, circulates through the various earth systems, including humans and the biosphere. Carbon and nitrogen together with oxygen and hydrogen are the basic elements of organic structures and metabolites of all living systems. However, seven other elements are essential to humans and other living beings: sodium, potassium, calcium, magnesium, phosphorus, sulfur, and chlorine. Deficiencies of these chemical elements in soils and crops adversely affect the health of humans (and other animals) because these elements are involved in metabolic activities related to growth, reproduction, and overall well-being (Shukla et al. 2014).

The symbolic-linguistic dimension is found in the origins of the Judeo-Christian tradition. In Genesis, the name of the first human being is Adam, which derives from the Hebrew *adamah*, which also means soil, top soil or humus (Callicott 1994). Both the name and the material origin of the first human being are associated with the soil, with nature: "Then the Lord God formed man [*adam*] from the dust of the ground [*adamah*], breathed into his nostrils the breath of life, and man became a living being" (Genesis 2: 7–8 NIV 2011). This conception was born from a peasant culture in the Middle East, and today is alive in the popular culture of peasant communities that maintain a daily relationship of co-inhabiting with the land in Latin America and other regions of the world (May 2015, 2017). In "Tierra Ajena" ("Alien Land"), the Chilean poet Oscar Castro (1940, p. 17) tells the story of a peasant who co-inhabits with each plant and the land that he works on a daily basis: "Lisandro feels the land. He kisses her with his eyes and with his feet... [he kisses] each humble [plant such as the] *yuyito*, each furrow of the earth." In South America, peasant cultures and poets evoke their everyday practices of coinhabiting the land.

Since the mid-twentieth century, however, the so-called Great Acceleration not only has amplified processes of ecosystem degradation and the extinction of a growing number of biological species on a global scale (McNeill and Engelke 2016) but has also stimulated a culture that has silenced its consciousness about the interconnections of biologically and culturally diverse lives. Today, we humans are humus that have forgotten our biospheric connections. Local peoples and cultures have been displaced from their original habitats and their life habits and relationships with local ecosystems and biological species have been replaced by non-local ones (Rozzi 2012). Interrelated losses of biological and cultural diversity have resulted in processes of biocultural homogenization worldwide. However, it is not the whole human race that is equally responsible for this anthropogenic ecocide (Rozzi 2007; Figueroa 2011; Dunn 2018). Identifying a diversity of cultures, languages, values, and practices in heterogeneous habitats of the planet could help us reorient trends of biocultural homogenization towards biocultural conservation. This is a necessary (although not sufficient) step for moving away from the current *Necrocene* (an era of death) toward a *Biocence* (an era of life).

The term Necrocene –an era in which death, oppression, and extinction of biological and cultural diversity prevail– was coined by environmental historian Justin McBrien (2016). Despite the revulsion that this term causes us, the reality is that we live immersed in an era of death of human beings and other-than-humans, most of which pass unnoticed by global society. In opposition to a Necrocene, I have introduced the term Biocene to denote an era in which life flourishes in its biological and cultural diversity (Rozzi 2019a). Death is the silence of love. Love is the pulse that animates life. Life is silenced by the lack of consciousness of this pulse that is grounded in the interconnections of biologically and culturally diverse lives. This silence is leading us into a Necrocene. Recovering the historic memory that the word "humans" is rooted in material, symbolic-linguistic, and civilizational origins should reawaken our consciousness about our inextricable biocultural connections. This awakening can guide us towards a Biocene that is founded on an ethic of caring for all life.

To revitalize consciousness about our biocultural connections with the biosphere, in this chapter I explore how to favor a Biocene with an ethic of caring for life in its biocultural diversity by examining the work of the Brazilian Franciscan Leonardo Boff. To do so I use my "3Hs" model of the biocultural ethic that values the vital links among (i) the well-being and identity of the *co-in-Habitants* (humans and other-than-humans), (ii) their life *Habits*, and (iii) the *Habitats* that they share (Rozzi 2012).

25.2 Co-inhabitants and Custodians of the Biosphere

> Shouldn't we all be custodians of the [bromelia] Chagual and the [Monkey puzzle-tree] Araucaria, Franciscan servants of the wild [herb] centaury, keepers of the maitén [tree], protectors of the [beech] coigüe and the larch? Only then can we be worthy of this kingdom of beauty and life that we spoil every day (Oyarzún 1973, p. 58).[1]

With this text, which he included in his posthumous book *Defense of the Earth* ("*Defensa de la Tierra*"), the Chilean thinker Luis Oyarzún urged us to be "custodians" of this "kingdom of beauty and life." Envisioning the future, Oyarzún, as a naturalist, artist, and philosopher, practiced the values of beauty and defended the diversity of life. Observing the herbs and trees, poeticizing about their beauty, Oyarzún summons us to take care of them and to cultivate a prototype of biospheric humanism.

A contemporary of the Chilean poet Nicanor Parra and the philosopher Jorge Millas, Oyarzún died prematurely at the age of 52 in 1972, but despite his early departure, he has become one of the precursors of Latin American environmental thought. In the above quotation, extracted from the book published in May 1973, Oyarzún encourages us to be "custodians," "Franciscan servants," "caretakers", and "protectors" of plant species that coinhabit with us. Only then can we fully achieve human dignity.

In the development of the biocultural ethic, I focused on how the word ethics comes from the Greek *ethos*, which in its most archaic form meant an animal den. The dwelling place of an animal is its habitat. A fox takes care of her cubs in a den. A pair of southern lapwings takes care of the nest, and when a predator or another danger approaches, they scream to distract the intruder. Humans, other mammals, birds, and also invertebrates risk their lives to protect their progeny, as well as to protect the dens, territories, and habitats that they inhabit. In *Defense of the Earth*, Luis Oyarzún stresses the need to care for the habitat in a way that converges with a biocultural ethic.

The cornerstone concept of the biocultural ethic is the idea of *co-inhabitant*. I coined this term as analogous to the concept of companion, which originally referred to sharing bread (from Latin, *cum* = with; *panis* = bread) (Rozzi 2019b). Just as life is shared with bread, it is also shared in a habitat inhabited by diverse co-inhabitants. The concept of co-inhabitants alludes descriptively and normatively to sharing the same habitat. We humans share local habitats with vertebrates, invertebrates, and a

---

[1] Original in Spanish. English translation by Ricardo Rozzi.

multitude of other living and ecological beings (such as rivers, mountains, rocks, or oceans), and this sharing has ontological, epistemological, and ethical implications.

Ontological, because human and other-than-human beings do not exist as isolated individuals, but rather exist together in co-inhabitation interrelationships.

Epistemological, because to understand human beings and other animals it is necessary to consider the co-inhabitation relationships that forge their identities and well-being.

Ethical, because human beings share a common habitat –the biosphere– which we must take care of for the welfare of all living beings by cultivating life habits that recover a sense of being co-inhabitants with myriads of ecological beings, most of which go unnoticed, but are critical for the health of humans and ecosystems (Rozzi 2019b). Undertaking a biocultural ethic also demands that we overcome epistemic injustices (*sensu* Fricker 2007). These injustices may derive from an arrogance of hegemonic forms of knowledge linked to Eurocentrism or scientism, which imply exclusion and oppression of other forms of knowledge such as ecological knowledge hosted by indigenous and other local communities (Grosfoguel 2007). This injustice and direct oppression of local communities also implies an indirect oppression of the biodiversity with which they co-inhabit. Epistemological arrogance "makes invisible" ways of knowing which are intimately linked to local forms of co-inhabitation with multiple biological species.

No South American theologian has been more forthright in highlighting the links between oppression of the poor and oppression of biodiversity than Leonardo Boff, a Brazilian Franciscan who has developed an eco-philosophy and an eco-theology of liberation. Boff's ethical turn arises from his demand to "listen" to others, diverse humans and diverse ecological beings. Such listening is required in order to reverse the silence and the inability of global society to hear the pulse that animates life in its diverse manifestations.

How did Boff develop his distinctive integration between social and ecological dimensions? How is it possible to achieve what he calls a reconnection with the Earth as a whole, a *dignitas terrae* that demands a material and spiritual reconnection with both the exterior and the interior nature of each human being and society? To explore these questions, I interviewed Leonardo Boff about his life and ideas.

25.3 A Dialogue with Leonardo Boff –Interview

On October 9–12, 2006, during a workshop that was part of the "*Patagonia sin Represas*" (Patagonia without Dams) resistance movement, we held a workshop with Leonardo Boff in Bariloche, Argentina. At the end of the workshop, Chilean sociologist Claudia Sepúlveda and I interviewed Boff (Fig. 25.1). His answers help us to understand how his work was inspired by different sources of knowledge and experiences that are woven throughout his thinking and life, and that have pedagogical value for members of global society. Below, I reproduce an extract of this interview that was recorded and, in agreement with Boff, was uploaded at the television

Fig. 25.1 Left. Leonardo Boff and Ricardo Rozzi during the interview conducted on October 11, 2006. Right. Group of participants in the workshop with Boff that was part of the "*Patagonia sin Represas*" (Patagonia without Dams) resistance movement in Bariloche, Argentina, October 9–12, 2006. (Photographs by Alfredo Agüero)

channel of the Universidad de Magallanes, Chile.[2] It is published for the first time as a written text because it can be of great support for researchers interested in eco-theology and, more specifically, because it undergirds the relationship between pantheism and ecology. It also offers a window into the life and ideas of one of the founders of Latin-American eco-theology.

Ricardo Rozzi Dear Professor Boff, we are very happy to meet you in Patagonia. Claudia is from Valdivia, Chile, and works with a civil organization that protects swans as well as their wetlands and rivers. I work in biocultural conservation at Omora Park in Puerto Williams in the Cape Horn Biosphere Reserve, where I pay special attention to the diversity, beauty, and values of little organisms such as mosses, lichens, and insects. Those of us who work in the South of Chile would greatly appreciate it if you would tell us about your life so that we may share the story with others when we return home. How, in your life, have you been able to cultivate yourself, reflect, read, write, and pray and at the same time maintain a social existence and be in demand in the community? We would enjoy a short history of your life.

Leonardo Boff For me, it is highly significant that my father was a school teacher. In our region, there was a colony of Germans, all Protestants, Polish, Italians, and small numbers of Mulattos, Blacks, and Indigenous people. The colonists were the first to arrive in the interior of the province of Santa Catalina [Brazil]. My father taught us to live with and not to discriminate against the marginalized people of our region, the Blacks and Natives. He obliged us to sit together with them at school because the Italians and the others greatly discriminated against them. I was born into a family of eleven siblings and thus with the virtue of accepting everybody's differences. My second most important influence was my studies in Germany. There I found a group of professors that were very famous at the time. They were the

[2] The recorded interview can be found at: https://youtu.be/b3m5t0jPrGQ

catholic theologian Karl Rahner, the protestant [Wolfhart] Panneberg, and I had the privilege of attending the very last lectures that Werner Heisenberg gave over the topic "God in Modern Physics". He was one of the founders of quantum physics and he obligated me to open space for dialogue with science and it's new perception of reality based more on chains of energy and related objects. These two points were important in my life. But how did I experience this subjectively? Here I must speak of the importance of the Franciscan spirit. I began seminary as a child when I was eleven years of age and throughout my college education philosophy and theology were marked by the Franciscan spirit of fraternity, love of nature, sensibility to the poor, and simplicity. This has, I believe, constructed my soul; my perception of reality. I today consider myself to be a Franciscan Apostolic Catholic, not so much Roman. Roman is based on laws and organization. Franciscan on the other hand is more chaotic but generative of surprising relationships. And all that has deeply defined me until today. The very place where I cherish the deepest religious experience is in nature, hugging trees, becoming like a tree: my feet become the roots, my body becomes the trunk, my hair becomes the branches, to feel the air on my skin, to feel like there is no duality with the environment. It seems to me that Saint Francis lived considering all others his brothers and sisters, not as a rhetorical approach or as a derivation of dogma. "God is father, all are sons and daughters" and not as an ordinary experience of fraternity either. And this I believe this is the biggest inheritance that I have had from Saint Francis. I carry it forward in my life.

Rozzi To live like this, do you have a daily structured schedule? How do you organize your everyday life to be able to accomplish all what you do, more or less?

Boff Well, my life changing experience was like that of Saint Paul [who was first known as] Saul. I came from Germany knowing critical theology, German, Protestant and Catholic, in February of 1970. In July I had to preach about spiritual exercises in Amazonia, and there I met missionaries who told me: "I have already travelled for a month, I work deep in the jungles one day, and the next day I work with towns along the riverbanks in very difficult situations." I preached to them the most modern, critical theology but I saw in their eyes that they did not understand it. This theology does not reach them! One day somebody came to me and asked me: "Boff how am I going to preach about Jesus… his resurrection, to my little town where the Indians are dying of the white man's diseases… how am I going to preach? How am I going to preach about Christ's cross to those that are working in the rivers looking for gold? And what I am going to say about this fight between Indians and the inhabitants of the rivers who do not understand each other, how am I going to reconcile them?" For that I had no words, none; as a matter of fact, I became so sick that my neck muscles hardened, and I could not speak. Then to alleviate my perplexity I created groups to discuss Biblical texts, and with these groups I would comment on the reflections. But there I realized that I had to do a theology adapted to this reality. This was in July. In August they begin the classes of the second semester. I begin to lecture on Christology, to give classes on Christology and to write all that. Out of this came my book which is one of the books that founded the theology of the liberation: *Jesus Christ Liberator* (Boff 1978). It was a painful experience, from which I had to dismantle my old vision and learn from this reality. And I began to work with the person who today is my partner, who had worked in the same area landfill for more than ten years. I started doing what I stipulate is the theology of the liberation: "A foot in Misery and a Foot in the Academy." So, then I accompanied this group of scavengers, living amongst them, constructing a school, preschool, a small factory to process garbage, creating communities, meditating together. This happened all over Brazil. We always connected with peace groups and those without land, with Black and Indigenous people, and with the poor, because it occurred to me that here are the true challenges. It produces an experience of faith that one doesn't have in the convent/monastery, because in the village faith is born into you. They don't believe in God, believing in God to them is "something for the dogs." They feel God, feel and live God; they feel it as a fundamental pillar in their lives. For a theologian, that is something very wholesome, the healing of their ideologies and theologies makes it simple to be honest with reality, and to coexist and share with them. I remember, and this I am a little discreet about, but I will say it anyways. After a tough fight within the holy office in Rome, when I basically wanted to punch the cardinal that laughed at my cynicism, I became extremely desperate. I told myself, this can't be possible, the Vatican is the symbol of our faith, and it has cardinals that wrongly represent the heritage of Jesus. And I fell into a deep emotional crisis, and I went to heal in the Amazon of Brazil on the outskirts of Bolivia. I immersed myself in the forest for two months along with Chico Mendes,[3] walking and visiting communities, absolutely lost inside the jungle. It was this village that gave me hope, desire to live. To them, Rome had no significance at all. For them, these theological arguments do not mean anything – true problems were something different. For me immersion in the village was very healthy, as it regenerated my faith and allowed me to put in perspective the problems with the Vatican and the Church. This makes me smile because what lives in that village is faith without any metaphors, true and transparent. I wish that all the theologians in crisis could go there. People who have doubts about faith go to the village because as the ancient fathers once said, "the poor are our teachers, they are our prophets."

Claudia Sepúlveda I wanted to ask you: this long road you have walked, that has been so beautiful, and being able to share it in some way at this moment in your life, has brought value and importance about the formulation of the "Liberation Theology" and the experience related to it. This also has allowed for encounters, as you told us, with very powerful people, such as company executives, presidents –

[3] Francisco Alves "Chico" Mendes Filho devoted his life to the conservation and defense of the Brazilian rainforests. In 1985, Chico Mendes introduced the concept of concept extractive reserves and implemented it as way to defend the Amazonian forests and the rights of sustainable use practices by Rubber Tapper communities. In 1988, Chico Mendes was murdered by ranchers. As a tribute to his invaluable work for conservation, in 2007 the Brazilian Ministry of the Environment created the *Chico Mendes Institute for Biodiversity Conservation*, which today manages federally protected areas. See Gross (1989), and Amos Nascimento and James Griffith (2012).

people who make important decisions nowadays and that will probably define the course of our future. In what way has this experience transformed you; having the contact and possibility of conversing openly, and perhaps aiming to disseminate your feelings and knowledge to these powerful people?

Boff I believe there are two fundamental points in my life. The first is when I discovered the world of the poor and recognized an option for the poor against poverty, favoring life. That is where Liberation Theology comes from, from listening to the shouting of the oppressed. The second one was my strife within the Vatican. I even had to sit on the same chair as Galileo Galilee, Giordano Bruno. I paid reverence to that chair, which infuriated, at that time, cardinal Ratzinger, nowadays Pope Benedict XVII. And when the conflict was finished, I accepted silence and the Pope wrote a letter to me saying two things: that I should focus on more important, more serious subjects, and that I personally had to show more seriousness. I consider myself a serious theologian... I didn't know how I was going to demonstrate it. So, I let my beard grow –that's a way to appear more serious– and then I realized that having a beard had advantages, because then you don't have to shave every day. I took a serious look at the Pope's petition and began to reflect: What are the bigger problems to consider? And I realized that none of the problems had to do with the church –none. They had to do with the poor, with ecology, with land... the systematic aggression that the capitalist system has over the nations, towns, and nature. From there I began to read all the literature that comes from the new ecology, quantum physics, and the new biology. For me, Maturana and Varela[4] are very important because they are part of "us." They are Latin Americans and have opened a new perspective. I am sad that I did not get to meet them, because in my opinion, it would have been more interesting to dialogue with them than Marx because they reframed all those problems within democracy, participation, and education as a process of interaction with reality. At this point I was completely confused. I lived in the old paradigm and now I entered the new one. I didn't understand anything, and for two years I lived completely lost. But slowly there was a decantation, and I began to notice the hard points, the fixed points, and I was able to construct, after 10 years of work, the book *Cry of the Earth, Cry of the Poor* (Boff 1997). I had realized that it is not only the poor people that cry out, but land cries, water cries... nature cries out for redemption. Therefore, an eco-theology of liberation is necessary, and that book was a synthesis of everything that I have learned from lectures, participation, and congregations, and from speaking to scientists. From there I

[4] Humberto Maturana (1928–2021) and Francisco Varela (1946–2001) were Chilean biologists who developed together the idea of *autopoiesis* (self- production) as the primary feature that distinguishes living things from non-living things. Based on this idea they developed an observer-dependent interpretation of cognition, language, and consciousness. They argued against any absolutely objective world. Instead, Maturana and Varela affirmed that humans bring forth a world with others through the process of our living in created worlds that arise through language and coordination of social interactions (see Hallowell 2009). Their perspectives have implications for education (Medina and Rozzi 2021), as well for media ecology and environmental philosophy (Holtmeier 2019).

changed focus, as this issue is an issue that is more global, which interests many more people than just churches. Today, I have an active involvement in the church –I accompany groups without land, peace groups, and Blacks, and that is something fundamental for me. I couldn't cite myself biographically without an option for the poor. I mean, to look at all different realities, ecology, quantum physics, anything from the eyes of the poor, on how to better their reality and broaden the extent of their liberties. From that point with the books I have written, some more popular than academic, I have penetrated into more non-professional territory. As I was a university professor in a state university, I became part of the lay or non-professional community within the university. Thus, I began to have a large presence in academia and in groups of executives who realized that the system does not work properly, that it is necessary to acquire social and environmental responsibility and be creative. Therefore, my book transcends these dimensions. There is no Brazilian executive that has not read, had discussions, because creativity is very important, to cross boundaries, be somehow heretical and daring. So, I received many invitations, the main one being to belong to the United Nation's commission to write the *Earth Charter*, which was composed of twenty-five participants with the intent of writing a document inspired by the poor. So, I talk with many scientists, cosmologists, astrophysicists, biologists, and that has greatly enlightened me by listening and by having to read and work in the field. I feel very well positioned in that project. I can read that literature, understand it, and have a dialogue with science. I am greatly sorry that there are not many theologians who dialogue with the phenomena of the earth, because such dialogue strengthens the faith. These dialogues could help much more than the other philosophies based on the Christian promise and could broaden the meaning of the fight against poverty. Strangely, many people consult me –they are company executives who are desperate, and I can see that well. They know that the so-called system does not even work for them. They own the main benefits but know it will soon come to an end. So, they want to prepare, want to try new ways to relate to their employees, other forms of production, to incorporate more human and spiritual dimensions. I work consistently on the theme of spirituality for laymen, physicians, psychologists, and managers. They ask me to address the topic of spirituality, so I underline that a lot of the spirituality is the anthropologic dimension of the human being. We all have a body that has a spirit the human spirit, but it is a spirit that not necessarily has to be religious or affiliated to a church to be spiritual. Religions are the natural realm of spirituality; however, religions have become substantive-like and with their rites sometimes kill spirituality. I say spirituality is that conscious moment in which a person feels part of something greater and discovers the majesty of the universe. That there exists a message behind everything and that there is a name for that realm, call it the original source, I don't know, or God, names don't matter. The only thing that matters is that it enlightens the human spirit. To have a deep silent dialogue of veneration which leads to more compassion, more sensibility, more love towards each other, love for nature, these are the good fruits of spirituality. And I try to incite these feelings in people. It is not that I am special, no, very often I am amazed by the large crowds that come to see me, how they fill up the auditoriums, and they even have to put up large screens to see me.

And I say, I am not a chicken thief, what do I have to say? But I feel it touches people deeply. The most important thing is for people to have more hope in humanity, for us to create a better world that we can't let go. We don't need to be discouraged, because there are a thousand reasons to keep fighting.

25.4 Boff's Ecotheology and the Biocultural Ethic

From our interview with Leonardo Boff and his methodological approach to liberation we can identify two necessary and complementary tasks.

First, *a de-constructive task*: to free ourselves of the myth of development and unlimited growth. This myth teaches that society is governed by the principle of economic growth. Axiologically, this myth implies that the value of economic growth is placed above the value of life and its diversity. Consequently, it triggers the structural oppression of the diversity of human cultures and the massive extermination of biological diversity with whom we share the planet. It also degrades and reduces the meaning of human existence to an economic one-dimensionality.

Second, freeing ourselves from this myth enables us to undertake *a re-constructive task*: to value and respect life in its biological and cultural diversity. Axiologically, this implies that the value of life is relocated above the value economic growth. Consequently, the primordial meaning of economy is recovered. Boff underlines that economics is not the unlimited accumulation of capital, but rather the sensible administration of the home. Economy and ecology share the same Greek root *oikos* (= home). One refers to the rules and laws (nomos) for the administration of the oikos, while the other refers to the study (logos) of the oikos. Boff does not simply use abstract philosophy to interweave economics and ecology. Rather he relies on practices of solidarity, resistance, and defense of the poor and marginalized human and other-than-human beings. For instance, in the interview he refers to "Chico" Mendes and during his life Boff has supported numerous grassroots movements such as the "Landless Rural Workers' Movement" (MST, *Movimento dos Trabalhadores Rurais Sem Terra*)[5] in Brazil and "Patagonia without Dams" in Chile. Although these two examples are from South America, there are similar cases in other world regions, as the development model underlying them has been globalized.

[5] The Landless Rural Workers Movement (MST) has been one of the largest and most active social movements in Brazil. The MST organizes unemployed laborers and landless farmworkers to challenge landowners and authorities by taking over absentee-owned farmland and agitate for a broad agrarian reform. It grew out of land occupations beginning in 1978 in Brazil's southernmost state of Rio Grande do Sul, led by activists from the Catholic Church's Christian base communities and some Protestant churches under the inspiration of liberation theology (Hammond and Rossi 2013). Formally founded in 1984 near the end of a 21-year military dictatorship, this Brazilian agrarian social movement has also introduced a vision of learning as territoriality and of a political ecology of education as part of a long-term strategy of transformation (Meek 2015). A central pedagogical practice is called *mística*, which draws on Christian mysticism as way of connecting with a transcendental reality (Hammond 2014).

Boff's eco-theology of liberation is relevant to twenty-first century philosophical thinking for at least two reasons:

(i) It re-establishes the philosophical unity of theoretical and practical knowledge to tackle the global scale from and towards local histories and reality.
(ii) It places environmental issues at the heart of our century's challenges, while also maintaining a transcendent sense of the cosmos that precedes and succeeds the human species.

In *Ecology and Liberation: A New Paradigm*, Boff (1995) stated that the Spirit sleeps in stones, dreams in flowers, awakens in animals and humans. He also affirmed that this intuition of the cosmic ubiquity of the Spirit is testified to by many mystics of various cultures, such as the Sioux Indians of North America or the Bororos of Brazil, Eastern Zen masters, and the Fathers of the Latin and Greek Church of the fourth and fifth centuries. Unlike pantheism that affirms that everything is God, panentheism maintains that God is incarnated, is immanent, in all the beings that make up the cosmos. Boff explains that "Everything is not God, but God is in everything. This is what the etymology of the word panentheism suggests: God is present in everything. He makes every reality his temple" (Boff 1995, p. 50).

In Boff, panentheism is not confined only and solely to the Christian God. Panentheistic worldviews guide forms of justice, prudence, and well-being in numerous peoples, cultures, and religions. Latin American panentheism is expressed both in the ingenious rogue who guides the people and in the wise woman of antiquity and indigenous peoples (Rojas-Salazar 2018). This plurality of panentheistic expressions is forged into three types of values that have oriented Boff's thinking and life and should also orient global society.

The Value of Non-discrimination and Openness to "Otherness" Boff's childhood experiences in Brazil were oriented by parents who taught him to live with members of Indigenous, Mulatto and Black communities (who were often marginalized by other European immigrants). As a member of a family of Italian immigrants, Boff clearly remembers the virtue of accepting differences and sitting together at school and around the table.

The Value of (Re)connection with Life in Its Biophysical and Cultural Diversity Boff's education was strongly influenced by the Franciscan spirit of fraternity, love of nature, sensitivity towards the poor, and simplicity. Religious and everyday life experiences are connected with nature. Feeling the trees, birds, rivers, the moon, and the diversity of beings with whom we share the planet leads us to overcome the duality between humans and others.

The Value of Transdisciplinarity Boff's higher education included the theory and practice of philosophy, theology, and science. This transdisciplinary education inspired his systemic and situated thinking in a way that converges with the biocultural ethic. Boff has proposed seven peaceful "pathways" or practices of ecology to articulate his holistic eco-theological approach. In my 3Hs biocultural ethic model

I have proposed that habitats, life habits, and co-inhabitants include biophysical, symbolic-linguistic, and institutional-social-political dimensions (Fig. 25.2).

Figure 25.2 illustrates that Rozzi's *biophysical dimension* matches Boff's *eco-ethics path* that demands responsibility for everything that exists and lives. The supreme good does not merely amount to the common good of humanity, but it entails earthly and cosmic integrity, care for and respect for biological and cultural diversity, aiming for the welfare of all beings or co-inhabitants (Rozzi 2013; Boff 1995).

Rozzi's *institutional-social-political dimension* matches Boff's eco-technology, eco-politics, and social-ecology paths that demand an urgent transformation of the instrumental and mechanistic view that allows a few humans, institutions, nations, and corporations to exploit without limits other humans, animals, plants, rivers,

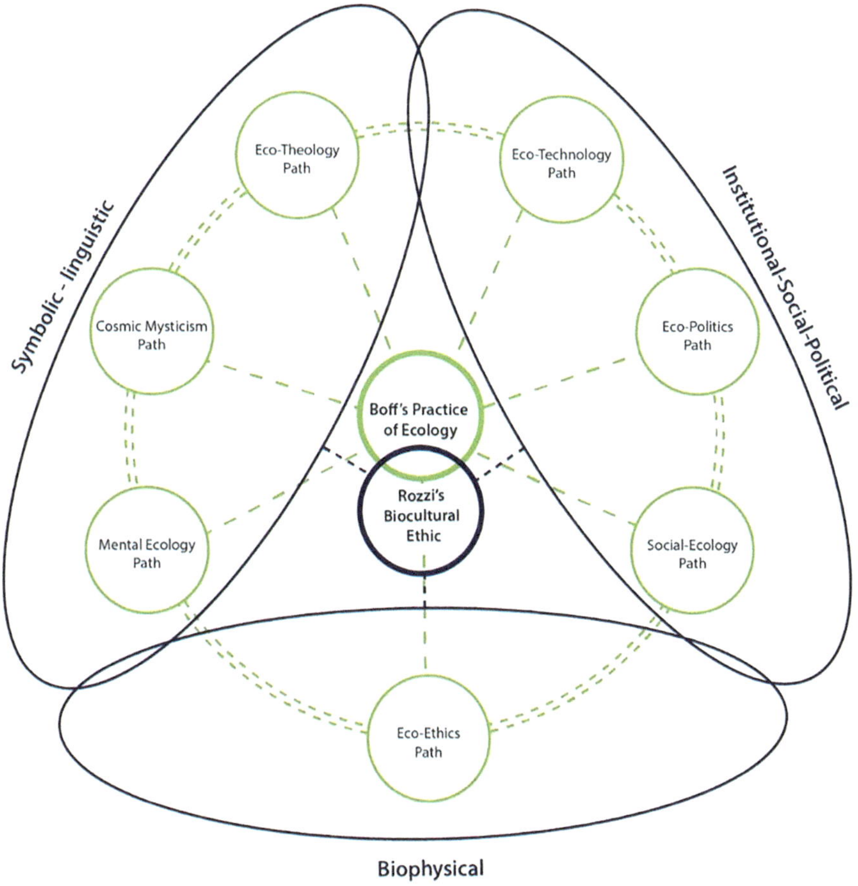

Fig. 25.2 Scheme of the intersection s between the seven paths of Leonardo Boff's ecotheology and the three *dimensions* of the 3Hs model (habitats, life habits, and co-in-Habitants) of the bio-cultural ethic (Rozzi 2013; Boff 1995)

oceans, minerals, and other beings. These beings are the majority of biophysical and cultural beings with whom we share our local ecosystems, the biosphere, and the cosmos. It is urgently necessary to stop stripping these of their autonomy and intrinsic value so that they are not reduced to mere market commodities. To counteract this trend, Rozzi proposes the concept of co-inhabitants that are valued as subjects with their own ends. Boff proposes an economy conceived as the administration of fair means necessary for life and well-being, and that embraces not only work but leisure, not only efficiency but gratuitousness, not only productivity but the absurd. The playful dimension must be encouraged for integral human existence. Toward this aim, practices of co-inhabitation involving imagination, fantasy, utopia, dreams, emotions, symbolism, poetry, and religion must be valued as much as production, organization, functionality, and rationality should be fostered (Boff 1995; Rozzi 2013).

Rozzi's *symbolic-linguistic dimension* matches Boff's mental ecology, cosmic mysticism, and eco-theology paths that require consideration of the diversity of beings inhabiting not only nature but also ourselves, as images, symbols, and values. Languages represent an essential cultural component that guide human habits. Hence, in order to reorient the relations of members of global society with other-than-human co-inhabitants it is necessary to consider the diverse interrelationships between the *biosphere* (*sensu* Vladimir Vernadsky; see Huggett 1999) and the *logosphere* (*sensu* Michael Krauss 2007; i.e., the diversity of languages that exists around the globe including a rich symbolic planetary web of logos or words, languages, and narratives) (Rozzi 2018). The water, plants, and animals that inhabit us are archetypes and figures filled with emotions. This understanding counteracts the modern fabrication of the "one-dimensional man" (*sensu* Marcuse 1964) and helps us to reintegrate the multiple forces of the universe that are present in our impulses, visions, intuitions, dreams, and creativity. This includes spirituality that captures a sense of communion among all beings.

Latin American spiritualities and epistemologies include diverse forms of knowledge that pass through the bodies and arise from experience, from knowledge acquired in daily life that rescues ancestral wisdom, experiences of the sacred, and relationships of Indigenous and Afro-American peoples who are the heirs of these spiritualities and epistemologies (Rojas-Salazar 2018). These spiritualities are marked by the presence of divinity in its liberating and immanent character both in history and on earth (*pachamama*) and in the entire cosmos. The heirs to this wisdom are the ones who uphold the practice of justice, the common good, equity, and the right of citizenship of every being that inhabits the planet.

Each being and the earth is conceived as a coinhabitant with rights (Rozzi 2022). In turn, for Boff (1995) each is inhabited by an expression of divinity. Spirit is the vital energy that permeates the cosmos and everything is filled with this vital energy (Conradie 2002). Latin American spiritualities are an ancestral force that remains latent in every human and other-than-human being. These spiritualities imply corporeality, sexuality in all its expressions and eroticism as a positive and necessary element, since eroticism is understood as the vital force and energy of the human being that leads them to the praxis of love (Rojas-Salazar 2018, p. 102).

We humans are humus. Recovering the ancestral meaning of the name of our species can (re)awaken the consciousness about our inextricable connections to earth and the cosmos. Boff's Franciscan panentheism is rooted in tenderness as the main attitude in the encounter with other beings, generating a cordial knowledge (cordial = from the heart) which does not distance itself from diverse cultural and biological realities. Instead, it makes possible communion with them, as was done by St. Francis for whom the moon and the sun, water and fire, the birds and the herbs are our sisters and brothers with whom we share the same divine genealogy. Our connections and impulses towards life in its diversity reside in love. Enthused by Latin-American panentheism, I want to paraphrase the sixteenth-century Spanish poet Francisco de Quevedo to conclude that *humans are humus and that humus unfolds vitally and cosmically as it releases its loving force.*[6]

Acknowledgments I thank Luca Valera for his stimulating invitation to be part of this volume; Roy May for his valuable English revisions; and the Cape Horn International Center (CHIC) – ANID PIA/BASAL (PFB210018) for its support.

References

Boff L (1978) Jesus Christ liberator: a critical Christology for our time. Orbis, New York

Boff L (1995) Ecology and liberation: a new paradigm. Orbis, New York

Boff L (1997) Cry of the earth, cry of the poor. Orbis, New York

Callicott JB (1994) Earth's insights: a multicultural survey of ecological ethics from the Mediterranean Basin to the Australian outback. University of California Press, Berkeley

Castro Ó (1940) Huellas en la Tierra. Editorial Andrés Bello, Santiago de Chile

Conradie EM (2002) Notes on Leonardo Boff and the filioque. Scriptura: J Contextual Hermeneut South Afr 79(1):14–24

Da Silva JJRF, Williams RJP (2001) The biological chemistry of the elements: the inorganic chemistry of life. Oxford University Press, New York

Dunn CP (2018) Climate change and its consequences for cultural and language endangerment. In: The Oxford handbook of endangered languages. Oxford University Press, New York, pp 720–738

Figueroa RM (2011) Indigenous peoples and cultural losses. In: The Oxford handbook of climate change and society. Oxford University Press, New York, pp 232–246

Fricker M (2007) Epistemic injustice, power and the ethics of knowing. Oxford University Press, New York

[6] I paraphrase the poem *Love That Endures Beyond Death* ("*Amor constante más allá de la muerte*") by Francisco de Quevedo (1580–1645): "Though my eyes be closed by the final/Shadow that sweeps me off on the blank white day/And thus my soul be rendered up... The body they will leave, though not its cares;/Ash they will be, but filled with meaning; Dust they will be, but dust in love." Quevedo's text in Spanish is: "Cerrar podrá mis ojos la postrera/sombra que me llevare el blanco día,/y podrá desatar esta alma mía ... su cuerpo dejará, no su cuidado,/serán ceniza, más tendrá sentido, polvo serán, mas polvo enamorado".

English translation taken from Margaret Elisabeth Jull Costa, British translator of Portuguese- and Spanish-language fiction and poetry; available at: https://english.duke.edu/news/poems-moment-00

Grosfoguel R (2007) The epistemic Decolonial turn: beyond political-economy paradigms. Cult Stud 21:211–223
Gross T (ed) (1989) Fight for the Forest –Chico Mendes in his own words. Latin America Bureau, London
Hallowell R (2009) Humberto Maturana and Francisco Varela's contribution to media ecology: Autopoiesis, the Santiago School of Cognition, and enactive cognitive science. Proc Media Ecol Assoc 10:143–158
Hammond JL (2014) Mística, meaning and popular education in the Brazilian landless workers movement. Interface: J Soc Mov 6(1):372–391
Hammond JL, Rossi FM (2013) Landless workers movement (MST) Brazil. In: The Wiley-Blackwell encyclopedia of social and political movements. Wiley-Blackwell, Oxford, pp 680–683
Holtmeier M (2019) Vital coasts, Mortal Oceans: The Pearl Button as media environmental philosophy. Video Essay. Film-Philosophy, Brighton, UK. Available at: https://www.film-philosophy.com/conference/public/conferences/2/schedConfs/13/program-en_US.pdf
Huggett RJ (1999) Ecosphere, biosphere, or Gaia? What to call the global ecosystem. Glob Ecol Biogeogr 8:425–431
Jéquier E, Constant F (2010) Water as an essential nutrient: the physiological basis of hydration. Eur J Clin Nutr 64(2):115–123
Krauss ME (2007) Mass language extinction and documentation: the race against time. In: The vanishing languages of the pacific rim. Oxford University Press, New York, pp 3–24
Marcuse H (1964) One-dimensional man: studies in ideology of advanced industrial society. Beacon Press, Boston
May RH Jr (2015) Andean llamas and earth stewardship. In: Earth stewardship: linking ecology and ethics in theory and practice. Springer, Dordrecht, pp 77–86
May RH Jr (2017) Pachasophy: landscape ethics in the Central Andes Mountains of South America. Environ Eth 39:301–331
McBrien J (2016) Accumulating extinction: planetary catastrophism in the Necrocene. In: Anthropocene or Capitalocene? Nature, history, and the crisis of capitalism. PM Press, Oakland, pp 116–137
McNeill JR, Engelke P (2016) The great acceleration. An environmental history of the Anthropocene since 1945. Harvard University Press, Cambridge
Medina L, Rozzi R (2021) Maturana y la educación como experiencia cotidiana multivocal y transformadora. Estudios Públicos 163:173–183
Meek D (2015) Learning as territoriality: the political ecology of education in the Brazilian landless workers' movement. J Peasant Stud 42(6):1179–1200
Nascimento A, Griffith JJ (2012) Environmental philosophy in Brazil: roots, intellectual culprits, and new directions. Environ Eth 34(4):379–397
NIV (2011) The Bible new international version. Available at: http://www.biblegateway.com/versions/
Ohashi Y, Sakai K, Hase H, Joki N (2018) Dry weight targeting: the art and science of conventional hemodialysis. Semin Dial 31(6):551–556
Oyarzún L (1973) Defensa de la Tierra (Edición 2020). Ediciones Universidad Austral de Chile, Valdivia
Rojas-Salazar M (2018) Descolonizando la teología: Espiritualidades panenteístas: Propuestas desde las epistemologías ecofeministas del sur. J Fem Stud Relig 34(2):98–104
Rozzi R (2007) Future environmental philosophies and their biocultural conservation interfaces. Eth Environ 12(2):142–145
Rozzi R (2012) Biocultural ethics: the vital links between the inhabitants, their habits and regional habitats. Environ Eth 34:27–50
Rozzi R (2013) Biocultural ethics: from biocultural homogenization toward biocultural conservation. In: Linking ecology and ethics for a changing world: values, philosophy, and action. Springer, Dordrecht, pp 9–32

Rozzi R (2018) Biocultural Homogenization: a wicked problem in the Anthropocene. In: Rozzi R, May RH Jr, Chapin FS III, Massardo F, Gavin M, Klaver I, Pauchard A, Núñez MA, Simberloff D (eds) From biocultural homogenization to biocultural conservation. Ecology and Ethics Book Series, Vol 3. Springer, Dordrecht, pp 21–47.
Rozzi R (2019a) Collaborative inter-continental dialogues: from a Necrocene to a Biocene. Environ Eth 41(4):291–292
Rozzi R (2019b) Taxonomic Chauvinism, no more! Antidotes from Hume, Darwin, and biocultural ethics. Environ Eth 41(3):253–288
Rozzi R (2022) An ethic of co-inhabitation for the biocultural conservation of rivers. Naturaleza y Sociedad. Desafíos Medioambientales 3:59–72
Schlesinger WH, Bernhardt ES (2013) Biogeochemistry: an analysis of global change. Academic, Waltham
Shukla AK, Tiwari PK, Prakash C (2014) Micronutrients deficiencies vis-a-vis food and nutritional security of India. Indian J Fertilizers 10(12):94–112
Szent-Gyorgyi A (1957) Bioenergetics. Academic, New York

Chapter 26
On the Compatibility Between Panentheism and Fragmentation: An Experimental Ecofeminist Loosening of the "in" in *Allingottlehre*

Casey Norrington Jackson

Abstract Traditionally, ecologists wishing to assert the spirituality of nature have turned to pantheism, a worldview that equates God with the world. More recently, however, the attention of some such ecologists has turned to panentheism, a worldview that suggests that God includes the world while also transcending it. By clarifying the key differences between pantheism and panentheism as they relate to ecology, this chapter aims to illuminate, over three sections, the pitfalls and opportunities that come with shifting from pantheism to panentheism-based ecotheologies. Section 26.2 explores the historical link between ecology and pantheism, explaining that pantheism has been deemed useful for environmental purposes on account of its attribution of intrinsic spiritual value to nature. Section 26.3 suggests that panentheism, as it is traditionally defined, is more problematic than pantheism to this end, as panentheism tends to consider nature as being of only instrumental value to God's transcendent aspect. Section 26.4 suggests that, ultimately, all holistic visions of the world's divinity – pantheistic and panentheistic – prove problematic in implying that historical atrocities, such as the Shoah, are included within divinity. From this, it is suggested that a new ecological role for panentheism is identifiable: if panentheism is remodelled to denote the inclusion of most (rather than all) of the world in God, then it could provide an underpinning for an environmental ethic that, while broadly holistic, refuses to encompass worldly atrocities like the Shoah. The benefits of this remodelling for feminist purposes are also considered, with Sect. 26.3 noting that holistic panentheisms can be guilty of reinforcing the problematic gendered immanence-transcendence binary, and Sect. 26.4 arguing that a fragmented panentheism would counter this.

Keywords Panentheism · Pantheism · Ecology · Feminism · Fragmentation

C. Norrington Jackson (✉)
School of History, Archaeology and Philosophy, University of Winchester, Winchester, UK
e-mail: C.NorringtonJacks.17@unimail.winchester.ac.uk

L. Valera (ed.), *Pantheism and Ecology*, Ecology and Ethics 6,
https://doi.org/10.1007/978-3-031-40040-7_26

26.1 Introduction

For a divine worldview intended to add nuance at the intersection between classical theism and pantheism, panentheism (German: *Allingottlehre*) is all too often reduced to a vague and oversimplified statement of its two main tenets: (a) that the world is in God, and (b) that God extends beyond the world. The former tenet has been the subject of particular scrutiny in recent decades, with proponents and detractors of panentheism alike seeking to uncover exactly what it means to suggest that the world is *in* God (Clayton 2004; Tabaczek 2021; Göcke 2022).

Taken literally and straightforwardly, the panentheistic adage of the world being in God has, in the vast majority of cases, been understood to suggest that the world *in its entirety* is a part of God. Considered such, panentheism is a totalizing divine worldview in which no aspect of the created world escapes God's being. Hegelian panentheism, in this vein, culminates in the dialectical sublation of *all* of human history into God's rationality, such that the world is entirely returned to God (Cooper 2006). Similarly, John Polkinghorne's (1994) eschatologically-focussed panentheism asserts that the world may only properly be considered to partake in a panentheistic relationship with God when salvation is complete. That is, in the words of Ward (2004, pp. 71–72), at a time when "all evil and imperfection will be eliminated, and we can become fully and unambiguously coworkers with God."

Contrarily, detectable in the panentheistic literature is a small wave of nuanced scholarship hinting at a loosening of the totalizing nature of its deity. Edwards (2004, p. 200), for instance, adds a disclaimer to the utility of its foundational language: "While the image and language of all-in-God is useful, [...] it is important that it is not taken in a literal sense. God is not literally any kind of container." Deane-Drummond (2004), moreover, proposes a Sophianic panentheism, based upon a friendship between God and the world, in which an ontological difference between these two entities is maintained. Accordingly, in his effort to consolidate panentheism as a paradigmatic research program, Göcke (2022, p. 53) finds that panentheisms can "either [...] assume that the relation between God and the world entails a genuine ontological distinction between both of them and thus is modelled as an external relation, or [...] suppose that everything belongs to the nature of God Himself."

Though the requirements of panentheistic totalization have been thus loosened, scholars have nevertheless remained reticent to *explicitly* affirm that it is possible for part of the world, no matter how small, to remain *outside* of God within a panentheistic framework. With a view to overcoming such reticence, this chapter seeks to demonstrate the necessity and possibility of a fragmented panentheism by bringing this divine worldview into a discussion with eco-theology and feminist philosophy of religion.

This discussion will be split into three sections. The first section will delve into the traditional relationship between holism and pantheism within environmentalism, outlining how pantheism, with its spiritualization of the entire natural world, is seen as a useful metaphysical underpinning for systems that seek to place humankind and nature on an even footing. The second section will argue that – contrary to the belief of some ecologists – panentheism does not, like pantheism from which it sprang, provide a suitable metaphysical basis for eco-holism, and is, moreover, problematic as pertains to feminism. This, it will explain, is because panentheisms have a tendency to lose sight of both the finite and the feminine in their plight to assimilate the *entire* world into an otherwise transcendent God. The final section will nevertheless critique (pantheistic) eco-holism itself, arguing that all-encompassing philosophical-ecological models prove problematic in view of their theoretically including the *Shoah*. This point, it will argue, might serve to carve out an important metaphysical role for panentheism within ecological ethics, should panentheism be able to loosen its concept of "in" so to allow for the world to mostly, *but not entirely*, reside within God. Loosening the panentheistic "in" as such will be shown to render panentheism of metaphysical utility to the development of eco-theologies that, while broadly holistic, are able to section themselves off from those most abominable instances of worldly destruction.

26.2 Pantheism and Holism in Environmentalism

With even a cursory survey of the seminal texts on environmental ethics, one will repeatedly encounter a certain train of thought common to numerous leading eco-philosophies. It goes something like this: the human tendency to separate oneself from nature, so to view oneself as over and above nature, has led to humankind justifying their destruction of the world on the basis that it is of only instrumental value to them. The antidote to this, the argument continues, is to attribute intrinsic value to the entire world, and to consider the human race as a part of this natural whole rather than as something transcending it.

Three particularly dominant iterations of this argument come from Aldo Leopold, James Lovelock, and Arne Næss. In "The Land Ethic," the final section of his esteemed *A Sand County Almanac: And Sketches Here and There*, Leopold (1968) laments that the point of departure for the human-animal/land relationship is currently monetary. Lamenting also that a suitable land ethic did not yet exist to negate this environmental profiteering, Leopold sketches one of his own:

> The land ethic simply enlarges the boundaries of the community to include soils, waters, plants, and animals, or collectively: the land. [...] In short, a land ethic changes the role of *Homo sapiens* from conqueror of the land-community to plain member and citizen of it. It implies respect for his fellow-members, and also respect for the community as such (1968, p. 204)

This is a sentiment shared by Lovelock (2000) in *Gaia: A New Look At Life on Earth*. Lovelock's Gaia hypothesis, as is by now well-known, suggests that earthly life forms synergistically conspire to regulate the planet such that it may continue to support their livelihood on its surface. Built within Lovelock's argument to this end is a strong rejection of Cartesian anthropocentrism, with Lovelock making his bewilderment known that an otherwise profound intellectual like Descartes would see fit to delineate between an ensouled humanity and a machine-like vision of animals. Lovelock provides no surer statement of Gaia's holism than the following:

> [T]he entire range of living matter on Earth, from whales to viruses, and from oaks to algae, could be regarded as constituting a single living entity, capable of manipulating the Earth's atmosphere to suit its overall needs and endowed with faculties and powers far beyond those of its constituent parts (2000, p. 9).

Næss (1973), moreover, positions holism at the centre of Deep Ecology, choosing to focus this movement upon a rejection of the notion that humankind subsists above its earthly co-inhabitants. Næss, in fact, goes as far as to suggest that abstractions of component parts of nature away from its whole are necessarily superficial, as per his following definition of the Deep Ecology movement:

> [It comprises the] [r]ejection of the man-in-environment image in favour of *the relational, total-field image*. Organisms as knots in the biospherical net or field of intrinsic relations. An intrinsic relation between two things *A* and *B* is such that the relation belongs to the definitions or basic constitutions of *A* and *B*, so that without the relation, *A* and *B* are no longer the same things. The total-field model dissolves not only the man-in-environment concept, but every compact thing-in-milieu concept — except when talking at a superficial or preliminary level of communication (Næss 1973, p. 95).

That pantheism might provide a fruitful metaphysical underpinning for this strain of eco-holism is picked up on by Mander (2022). A common trait of pantheism, notes Mander, is the attribution of such intrinsic value to the natural world that it may be considered the beholder of divinity. This position provides a theological underscoring for attempts to shift the balance of value away from humankind toward the natural world, with Mander quipping in this regard that "[t]he pantheist finds God more in the waterfall or the rainforest than in the car park or the gasworks".

Levine (1994, p. 121) sees it as "no accident" that pantheism is a forerunning metaphysic within the environmental sphere. He focusses heavily upon the pantheistic principle of a "divine Unity," which is considered an anti-anthropocentric principle on account of its recognition of an "important commonality" between human beings and nature (1994, p. 132). For Levine, the situation of a natural object within the pantheistic divine unity suffices for it to be considered an inherently valuable agent within the "moral community" (1994, p. 132); a community upon which humans do not have a monopoly.

Perhaps the most interesting interaction between pantheism and ecology in the literature, however, is Næss's own interaction with Spinozism in "Spinoza and Ecology" (1977). Here, Næss lists sixteen possible points of accordance between Spinozism and the ecological movement, though one need look no further than the first three to find evidence of Næss's recognition of the continuation between pantheistic unity and eco-holism. Point one sees Næss praise Spinoza's God, or Nature (Latin: *Deus sive Natura*) for its "[a]ll-inclusive[ness]" (1977, p. 46). The second point, moreover, sees Næss praise Spinoza for his rejection of Cartesian dualism, while the third sees Næss affirm Spinoza's rejection of teleology in his assertion that God, or Nature is, in Næss's words, "an *absolutely* all-embracing reality" (1977, p. 47).

While modern philosophy has traditionally been content to, on account of Spinoza's role as a founder of modern pantheism, utilise Spinozism as a synonym for pantheism, much attention in recent years has been paid to the question of whether Spinoza was in fact a panentheist rather than a straightforward pantheist. This, it ought be noted, is no mere moot point: where Spinoza is read as a panentheist, it is perhaps fair to say that panentheism, rather than pantheism, is the superlative metaphysical basis for the afore-outlined holistic environmental ethic (Valera and Vidal 2022). By contrast, this paper will now proceed to argue that panentheism cannot act as a suitable basis for the eco-holism of such thinkers as Leopold, Lovelock, and Næss.

26.3 Panentheism and Holism in Environmentalism and Feminism

Whether Spinoza is better understood as a pantheist or a panentheist is a question that remains open for debate. It is, furthermore, a question for which an attempt at a definitive answer lies outside of the scope of this paper. A point that does not receive due attention in the debate on panentheism and ecology, however, and which this paper accordingly wishes to underscore, is that even if Spinozism is accepted as a panentheism, one cannot deduce from this that panentheism more generally is a suitable basis for a holistic environmental ethic. That is, Spinozism, if a panentheism, is an atypical panentheism to the extent that its notion of transcendence is more limited than in other panentheisms.

Spinoza's God, or Nature, as Næss (1977, p. 51) points out, accords with ecological thought in that neither "lead away from the singular and finite." To this end, Spinozism is able to treat nature as an end in and of itself. Where it can be said to panentheistically contain the world within a transcendent being, this being is, indeed, itself essentially local to nature – hence the maxim God, or Nature. As Nadler (2022) states: "For Spinoza, there is nothing but Nature and its attributes and modes. And within Nature there can certainly be nothing that is supernatural."

Conversely, the overarching problem with panentheism more generally, particularly as it pertains to ecology, is as follows: in placing the world within God, panentheism does not assign value to nature in and of itself, but rather only insofar as it relates to a transcendent being. While certain panentheisms prove more nuanced than others in this regard, nature is frequently posited as secondary by panentheists; so much so, in fact, that Peterson (2001) and Richardson (2010) each fear that totalizing panentheisms might lose sight of finitude's distinctiveness. The instrumentality of nature, indeed, may be traced throughout Gregersen's typology of panentheisms: "*soteriological panentheism*," "*expressivist panentheism*," and "*dipolar panentheism*" (2004, p. 21).

Soteriological panentheism is synonymous with eschatological panentheism, and denotes the belief that, though evil contradicts God's perfection, and its presence thus demonstrates that all is not currently within God, all will eventually return to God in the end times (Gregersen 2004). John Polkinghorne, the primary exponent of eschatological panentheism, does not specify exactly what will come of the planet in the aftermath of its return to God. He is, nevertheless, clear that it will look markedly different than it does in its natural state, affirming that it will be rid of suffering: "[T]he 'matter' of that world-to-come must be such that it will not enforce recapitulation of the deadly raggednesses and malfunctions of the present universe" (Polkinghorne 1994, p. 166). Central to Polkinghorne's panentheism, then, is a belief that the world is home to an unholiness that needs saving by an otherworldly agent – a view that contrasts eco-holism's proffering of intrinsic value to nature. Eschatological panentheism, moreover, sits uncomfortably alongside studies suggesting that there is a link between belief in Christian eschatology and a lack of support for global warming measures (see, for instance, Barker and Bearce 2013).

Expressivist panentheism is an umbrella term for models of God that were part of, or follow on from, the German idealism of the early 1800s. Expressivist panentheists contend that an original logical divine principle, in need of determinate content, posits the physical world as its own contradiction within itself, with a view to overcoming this contradiction by raising the world into its divine perfection (Gregersen 2004). The world is thus a means by which the transcendent divine can actualise its desire for determinacy and perfect itself, which, again, raises questions of incompatibility with eco-holisms that intend the divinization of nature to be an end in itself. Even more concerning than that, however, is the tendency of expressivist panentheists, including Hegel, to claim that the world's rise to divinity is achieved by human beings alone, with nature essentially acting as a barrier to this human achievement. Such is one of the key messages of Hegel's introduction to his *Aesthetics*: "Even a useless notion that enters a man's head is higher than any product of nature, because in such a notion spirituality and freedom are always present" (1975, p. 2). Hegel's reflections here are indicative of a considerable distance between expressivist panentheism and eco-holism, which generally looks to level the playing ground between humanity and nature.

Instrumentality in Hegelianism, furthermore, is not reserved for nature; Hegel also utilises femininity as a springboard via which masculinity (alone) can move beyond sensible contingency (Irigaray 1985). As Oliver (1996) aptly points out, woman is therein essentially abandoned by the dialectic, and left unconceptualized, which makes nonsense of Hegel's claim to have reconciled *all* in his grand dialectical narrative.

Dipolar panentheism, the last in Gregersen's typology, is another term for process theology. Following in the tradition of Whitehead's philosophy of organism, process theologians, though a broad church in themselves, generally hold that there are two aspects of God. These are God's "*primordial*" and "*consequent*" natures, with the former referring to God's knowledge of the possible courses of action that created beings might take, and the latter referring to His/Her knowledge of the actions that creation has already partaken in (Viney 2022). For process theologians, all worldly events are grounded in God's primordial nature and enter into God's consequent nature, with God continually acting to persuade created beings to take the best courses of action available to them (Viney 2022). Central to process theology, furthermore, is an all-encompassing notion of relationality, which Barbour (1990, p. 222) explains in terms of Whitehead's organism metaphor: "The basic analogy for interpreting the world is not a machine but an organism, which is a highly integrated and dynamic pattern of interdependent events. The parts contribute to and are also modified by the unified activity of the whole. Each level of organization – atom, molecule, cell, organ, organism, community – receives from and in turn influences the patterns of activity at other levels."

As with eschatological and expressivist panentheism, process theism, too, is problematic in relation to eco-holism wherein it posits the natural world as a prerequisite to a grand divine reconciliation (though some process theisms are not eschatologically-focussed, this remains a reasonably common feature of them: see Cobb 2004). That notwithstanding, Barbour's reflections on relationality here are reminiscent of Næss's "*relational, total-field image*" (1973, p. 95), and to this end, process theism differs less from eco-holism than the other panentheisms in Gregersen's typology.

Worth noting here, however, is that process relationality has a troubled relationship with feminism. By way of explanation, Frankenberry (1993) fears that panentheistic process theologies sustain problematic binary notions of God by asserting the importance of God's transcendent aspect (an all-embracing spiritual totality, patriarchally associated with masculinity) over and above the natural world (patriarchally associated with femininity) that inheres in Him/Her. Process theology's focus on relationality plays into this structure, and Frankenberry's (1993, p. 36) comments on this are worth quoting at length:

> "From the standpoint of gender politics, is the time perhaps ripe for reevaluating the premium currently being placed on wholeness, unity, the holistic, and the organic, and for remembering that, historically, women have suffered not so much from fragmentation in their lives as from repressive unity? [...] [T]hose who consider it the task of feminism to tear away at deep social cleavages and pervasive webs of dominance and exploitation, precisely to shatter the unity of the patriarchal cosmos, will need plenty of elbow room. Do we get that, or enough of that, from a panentheistic understanding of the relation of whole and parts?"

Thus rejecting panentheism, Frankenberry suggests that pantheism is a more suitable basis for a feminist philosophy of religion. Jantzen (1997), too, opts for pantheism in her construction of a feminist religious symbolic.

26.4 An Eco-Feminist Fragmented Panentheism?

All in all, then, each of the variations of panentheism identified in Gregersen's typology prove incompatible with, or otherwise problematic with relation to, eco-holism. The most pressing issue here is the totalizing placement of the world within a transcendent God, which serves to denigrate the finite and results in an inability to treat the world as a divine entity in and of itself. That totalizing panentheisms do not provide a suitable metaphysical basis for eco-holism, however, does not imply panentheism's absolute lack of utility within the ecological field. Rather, it merely implies that a tweaking of panentheism's doctrines and role is in order. In attempting to carve out a new role for panentheism within ecology, it is interesting to note that a key critique of totalizing panentheisms may also be applied to pantheistic eco-holisms. To be specific, totalizing panentheisms have been criticised for placing within God world-historical events that simply do not belong in a superlatively wonderful being. Arguably the most authoritative instance of this argument comes in the form of Emil Fackenheim's suggestion that the Hegelian Absolute fails insofar as, while it is supposed to be all-encompassing, it would not be able to stomach Auschwitz (though Fackenheim does not explicitly use the term "panentheism"):

> Hegel's own countrymen – in his view permanently raised above their original barbarism by a Christian culture originally alien to them – have shown at Auschwitz a depravity unequalled in all history. [...] Hegel, were he alive today, [could not] remain with his own nineteenth-century synthesis. For if a truly modern philosophic thought must stay with the world rather than flee from it, then a twentieth-century Hegelianism would have to stay with a fragmented world (Fackenheim 1982, p. 12).

How, then, does Fackenheim's criticism of Hegelian panentheism relate to pantheism-underpinned eco-holisms? A re-quoting of Lovelock's Gaia hypothesis might prove enlightening here: "[T]he entire range of living matter on Earth, from whales to viruses, and from oaks to algae, could be regarded as constituting a single living entity, capable of manipulating the Earth's atmosphere to suit its overall needs" (Lovelock 2000, p. 9). As might another look at Næss's praise for Spinozism as a fruitful metaphysical underpinning for Deep Ecology:

> The nature conceived by field ecologists is not the passive, dead, value-neutral nature of mechanistic science, but akin to the *Deus sive Natura* of Spinoza. All-inclusive, creative (as *natura naturans),* infinitely divers, and alive in the broad sense of panpsychism, but also manifesting a structure, the so-called laws of nature (1977, p. 46).

Following on from Fackenheim, one questions whether organismic metaphors in which the natural world works *holistically*, as per Lovelock, or claims that *all* of nature is underpinned by divinity in a pantheistic sense, as per Næss, are true to experience. Might it be more suitable to reluctantly concede that some atrocities are so barbarous that they prevent the world from being thought about as a unified whole? That is to say, are there events sufficiently destructive as to be fundamentally irreconcilable with notions of a divine natural unity and continuous community? It is worth again quoting Fackenheim with these questions in mind:

> [Auschwitz was] a "world" in its own right. [...] The logic of normal worlds is, if not creation, at any rate preservation, and this is what holds them together. The logic of the Auschwitz world, in contrast, was destruction, and this was what held *it* together. Auschwitz was therefore an *anti*-world, indeed, the anti-world *par excellence*, for its like had never existed before (1994, p. xxxvii).

Fackenheim's reflections here, one might fairly believe, are sufficient for the aforementioned questions to be answered in the affirmative. Accordingly, it is worth considering the development of a fragmented ecological worldview in which the most extreme instances of destruction are admitted to *transcend* and work *against* the world, rather than as part of it. This chapter will now look to outline a fragmented panentheism, which might act as an interesting underpinning for such a fragmented ecology.

A fragmented panentheism would hold that the vast majority of the world lies within God, but that created beings have the potential to undertake actions that fall outside of God's sphere of divinity should they prove especially unethical, destructive, and heinous. Though any precise lines of demarcation would be arbitrary, those actions worthy of falling outside of God should be thought of as being limited to only the most extreme acts of evil. Unfortunately, evil is a normal and characteristic part of nature, and, as such, it would essentially negate nature to suggest that everyday instances of sinfulness constitute a move outside of God. To the contrary, the *Shoah*, for instance, brought evil to a decidedly *unnatural* level – its unique unnaturalness being attested to by Fackenheim. The *Shoah* would, accordingly, be an example of a worldly event that explicitly falls outside of God's being– indeed, as per Fackenheim, it is almost improper to describe it as a *worldly* event.

Instead of picturing the world as the smaller and God as the larger of two concentric circles, as is the norm in panentheism, a fragmented panentheism would more closely resemble a Venn diagram, in which the majority of the circle representing the world overlaps with that representing God, but in small part falls outside of it. Such fragmentation is not reserved for one strain of panentheism, but is rather, with some work, compatible with them all. It might be introduced into eschatological panentheism, for instance, by saying that, in the end times, God recovers most of the world, but not all of it. Equally, expressivist panentheisms could loosen their notion of deity so to hold that most, but not all, of the world eventually ascends into God's divine perfection. Process theologians, lastly, could fragment their accounts of relationality such that the events of the *Shoah*, and their like, may be removed from the divine organism.

Now to address the elephant in the room: the phrase "fragmented panentheism" appears immediately as an oxymoron. If the prefix "fragmented" is taken to suggest that part of the world lies outside of God, then how can this be compatible with panentheism – the doctrine that all is in God, no less? In justifying a fragmented panentheism, it is first necessary to question what the word "all" is taken to mean. By definition, "all" can refer to the collective of *absolutely* everything, or, more humbly, to the entirety of a particular group. While totalizing panentheisms think of "all" in the first sense, fragmented panentheism utilises the second, taking "all" to denote everything that is included in God's being. To elaborate, with a view to affirming the importance of responsibility, it holds that God creates every being with an intrinsic moral dignity, but also with absolute free will such that they can (in very extreme cases) transcend this dignity by committing heinous acts. Wherein this is the case, their problematic actions cease to be included in the "all" (God) –rather, they are seen to transcend and run counter to it. All, thus, may be said to reside in God according to fragmented panentheism, with the caveat that an action's being underpinned by a minimal level of responsibility is its price of entry into this "all."

The focus upon *actions* as opposed to *actors* is important here; to suggest that certain actors may be deemed to irremediably fall outside of an "all" would sit uncomfortably alongside narratives associating an "evil" otherness with those who are of specific races, religions, genders, sexualities, levels of physical ability, etc. All too easily, such an "all" could become patriarchal, exclusive, and dominating, and lead to unnecessary violence against those actors who are deemed to run counter to the all. The condemnation of actions, however, can be measured against (at least somewhat) objective criteria relating to levels of destruction, which would – instead of nurturing them – condemn prejudiced actions against certain peoples (as is the case with the actions of the *Shoah* perpetrators being deemed absolutely irreconcilable with God's divinity).

Further justification for its reconcilability with fragmentation lies in the fact that panentheism is often quite different in essence than in definition. As has been demonstrated, for instance, those panentheisms which stick to the notion of "all-in-God" in the strict sense often actually end up doing little justice to the world, insofar as the finite is lost sight of and transcendence takes precedence. Brierley (2004, p. 6), moreover, lists eight ideas that are frequently adhered to by panentheists, and it is interesting to note that a fragmented panentheism would score as high as most typical panentheisms on this count. Brierley's eight criteria are as follows:

> [1] the cosmos as God's body; [2] language of 'in and through;' [3] the cosmos as sacrament; [4] language of 'inextricable intertwining;' [5] the dependence of God on the cosmos; [6] the intrinsic, positive value of the cosmos; [7] passibility; and [8] degree Christology.

Though it is not possible to go through them all individually here, there is a case to be made that a fragmented panentheism accords with six of these criteria (specifically: 2, 3, 4, 5, 7, and 8). This puts fragmented panentheism at least on a par with Philip Clayton's panentheism, which Brierley notes falls afoul of criterions five and six.

It remains for this paper to further explain why a fragmented panentheism would be of greater utility for feminist and ecological purposes than a regular, totalizing panentheism. As a discipline, feminist philosophy of religion takes issue with God being equated with masculinity. The latter generally manifests itself in the form of a straightforward binary pitting of an all-powerful spiritual being (problematically associated with masculinity) against a passive world (problematically associated with femininity). Though they intend to secure the divinity of the world by placing it within God, it has been noted throughout this chapter that totalizing panentheisms often actually serve to reinforce this gendered binary that prioritises transcendence above worldliness. This is true insofar as the placement of *everything* in God validates the classical gendered notion of God as an all-powerful and dominating force; which is attested to by the aforementioned eschatological allusions to God's ultimate transformation of nature, the treatment of the world as a means to an end via which God can glorify *Him*self, the absorption of the world into God without much of a trace, and the suggestion that *He* must be viewed as a holy organism within which all created beings ineluctably partake. Acknowledging that aspects of worldly finitude fall outside of God's sphere of divinity would paint a more complicated and chaotic picture of the relationship between the finite and the infinite; one which would serve to emasculate God by limiting *His* ability to utilise the world as a means to *His* own glorification in making *His* creation, in part, permanently recalcitrant to *His* purposes. According to this view, God would become less of an all-powerful and all-encompassing agent of (masculine) dominance, and more of an ideal of responsible behaviour that people of any gender can choose to reflect or fall short of.

Asserting that it is possible for part of the world to fall outside of divinity would, furthermore, serve the ecological purpose of making it known that there is no guarantee of a holistic worldly salvation. When God is posited as wholly containing the world, there is a confidence that God has the power to, and inevitably will, transform the world into a beacon of His/Her glory. Such guarantees lend themselves to the shirking of responsibility among human beings, for if God will ensure that all is fine in the end anyway, humans may wonder whether it really matters what course of action they take in the present. Allowing for the possibility of genuine and irreversible worldly destruction that takes place outside of God and is thus beyond His/Her sphere of influence might, to the contrary, give humans pause for thought concerning their treatment of the environment. This is a particularly pertinent message in a nuclear age where natural destruction of unprecedented proportions remains a constant possibility. In a word, whereas totalizing panentheisms gaze at the heavens, a fragmented panentheism would keep a watchful eye on the world at all times.

26.5 Conclusion

In sum, this chapter has argued that panentheism is an unsuitable basis for environmental holism, as there is a tendency for panentheisms to contain the world within a transcendent God who is considered superior to nature, and for whom nature is only instrumental. This God, it has been shown, is also sometimes problematically framed as a masculine agent encompassing a feminine world. Nevertheless, this chapter noted that eco-holism, taken as being absolutely all-encompassing and underpinned by pantheism, is itself problematic insofar as it must extend even to those most extreme instances of worldly destruction. It noted in this point an opportunity for panentheism to cement itself as an important foundation of environmentalism; that is, provided panentheism is able to loosen its strict criteria of *all* of the world being contained within God, and thus allow for part of the world to lie outside of God. The affirmation of such a fragmented panentheism, it argued, could serve to transform panentheism into a useful metaphysical tool for the development of an environmental ethic that, while broadly holistic, is able to separate itself off from extreme, and decidedly unnatural, instances of worldly destruction. This panentheism was, moreover, argued to prevent the straightforward association of spirituality with masculinity and naturalness with femininity, as has been a problem throughout the history of philosophy.

References

Barbour I (1990) Religion in an age of science: the Gifford lectures –volume one. Harper & Row, New York

Barker DC, Bearce DH (2013) End-times theology, the shadow of the future, and public resistance to addressing global climate change. Polit Res Q 66(2):267–279

Brierley MW (2004) Naming a quiet revolution: the panentheistic turn in modern theology. In: Clayton P, Peacocke A (eds) In whom we live and move and have our being: panentheistic reflections on God's presence in a scientific world. William. B. Eerdmans Publishing Company, Grand Rapids, pp 1–15

Clayton P (2004) Panentheism today: a constructive systematic evaluation. In: Clayton P, Peacocke A (eds) In whom we live and move and have our being: panentheistic reflections on God's presence in a scientific world. William. B. Eerdmans Publishing Company, Grand Rapids, pp 249–264

Cobb J (2004) Is there a process eschatology? Process and faith. Available at: https://processandfaith.org/ask-dr-cobb/is-there-a-process-eschatology-december-2004/

Cooper JW (2006) Panentheism: the other god of the philosophers, from Plato to the present. Baker Academic, Grand Rapids

Deane-Drummond C (2004) The logos as wisdom: a starting point for a Sophianic theology of creation. In: Clayton P, Peacocke A (eds) In whom we live and move and have our being: panentheistic reflections on God's presence in a scientific world. William. B. Eerdmans Publishing Company, Grand Rapids, pp 233–245

Edwards D (2004) A relational and evolving universe unfolding within the dynamism of the divine communion. In: Clayton P, Peacocke A (eds) In whom we live and move and have our being: panentheistic reflections on God's presence in a scientific world. William. B. Eerdmans Publishing Company, Grand Rapids, pp 199–220

Fackenheim EL (1982) The religious dimension in Hegel's thought. University of Chicago Press, Chicago

Fackenheim EL (1994) To mend the world: foundations of post-holocaust Jewish thought. Indiana University Press, Bloomington

Frankenberry N (1993) Classical theism, panentheism, and pantheism: on the relation between god construction and gender construction. Zygon 28(1):29–46

Göcke BP (2022) There is no panentheistic paradigm. Heythrop J 63(1):49–56

Gregersen NH (2004) Three varieties of panentheism. In: Clayton P, Peacocke A (eds) In whom we live and move and have our being: panentheistic reflections on God's presence in a scientific world. William. B. Eerdmans Publishing Company, Grand Rapids, pp 19–35

Hegel GWF (1975) Hegel's aesthetics: lectures on fine art, volume one. Clarendon Press, Oxford

Irigaray L (1985) Speculum of the other woman. Cornell University Press, Ithaca

Jantzen GM (1997) Feminism and pantheism. Monist 80(2):266–285

Leopold A (1968) A Sand County Almanac: and sketches here and there. Oxford University Press, Oxford

Levine M (1994) Pantheism, ethics and ecology. Environ Values 3(2):121–138

Lovelock J (2000) Gaia: a new look at life on earth. Oxford University Press, Oxford

Mander W (2022) Pantheism. In: The Stanford Encyclopedia of Philosophy. Available at: https://plato.stanford.edu/archives/spr2022/entries/pantheism/

Nadler S (2022) Baruch Spinoza. In: The Stanford Encyclopedia of Philosophy. Available at: https://plato.stanford.edu/archives/sum2022/entries/spinoza/

Næss A (1973) The shallow and the deep, long-range ecology movement. A Summ Inq 16(1–4):95–100

Næss A (1977) Spinoza and ecology. Philosophia 7(1):45–54

Oliver K (1996) Antigone's ghost: undoing Hegel's phenomenology of Spirit. Hypatia 11(1):67–90

Peterson GR (2001) Whither Panentheism? Zygon 36(3):395–405

Polkinghorne J (1994) The faith of a physicist: reflections of a bottom-up thinker. Princeton University Press, Princeton

Richardson WM (2010) Evolutionary-emergent worldview and Anglican theological revision: case studies from the 1920s. Angl Theol Rev 92(2):321–345

Tabaczek M (2021) Divine action and emergence: an alternative to panentheism. University of Notre Dame Press, South Bend

Valera L, Vidal G (2022) Pantheism, panentheism, and ecosophy: getting back to Spinoza? Zygon 57(3):545–563

Viney D (2022) Process theism. In: The Stanford Encyclopedia of Philosophy. Available at: https://plato.stanford.edu/archives/sum2022/entries/process-theism/

Ward K (2004) The world as the body of god: a panentheistic metaphor. In: Clayton P, Peacocke A (eds) In whom we live and move and have our being: panentheistic reflections on God's presence in a scientific world. William. B. Eerdmans Publishing Company, Grand Rapids, pp 62–72

Chapter 27
Hossein Nasr on the Environmental Crisis

João Almeida Loureiro

Abstract In this article I aim, on the one hand, to explore, in general terms, the complex relationship that religion has had and continues to have with nature and the environment, and, on the other hand, to assess what the Muslim philosopher Hossein Nasr considers to be the only solution to the environmental crisis. Toward this aim, I will briefly trace the history of what Hossein Nasr designates as the "desacralization of nature" and try to understand how what for the Stoic was a κόσμος ("kosmos"), for Heraclitus a λόγος ("logos"), or for Ephrem of Syria a book *scriptus digito Dei* ("written by the hand of God"), could become a mere object, neutralized of all that animated dignity. As we will realize, a careful reading of the history of philosophy and scientific thought will reveal the conditions for the possibility of such a change of point of view. Hossein Nasr's proposal focuses, precisely, on the need to change this paradigm, postulating that only a return to the integrality of cosmological thought –which is not merely cosmographical but strives to know things *in divinis* ("in divine entities")–, while rejecting scientism, will allow us to restore nature's noble and sacred character.

Keywords Nasr · Religion · Cosmology · Nature · Ecology

27.1 Introduction

Seyyed Hossein Nasr is a most important figure in contemporary Islamic thought and religious studies. His role in the revival of *philosophia perennis*, understood as a necessary adjunct to a *prisca theologia*, is well known. Guided by an originality that is not afraid to clash with the most rooted assumptions of Western culture, in 1966, in the Rockefeller Foundation Lectures, given at the University of Chicago (Nasr 1990), Hossein Nasr became one of the first figures in theology to explicitly

J. A. Loureiro (✉)
Department of Philosophy, Universidade de Lisboa, Lisbon, Portugal
e-mail: jloureiro2@edu.ulisboa.pt

L. Valera (ed.), *Pantheism and Ecology*, Ecology and Ethics 6,
https://doi.org/10.1007/978-3-031-40040-7_27

affirm the existence of an ecological crisis, and, above all, to call attention to its religious aspect (Deniz and Borgerding 2018). Since then, both the existence of some kind of nexus between religion and ecology (and not in the sense of the famous, but already archaic, thesis of Lynn White (1967)), and the need for inter-religious collaboration in the response to the environmental crisis have become clear (Gottlieb 2006).

In this way, our aim will be, first, to outline the diagnosis that the Muslim philosopher makes of the ecological crisis when invoking the history of science –which, gradually, led the West to the desacralization of nature and the relationship that humanity has with it–, and his response, his antidote to fight the ecological crisis from its very pillars.

27.2 The Spiritual Dimension of the Ecological Crisis: Humanity, Spirit, and World

The existence of something like an ecological crisis is today a fact that belongs to common sense. It is, precisely, a crisis in the Greek sense of the word: κρῐσῐς refers, first of all, to the ability of a subject to make decisions and make judgments, especially in relation to the field of justice. It is originally a legal term. In a medical context, it means, on the one hand, the decisive moment in the course of a certain illness, which can lead the patient to recovery or death, and, on the other hand, to the very appreciation that the doctor (ῐᾰτρός; φῠσῐκός) must make of the patient. For Galen, a symptom is a κρῐ́σῐς because, depending on the will of the gods, it can dictate the patient's salvation or ruin (Cooper 2004).

Thus, to speak of an ecological or environmental crisis is to speak of a decisive event that demands an immediate response and whose outcome can dictate the continuity or the end of Humanity. The answer becomes more urgent when it is verified that its cause is human, and, therefore, the result of cosmovisions, deliberations and choices. Indeed, that the pollution of the atmosphere due to fossil fuels, the contamination of the oceans with industrial chemical residues, the deforestation, desertification and other attacks on the survival of countless species, are factors of human cause is a truism. The brutally negative impact they have on the environment is recognized by virtually everyone (Oreskes 2004). Likewise, the nefarious role of technology in the course of the modern economy is something that finds similar wide acceptance. What Hossein Nasr believes it is necessary to clarify is, precisely, the foundations, the conditions that made it possible to get where we are.

In this sense, the response to the ecological crisis requires the understanding of a series of phenomena of a historical and conceptual nature that allowed Humanity – mainly, or almost exclusively, in and from the West– to dissociate itself from Nature, to face it, to use Hossein Nasr's metaphor, as a prostitute, as something that one merely uses, instead of taking it as a wife, which the man naturally enjoys, but towards whom he has responsibilities and obligations (Nasr 1990). Thus, there is a necessary propaedeutic to the approach of the environmental problem, which consists, on the one hand, of a set of anthropological considerations that allow identifying the moment and the historical conjuncture that led Humanity to the possibility of ecocide, and, on the other hand, a philosophy of nature that locates and elucidates Humanity's place in the world. In this sense, promoting the reduction of the use of fossil fuels and the alteration of the means of transport as a total or partial solution to an ecological problem, for the Iranian philosopher, means not even realizing the problem itself, because what matters to investigate is, first of all, *what drives the men and women of modernity to this incessant desire to travel –to travel constantly– and to want to escape the area where they live* (recalling that, for example, in the USA, in South America, and in Europe, more than 80% of the population resides in urban areas (United Nations 2019)); why, as Ralph Waldo Emerson (1983, p. 542) said, "cities give not the human senses room enough?" To understand these and other questions is to understand that the environmental crisis has, to a large extent, a spiritual dimension.

This spiritual dimension can be understood in two senses. In fact, contrary to what the naturalist movements that emerged in Modernity advocate, people cannot be reduced to a mere historical moment in the cosmos, the vertex of an evolutionary process that, in the end, would not distinguish them, except by degree of complexity, and never by nature, from any other natural organism. On the contrary, a human being is a person. Being a person and being an individual belonging to the biological species *homo sapiens sapiens* are radically different things. In Boethius' formulation, the person is *rationalis naturae individua substantia* (equivalent to Chrysippus' formulation, λογικὸν ζῷον), that is, the personality results from the individuality of a body animated by a spirit (Galen 1981). While non-human animals certainly possess some kind of soul in the Brentanian sense, that is, a substance with intentional states, they cannot be said to have personality in this very strict sense. For this reason, humans are the only entity that are said to have free will, because we can be governed by axiological-normative principles and, therefore, self-determine; that can be free, because we can materialize and become who we are; who, finally, are said to be rational, because we are the only animal that has language and that, through it, manifests the truth. Heidegger expresses this idea when he affirms the status of Man as a shepherd of Being (*Hirt des Seins*) –because, precisely, he is the only entity that, due to his characteristics, is capable of configuring a world (Heidegger 2000).

In this sense, this personality from which the I emerges, irreducible to atoms or neurons, is truly, as Bergson first glimpsed and later Merleau-Ponty envisioned, an incarnate consciousness. Not even the staunchest nephelibate actually lives in the clouds. A person lives in a body vivified by a conscience, so that she is not a mere *res extensa*, but a *corps vécu*, a *Leib* in the Husserlian sense. The body is an organ of movement and orientation in the world (Vydrová 2016; Cusinato 2018). And just as the body is not a mere *res extensa*, so the world and space. It was perhaps Jakob von Uexküll who saw it most clearly: the world is always an environment-world (*Umwelt*), that is, a place of installation. From the outset, a place is not a mere geographic location that can be defined by a set of coordinates, like a point in a

Euclidean space, but a determination of meaning of a factical or affective nature – remembering that the Latin etymon *affectus*, derivation of the verb *ad-ficere*, has as its primary meaning "clash," that is, what results from the exercise of an action on a given object, properly, from contact. People access the world, are in contact with it, and know it in a peculiar way. It is Pascal's *cœur*, Bergson's *instinct*, Heidegger's *vorontologischen Seinsverstandnisses*, or, *mutatis mutandis*, the Augustinian *visio intellectualis*. Just as the yellow-winged digger wasp knows where to sting the cricket to immobilize its three pairs of legs (Bergson 1908), so Man finds himself, in the immediate of his existence, situated in a place of affective familiarity, having a point of view that is effective regarding navigation through life.

Thus, people are not a mere thing, like an object, but, from what we have just seen, a radically privileged entity that stands in a peculiar relationship of epistemological harmony with its world-environment. This is the first step towards understanding the spiritual dimension of the ecological crisis: a human is irreducible to her physiological particularities and participates in a fundamentally spiritual life, in such a way that what is essential to her cannot be captured by science in the modern sense of the word. Think, just to provide an example, of the notions of territory and habitation –including in the case of non-human animals. When a black bear rubs its back against the trunk of a pine tree to mark its territory, nothing changes, in scientific terms, in the landscape. Both the pine and the forest remain precisely the same. However, for the black bear, everything changes. In the same way, to describe a subject's childhood room as a polygonal structure of concrete is to miss all the life that is there. Science doesn't capture atmospheres – that's not what it is for. But life is fundamentally atmospheric, it is made up of nuances, insinuations, and hues.

27.3 Nature Throughout History: Modernity and the West

That harmony between humanity and its world-environment, however, is not only epistemological. It is also a harmony of meaning. So it was, at least, for the pre-Modern Man. The culture of classical antiquity had a cosmology that expressed it very clearly. For the Stoic and the Presocratic, the κόσμος, as opposed to χἄος, is that which is in order, that whose parts are intrinsically organized with a view to achieving an end (τέλος), and yet there is no *realität* –something entirely objective– but only a *Wirklichkeit*, a reality that is defined by its effectiveness, by its affective character –this is also the κόσμος of the Johannine epistles and the *mundus* of Augustine. The relationship between the subject and this world is not of a theoretical-cognitive character. On the contrary, the cosmos itself is understood as a rational animal (Von Arnim 1903). There is a relationship of homology between nature (Φύσις) and humanity itself. Like it, nature, understood as *natura naturans*, is something that is continually made, reinvigorated, as happens with humans. The root of Φύσις, φὔω, indicates precisely this: nature was and has been, it has a continuity that corresponds to a concept of diachronic identity. There is, therefore, a relationship of dialectical assimilation between the subject and the universe that makes it possible for humanity to find itself in an *acoustic* relationship with nature. Therefore, "the water of Thales is not what flows in rivers and streams but is the psycho-spiritual substratum and principle of the physical world" (Nasr 1990, p. 54). It is not yet a matter of a material world that is a symbol of a spiritual reality, since such a thing, for the pre-Socratic, makes no sense. Rather, the corporeal and the spirit are completely identified –the natural world is truly *animated.* This does not imply the inexistence of laws (in the sense of the term as used in modern sciences), including legal ones, as long as it is understood that they are an integral part of a philosophy of nature that involves a theology (Jaeger 1948).

As Jaeger (1948, p. 116) says,

> the "law of nature" is merely a general descriptive formula for referring to some specific complex of observed facts, while Heraclitus' divine law is something genuinely normative. It is the highest norm of the cosmic process, and the thing which gives that process its significance and worth.

The Heraclitean λόγος, in addition to being a jurisprudential concept, is a cosmological principle of intelligibility and meaning. *The concept itself is seen as an expression of the supreme principle that governs all things.* In the same way, the concept of νόμος, gradually stripped of its cosmological content and today used as a mere equivalent of "law," had a much richer meaning for a Greek that escapes the usual translation. If it is true that νόμος can have the meaning of "law," still it is a law that expresses the way of life of a people, and not, as a mere θεσμός, a legal norm promulgated by a legislator (Ostwald 1969). In this sense, νόμος is an expression of the πολίτευμα, and this does not consist only of "laws," but involves a set of notions of justice, *communitas*, *pietas*, harmony, etc. Again, Greek thinking is holistic. Humanity, civilization, nature and the world are always inseparable. For this reason, Josephus still equates the way of being of the Jew (Ἰουδαϊσμός) as a νόμος (making it equivalent to the Hebrew *Torah* and the Aramaic *'orayta* (Boyarin 2019; Barton and Boyarin 2016)).

Thus, Greek thought, as Mircea Eliade recognized in the figure of Pythagoras, looking at the cosmos holistically, always sought the constitution of a true *scientia sacra*, in which knowledge "had a function that was at once gnoseological, existential, and soteriological. It is the "total science" of the traditional type" (Eliade 2001, p. 195). It is only the decline of Olympic religiosity in its specificity that allows the emergence of a physics that tends towards naturalism. From Aristotle onwards, philosophy acquires a rationalist bent that strips science of its theological significance. In a way, there begins the movement that, gradually, would lead to the desacralization of nature and Man.

It is in this context that Christianity emerges, at a time when the *lingua franca* of intellectual speculation is, precisely, Greek philosophy. Indeed, without Greek philosophy and Hellenic culture there would be no Christian theology (Artemi 2016) and it is in Patristics that one finds the first systematic reaction against the growing naturalism of the Epicurean and Peripatetic thinkers of the classical period –a reaction that takes place through the radicalization of the distance between the supernatural and the natural. Thus, Charles E. Raven, biologist and eminent theologian born in the nineteenth century, who Panu Pihkala (2017) describes as an *ecotheologist*, considers that the patristic thesis that nature is *massa perditionis* inevitably led Christian theology, from the first centuries, to postulate a drastic dualism between nature, understood as the scope of the profane and the secular, and the place of the sacred, inherently supralunar (Raven 2010). In the words of H. F. Stander (2000, p. 169), "we can almost say that the early theologians 'de-divinized' nature." And this happens from a very early age. Lactantius, in his *Institutiones Divinae*, one of his most important treatises written in the first decade of the fourth century with the intention of establishing the falsity of non-Christian worldviews, goes on to explain that Stoic doctrines are wrong and that the idea of an animate, living, and rational cosmos makes no sense. The world, *terra quam calcamus*, is not only not God, nor divine, but it is not even a living thing *–nec mundus Deus est, nec animans*. The type of reasoning followed by Lactantius leads him to conclude that man is not even part of the world: *neque mundus generat hominem, neque homo mundi pars est* (*Institutiones Divinae*, II. 6).

Interestingly, as Raven and Hossein Nasr are careful to point out, this polarizing movement is almost exclusively characteristic of Western Patristics. The Fathers of Cappadocia, as well as Ephrem of Syria, among others, share the idea, naturally incipient, that the natural world can be understood as *liber natura*, and therefore as a source of revelation whose study can generate knowledge of God (Tanzella-Nitti 2005). However, as the process of Christianization of Europe advanced, some elements of this primordial Christianity from the East managed to penetrate the rapidly rising civilization. Scholasticism marks the zenith of this gnoseological ecumenism that integrated the esoteric dimensions of Christian knowledge prevalent in Eastern schools of thought. In this sense, the doctrine of the Book of Nature could be clearly articulated and developed by Hugh of Saint Victor. For him, the visible world constitutes a book written by the finger of God (*scriptus digito Dei*), and is, therefore, an expression of divine power (*virtute divina*). As a product of God's authority (*divino arbitrio*), its study can reveal knowledge of hidden aspects of the Creator (*Eruditiones Didascalicae*, Sept, IV). Bonaventure would later elaborate the same idea with greater delicacy and elegance. In his *Itinerarium mentis in Deum* (I, 14), he asserts that it is possible to see manifestations of God in the seven properties of all creatures: origin (*origo*), magnitude (*magnitudo*), multitude (*multitudo*), beauty (*pulcritudo*), fullness (*plenitudo*), activity (*operatio*), order (*ordo*). In this way, for example, the continuous effectiveness of the activity of fire "clearly manifests the immensity of the power, wisdom and goodness of the Trinitarian God who, by his power, presence and essence, exists uncircumscribed in all things" (*manifeste indicat immensitatem potentiae, sapientiae et bonitotis trini Dei qui in cunctis rebus per potentiam, praesentiam et essentiam incircumscriptus existit*).

Medieval Christianity in the first centuries of the first millennium manages to stand out from the preceding period through the more or less generalized adoption of an essentially Pythagorean gnoseological cosmovision. "The Pythagorean science of harmony, of numbers, geometric forms and colours, pervaded the science and art of the Middle Ages," says Hossein Nasr (1990, p. 59). In this sense, the scientific enterprise regained its holistic content, of a *scientia* of totality, in which the scientist is simultaneously a theopath and nature a theophany. Thus, an ancestral trend reappears in the West which, on the other side of the world, had already been very clearly consolidated in traditional Chinese medicine. In effect, traditional Chinese medicine, whose origins date back to the Neolithic period, has always seen humans and their bodies as an integral part of nature, with these being in a relationship of primordial harmony, and having as a fundamental principle that the mechanisms that regulate nature are the same ones that regulate human health (Schiffeler 1976) –their primary purpose has always been, therefore, the maintenance of this harmony. And just as in that civilization natural phenomena were seen as manifestations of divinities which one should aspire to understand, *mutatis mutandis*, the Christianity of the first Scholastic, with a hermetic-Pythagorean background, had as its pretension to ascend from *materia prima* to the celestial intelligible world.

However, the massive work of translation of works from Arabic into Latin that took place during this same period would gradually lead, from the thirteenth and fourteenth centuries onwards, to the "Aristotelianization" of Christian theology. Very quickly, a rationalism with a skeptical background would replace a Gnostic-oriented philosophy which had its zenith in the thirteenth century. Ockham's nominalism, by postulating that the universal is a mere *conceptus mentis*, represents the beginning of this trend that will forcefully reject any idea of the transcendent. In fact, for Ockham, the real is reduced to what is apprehensible by *cognitio intuitiva*, that is, by the senses – defining everything else as qualities (De Angelis 2019). Buridan's school, mainly through its most important disciple, Nicholas of Oresme, will continue its legacy, laying the foundations for the formation of modern science. Observation and experimentation become the touchstone of the validity of explanations of reality, relegating God and the intelligible world of angels to the domain of faith and, therefore, of uncertainty and doubt.

The Renaissance, notwithstanding the peripheral popularization of certain esoteric cosmologies of hermetic and kabbalistic origin –by such notable personalities as Marsilio Ficino and Giovanni Pico della Mirandola–, even so, following the trend, marks the beginning of the integral naturalization of Man and the solidification of the concept of science. In this sense, cosmology, devoid of any metaphysical background, becomes mere cosmography, an external description, only concerned with form and ignoring content. The Copernican Revolution –which does not make Copernicus a revolutionary (Africa 1961; Saliba 2002)– came at a crucial moment in that it was not accompanied by a new spiritual cosmovision. The result was, as Hossein Nasr says, the displacement of humanity in the cosmos. The geocentric model preserved a symbolism that invoked the divine nature of humanity, understood as *magnum miraculum* and, therefore, center of the universe, and, even so, as an inhabitant of a lower level of existence, with the purpose of ascending to the heights of the intelligible worlds, which flew over him. Thus, traditional cosmologies mirrored what was the metaphysical reality underlying the *sophia perennis* of all the great theologies and soteriological conceptions: the possibility that humanity has, and which appears to her as gradual, of ascending and approaching the divine through gnosis. The subtraction and annulment of higher levels of reality, that is, of a hierarchy of the real, explicit in the very notion of gnosis, implies not only the annihilation of an entire gnoseological field and a set of representations, but, above all, that vital form of intelligent realization which is a key to realize Life. Despite the symbolism associated with the heliocentric model –the Apollonian centrality of the light source, like Christ in Michelangelo's *Giudizio Universale*, etc.–, the new astronomical interpretation was not accompanied by a new spiritual conception of humanity, which had the effect of terrestrializing him, thus paving the way for the Scientific Revolution.

In this sense, what was the fundamental aspect of Cartesian philosophy –namely, the reduction of reality to quantities, mere *res extensæ* (therefore, fundamentally *res mortuæ*)– marked the radical paradigm shift that occurred with Modernity. In fact, many of Descartes' theories, such as those referring to zoology or human anatomy, have been peremptorily rejected, except for his conception of material reality. That novelty which Galileo –his contemporary– introduced in the history of scientific thought, namely, the application of mathematics to the study of movement, finds in Descartes a theoretical edifice that sustains and justifies it. Henceforth, reality *–true reality–* neutralized of any affective content, will be reduced to a complex of relationships between pure quantities, and access to it will be done by mathematics. It will be precisely what will result from the triumph of Newton's physics: a mechanism that will reject any holistic worldview that encompasses knowledge, nature, and the sacred in the same sphere of understanding. Science, in its modern sense, will come to be seen as the only means of accessing and knowing reality. Until today, so-called "scientific methods" (Harrison et al. 2011) –from Baconian inductivism to Popperian falsificationism– prevail, almost unscathed, as the only valid forms of *scientia* and valid knowledge. Science becomes, from Modernity onwards, to use the expression that gives the title to the famous book by the chemist Anthony Standen, an authentic sacred cow (Standen 1950).

27.4 A Paradigm Shift

The implications of the scientific paradigm prevailing in Modernity are evident. A mathematized nature, reduced to relations between quantities, and a Humanity dissociated from the natural world, are the ideal formula for the continuation of the systematic ecological abuse. Where the pre-Socratic Greek saw an immanent principle of divinity, the modern subject, under the lens of capitalism, sees an infinite constellation of resources that she can use for her profit (Williams 2010); where the scholastic Man marveled at the reading of the Book of God, Modernity sees quantities of force that it can dominate –why not? – for her benefit. To combat the ecological crisis, it is necessary, first of all, to realize that the crisis is itself due to such a paradigm shift that culminates in Modernity. In this sense, *fighting the ecological crisis necessarily implies fighting this paradigm.* This is Hossein Nasr's fundamental proposal. What is urgent is not the application of technology to problems technology itself has created –a nugatory and senseless project– but a change of point of view that generates the conditions of possibility for the promotion of a new relationship between humanity and nature.

Curiously, with very few exceptions, philosophers and scientists, expelled from life, are usually the last ones to see what is expressed in the great religious and sapiental traditions of Humanity. The Sioux people themselves, one of the subjects of Durkheim's famous study on the forms of religious life (Durkheim 2013), now almost extinct, had a cosmology marked by a principle of universal explanation called *wakan*. Like the *orenda* of the Iroquois, the *wakan* constitutes the efficient cause of all phenomena –it is what makes the wind blow, the sun illuminate the Earth, plants grow. The same can be seen in the ideas of the Australian aboriginal tribes. The totem is more than an object or sacred symbol, since, in addition to functioning as a kind of heraldic sign, it is something in which all of nature participates in: animals because they evoke it by their resemblance, plants because they serve as food, stones and rocks because they allow commemorative ceremonies and rituals to be carried out. The nature of so-called "primitive" peoples –those of traditional civilizations– is an animated nature, a living thing in the most radical sense of the word. The great stain of Modernity is the death of this historical nature, the cause of wonder and fearful admiration. The subtraction of this primordial understanding from humanity's reality not only impoverishes her but vilifies her, making her arrogant, allowing her to create a system and adopt practices that are based on the view of nature as a mere thing at her disposal.

Finally, Hossein Nasr points to the fact that modern science also fails in two crucial senses: on the one hand, (1) it fails to account for the totality of reality, and, on the other hand, (2) for not being integrated into a hierarchy of knowledge that has something sacred as its axis. The first is suppressed, according to Hossein Nasr, through the reintroduction of metaphysics in the cast of traditional sciences. By metaphysics, the philosopher understands the science of the Real, or, rather, the knowledge that allows us to discern the illusory, reaching the reality of things, knowing them in their essence, which means, in the last analysis, knowing them *in divinis* (Nasr 1989). In this discipline are contained the principles of the cosmological sciences themselves –not merely of cosmographies– and, above all, of an anthropology and a psychology that must have the cosmic Principle as a central element. What may seem an absurd attempt –after all, one of the most basic dogmas of Modernity is, precisely, the relegation of the scope of the sacred to the private domain of "belief" or "faith," understood as adherence to a set of theses, most often taken as unverifiable and irrational– however, for Hossein Nasr, it is not. And he says it very clearly: "There would in fact be no agnostics around if only it were possible to teach metaphysics to everyone" (Nasr 1993, p. 6). It is just that, just as it is unreasonable to expect everyone to understand quantum physics, one should not expect everyone to understand metaphysics. That contempt for Philosophy which leads everyone to believe they are capable and entitled to discuss it and to which Hegel drew attention (Hegel 1986), has only become more acute. The mass-man that Ortega y Gasset spoke of considers himself prepared to face the most perennial metaphysical questions without any kind of training.

In this sense, a first step involves the rejection of scientism, that is, the ideology that takes modern science as the only valid source of acquiring knowledge. The desacralization of nature and the cosmos, as well as the postulate of the scientific supremacy of mathematical knowledge that deals with pure quantities –which, as we have seen, arises along with Renaissance humanism– is not only a gnoseological error but ends up threatening the very status of humanity. This Promethean Man, conceived in rebellion against God, holder of absolute power over nature and the environment, total master of his destiny, is nothing but a truncated Man destined to externalize his inner chaos (Nasr 1993). Thus, the ecological and environmental reform has to pass, necessarily, through a particular reform of the modern Man. This is, once again, a reform of Modernity's point of view.

If what made possible and continues to aggravate the present ecocide is the desacralization of nature and humanity and the consideration that the ultimate aim of humanity is, as in the Baconian utopia (today turned dystopia), infinite material progress, the answer cannot fail to pass –as Hossein Nasr proposes– through the restoration of the sacred character of the natural world: "There is no alternative but to change our whole world-view" (Nasr 2000, p. 20). The Iranian philosopher believes that this can be done through education –namely, the teaching of metaphysics understood as *scientia sacra*, an undertaking that must be preceded by a critique of modern science and an understanding of its radical limitations. We don't know if this will be enough, or even feasible, but we are convinced of the benevolence of a wisdom that, in the warble of the nightingale and in the scent of the orchid, as much as in the snowy meadow, manages to glimpse the tone and the zephyr of God; and, above all, that if this ancestral wisdom were cultivated and generalized, there would be no ecological crisis.

References

Africa TW (1961) Copernicus' relation to Aristarchus and Pythagoras. Isis 52(3):403–409

Artemi E (2016) Embracing Greek philosophical thinking in the fathers of the 2nd–5th centuries. Vox Patrum 65:31–47

Barton CA, Boyarin D (2016) Imagine no religion: how modern abstractions hide ancient realities. Fordham University Press, New York

Bergson H (1908) L'Évolution Créatrice. Félix Alcan, Paris

Boyarin D (2019) Judaism: the genealogy of a modern notion. Rutgers University Press, New Brunswick

Cooper GM (2004) Numbers, prognosis, and healing: Galen on medical theory. J Wash Acad Sci 90(2):45–60

Cusinato G (2018) La distinzione Leib-Körper in Husserl e Scheler. In: Biosemiotica e psicopatologia dell'ordo amoris: in dialogo con Max Scheler. FrancoAngeli, Milan, pp 81–84

De Angelis L (2019) The "real" scientific revolution: 14th century England's contribution to modern scientific thought. Histories 10(2):3–22

Deniz H, Borgerding LA (2018) Evolution education around the globe. Springer, Cham

Durkheim É (2013) Les formes élémentaires de la vie religieuse: Le système totémique en Australie. Presses universitaires de France, Paris

Eliade M (2001) A history of religious ideas: from Gautama Buddha to the triumph of Christianity. University of Chicago Press, Chicago
Emerson R (1983) Essays & lectures. Literary Classics of the United States, New York
Galen (1981) On the doctrines of Hippocrates and Plato. Akademie-Verlag, Berlin
Gottlieb RS (2006) The Oxford handbook of religion and ecology. Oxford University Press, Oxford
Harrison P, Numbers RL, Shank MH (2011) Wrestling with nature: from omens to science. University of Chicago Press, Chicago
Hegel GWF (1986) Grundlinien der Philosophie des Rechts oder Naturrecht und Staatswissenschaft im Grundrisse Mit Hegels eigenhändigen Notizen und den mündlichen Zusätzen. Suhrkamp, Berlin
Heidegger M (2000) Über den Humanismus. Vittorio Klostermann, Frankfurt
Jaeger W (1948) The theology of the early Greek philosophers. Oxford University Press, Oxford
Nasr H (1989) Knowledge and the sacred. SUNY Press, Albany
Nasr H (1990) Man and nature. Unwin Paperbacks, Sydney
Nasr H (1993) The need for a sacred science. SUNY Press, Albany
Nasr H (2000) The spiritual and religious dimensions of the environmental crisis. Ecologist 30(1):18–20
Nasr H (2015) A religious nature: philosopher Seyyed Hossein Nasr on Islam and the environment. Bull At Sci 71(5):13–18
Oreskes N (2004) The scientific consensus on climate change. Science 306(5702):1686–1686
Ostwald M (1969) Nomos and the beginnings of the Athenian democracy. Clarendon Press, Oxford
Pihkala P (2017) Early ecotheology and Joseph Sittler. Lit, Zürich
Raven CE (2010) Natural religion and Christian theology: the Gifford lectures 1951. Cambridge University Press, Cambridge
Saliba G (2002) Greek astronomy and the medieval Arabic tradition. Am Sci 90(4):360–367
Schiffeler JW (1976) The origin of Chinese folk medicine. Asian Folklore Stud 35(1):17–35
Standen A (1950) Science is a sacred cow. EP Dutton, Boston
Stander HF (2000) Ecology and the church fathers. Acta Patristica et Byzantina 11(1):167–176
Tanzella-Nitti G (2005) The two books prior to the scientific revolution. Perspect Sci Christ Faith 57(3):235–248
United Nations (2019) World urbanization prospects: the 2018 revision. Available at: https://www.un-ilibrary.org/content/books/9789210043144
Von Arnim HF (1903) Stoicorum Veterum Fragmenta (Vol. 1). Teubner, Leipzig
Vydrová J (2016) The intertwining of phenomenology and cubism –in the analyses and works of art of Czech artists and theoreticians. Horiz Stud Phenomenol 5(1):214–231
White L (1967) The historical roots of our ecologic crisis. Science 155(3767):1203–1207
Williams C (2010) Ecology and socialism: solutions to capitalist ecological crisis. Haymarket Books, Chicago

Chapter 28
Francis Hallé's Project for a Large Primary Forest in Western Europe and a New Understanding of Our Relationship with the Biosphere

Fernando Calderón and Teresa Calderón

Abstract Dean of French botany Francis Hallé (1938) has been promoting a utopian project since 2019: to return to Western Europe a portion of its primitive jungle in the form of a primary forest (70,000 hectares). The project is neither romantic nor lacking in utility. Science, in fact, has already pointed out its benefits. And so has philosophy, a discipline that long has been concerned about the issue of deforestation and whose collaboration in the development of this initiative seems more than desirable. The project has an important philosophical background since, were it ever carried out, would be the most unequivocal material expression of a new understanding of humanity's relationship with the biosphere and, as such, a response undermining Cartesian dualism. In this chapter, we show that the project for a primary forest in Western Europe contains, in the formulation itself, a revision of the values that have governed human relationships with other species, and implies a new way of inhabiting the world, which is currently harassed by policies of domination of nature and beliefs of human superiority encouraged by certain forms of thinking.

Keywords Francis Hallé · Dualism · Primary forest · Biosphere · Rewilding

On 26 February 2019, the *Association Francis Hallé pour la Forêt primaire* (AFH) made its first public appearance. Since then, important steps have been taken towards its goal to return a portion of European lowlands to their original condition as *primary forest* (Hallé 2021). The project is not an isolated one. All over the world, wilderness policies promote the stewardship of certain territories to protect them from unsustainable human interventionism. The scientific community agrees that

F. Calderón (✉)
Department of Philosophy, Universidad de Valladolid, Valladolid, Spain
e-mail: quindos@uva.es

T. Calderón
Department of English Philology, Universidad de Valladolid, Valladolid, Spain

L. Valera (ed.), *Pantheism and Ecology*, Ecology and Ethics 6,
https://doi.org/10.1007/978-3-031-40040-7_28

easing pressure on ecosystems will improve their resilience and contribute to the preservation of species. At a time of climate emergency, resource depletion, and environmental crisis, it seems appropriate to entrust nature to take care of itself. However, it is not a question of abandoning it to its fate, nor of excluding humans from any relationship with wildlife. Rather, it is a question of renouncing a model of oppressive management of natural habitats in favor of one that gives nature greater freedom and counteracts the impoverishment of ecosystems and loss of biodiversity. It is also a question of reviewing human relations with the biosphere, which are currently prevailingly understood according to the logic of the other, as an inanimate reality whose value would ultimately depend on our gentility as a gifted species (Morizot 2020; Tafalla 2022).

Understanding the magnitude of the project conceived by Francis Hallé and pointing out its uniqueness compared to other rewilding projects requires saying something about the history of European forests. To do so, it is important to draw attention to the early days of the Holocene, after the last episode of continental glaciation. When the glaciers began to retreat, the forests settled in the southernmost regions of Europe took advantage of the new climatic conditions and began colonizing the territories abandoned by the ice in central and northern Europe. In a slow but inexorable process, the bare ground was covered with splendid forests leading to an explosion of life, a circumstance which our Mesolithic human relatives took advantage of to satisfy their subsistence needs. Like all other living forms, these human populations also took part in the balance and maintenance of the natural dynamics of the forest, modelled the landscape according to their own skills, and organized their existence in a territory with closed –almost non-existent– horizons without causing significant damage.

The transition from nomadic gathering to sedentary farming and livestock rearing drastically changed humanity's relationship with nature. The landscape also underwent a radical change. It was now the forests that retreated in the face of human industry, which needed pasture and arable land. Forests were an inconvenience, which led to their destruction, often by the use of fire. Indeed, fire was decisive in this work of appropriation and transformation of the landscape. The growth of human population due sedentary farming and the continued extraction of natural resources that new lifeways required, ruined the primeval European forest. It was eventually broken up and unevenly distributed from the Ural Mountains in the east to the lands ending at the Atlantic coasts in the west.

Since initially nature was able to metabolize the effects of human behavior, it hardly suffered any damage. Its resilient condition successfully mitigated the harmful behavior of our species. It was only later that the multiple and constant aggressions –together with their combined effects and the increasing variation in their velocity– provoked a fracture between human beings and nature in the territories that cover Europe today. Nature became, therefore, exposed to a situation of unique abuse and domination. European *homo faber* finally altered the common flow of natural dynamics and upset the evolutionary concert of species. *Homo faber*'s technical and cognitive faculties became a source of *disruption* (Morizot and Mantovani 2022), a concept borrowed from the philosophy of technology to illustrate the increasingly pronounced mismatch between natural rhythms –which were often invariable– and those rhythms caused by the protean action of human beings at varying rates of velocity and acceleration. According to some studies, this ecological disruption is nowadays about to provoke a collapse of ecosystems and lifeways dependent on them. It has already been responsible for the deterioration of the biosphere and a new episode of mass extinction, the only one caused by direct human action (Servigne and Stevens 2015). This is the scenario which the AFH project is responding to.

Recreating a primary forest in Western Europe is tantamount to restoring a lost ecosystem to the continent. However, this work of recreation should never be interpreted as a form of nostalgic expression or refined primitivism. The work of the AFH does not consist in setting aside a territory and declaring it sacred, immaculate, akin to paradise. The project does not adhere to the Cartesian dualism that has pitted human beings against nature, and forests in particular. Due to the geography of Europe, the absence or scarcity of horizon and perspective, of light and order, of always available and reliable points of reference in the forested habitats may have generated fear, suspicion, and unease in early human populations. In contrast to the landscapes designed by pastoral societies that are flooded with light, openness, and planned for human convenience, forests, with trees densely enclosing space and shutting out light, represent a place without us. And yet we originally belonged to the biotic community of forests.

The sinuous lines drawn by the perimeter trees of a forest have the ancestral value of *limes*. They are the frontier on whose sides separate existences are arranged. On one side, the governance of the human being; on the other, coarse and uncontrolled nature. Cartesian philosophy did not create this fractured space, but strengthened it, since it removed from the domain of moral considerations everything that could bring one world into relation with the other. To make the soul the seat of thought, and then say that to think is to feel, to doubt, or to will, is to declare that the soul is the exclusive heritage of our species. This promotes an ontology of annulment which consists in equating the other forms of existence by excluding the intermediate elements, those which –like pain or pleasure, intelligence or sensibility– intervene as factors of increasing gradation between the abiotic components of the biosphere –air, water, temperature, the acidity or alkalinity of the soil– and the whole of living organisms. This Cartesian ontology is monochrome, grey. Strictly speaking, bodies are reduced to extension, so all material existence is measurable. Farewell, therefore, to the elusive properties of bodies this reduction of nature to number and measure implies. It follows that the proclamation of Cartesian dualism not only means the expulsion of biological attributes –homeostasis, metabolism, adaptation, etc.–, but it also means that qualitative attributes of matter –color, taste, or smell– disappear or become irrelevant.

In his *Discourse on Method,* Descartes points out that, in the case of straying in the middle of a forest, the best alternative is to advance in a straight line (Descartes, in Adam and Tannery 1996, VI, pp. 22–27). He does not realize that a rectilinear trajectory is impossible there and that he cannot project his ideal onto a medium that escapes the conventions of geometry. A forest is not the same as a city. As is evident, the forest responds to natural dynamics in its spontaneity and freedom of growth. It does not need our help, which it can very well do without. The city, on the contrary, is a human artefact and can adapt to Cartesian preferences, although it does not always do so. Descartes himself distinguishes between the "*places regulières qu'un ingénieur trace à sa fantaisie dans une plaine*" ["regular places that an engineer draws at his whim on a plain"] (Descartes, in Adam and Tannery 1996, VI, pp. 11, 25–26) and the badly arranged cities. Of the two types of city, only the former meets Cartesian requirements, although what is interesting for our purposes is that both offer a certain relationship of similarity with two equally opposed ecosystem realities: the plantation and the forest, the former understood as the work of human beings and the latter as a genuine expression of free nature. It is worth examining the differences between these two types of cities to gain a better understanding of the project undertaken by the AFH. In other words, it seems to us that the metaphorical language devised by the main proponent of modern dualism could shed light on the uniqueness of this initiative.

We could say that the first type of city appears *ex novo*, in the way Minerva would emerge from Jupiter's head. It is the properly Cartesian city, a city without history, built on a grid in which the streets are arranged like lines drawn on a chessboard. It obeys the design of a single engineer, which guarantees a uniform and rational distribution of its premises. There is no use of "*vieilles murailles*" or old layouts, because to use them would compromise the success of the result and would give a disorderly and heterogeneous air to the whole. Cartesian urban planning is related to Hippodamian design and is present in many cities, especially the modern ones. One example is the work of Girolamo Marini, an Italian engineer in the service of Francis I. After being ordered to design a town near Vitry en Perthois, which had been devastated in 1544 by the troops of Charles I of Spain, Marini was given a blank sheet of paper. He then drew a rectangle on which he laid out squares, buildings, and streets, all perpendicular and equally rectilinear. The town was baptized Vitry-le-François. Of the old town, only part of its original name remains. What interests us about this example is that this type of town resembles a plantation, i.e., a farm or land used for growing trees, generally of the same species and always at equal distances.

The second type of city, on the other hand, bears some resemblance to a forest. Indeed, from a Cartesian perspective, we could identify the features of a poorly arranged city with the peculiarities of a primary forest. In this city, the streets are *courbées et inégales* (curved and uneven); the houses are built on *vieilles murailles*; the buildings form a motley mixture of ages and sizes. Medieval architectural elements –the remote origin of the great city– coexist with modern works. It is a waste-free place where old building materials are constantly recycled. The appearance is one of disorder, and one is tempted to attribute this appearance to the intervention of whim or chance. Undoubtedly, our forest can be seen through the metaphor provided by this type of city in which time generates an accumulative, never eliminative, effect of heterogeneous pieces.

Retrace now your steps and dwell on the Cartesian image of the plain as an element associated with a well-ordered city. It can also be transferred to our subject-matter. It is not for nothing that the project for a primary forest in Western Europe is to be located on a plain, an area of 70,000-hectares, which is equivalent, notes Hallé in his Manifesto (Hallé 2021, p. 29), to the surface of Menorca Island (Balearic Islands). If we map the primary forest on the Cartesian perception of nature, we will soon understand that the reductionist obsession of this model of rationality bears no relation to the forest ecosystem. Plant trees on your 70,000 hectares and wait. Give yourself time to watch them grow. Some trees will collapse in old age and open clearings in the forest; others will reach maturity and soar to incredible heights. Some tree species will give way to others. First will come the pioneers, such as birches, poplars, and ailanthus; last will come the yews, beeches, and oaks. Hundreds of years have passed, enough time for life to have settled everywhere, taking advantage of the scaffolding of the trees, from the apical tips of their deepest roots to the branches of the highest canopies. The stillness of the trees might give the viewer the idea of a dormant, dying world even where the trees have fallen. This is a false, hasty impression. In the flaky soil there are communities of microscopic creatures, legions of bacteria and fungi working to decompose organic matter. A step higher, on the surface horizon of the soil, lie the fallen leaves, the wood cut down from trees, the dead animals, a feast for decomposing organisms, many of which do not exist in the forests managed and groomed by our beneficent hand. There, life owes much to offal. The complex structure of this primary forest is generous in the formation of habitats. Its heterogeneous character, to which trees contribute as landscape engineers, favors the appearance of an extraordinary number of species. It is a vibrant world that hosts types of life that only here manage to thrive. The cavernous aspect of the trunks, a rarity in the bland groomed forests, offers shelter to birds that would not know what to do without them. Their barks are covered with mosses and lichens, organisms that are unfriendly to the timber industry. European megafauna finds their paradise here. The processes of biological interaction multiply; they shape and structure the form of the forest and optimize its multifunctional character. The interweaving of interspecific relationships has made the forest an exceptionally complex biological community and has brought together more species than any other ecosystem in the temperate regions of the globe could support. It is the work of nature working on its own.

French philosopher and writer Baptiste Morizot has adopted the concept of "heterochrony," a term coined by Ernst Haeckel for quite different purposes, to point out that in any primary forest the present can be described as the concurrent expression of multiple pasts. "*J'appelle hétérochronie*," he notes in *S'enforester*, "*la coprésence de plusieurs passés à la surface du même présent, et leur capacité à communiquer, à intéragir pour inventer des réponses aux sollicitations de l'avenir*" ["I call heterochrony the co-presence of several pasts on the surface of the same present, and their capacity to communicate, to interact in order to invent responses to the demands of the future"] (Morizot and Mantovani 2022, p. 98). This phenomenon of heterochrony, Morizot continues, unfolds in three ways. First, as cohabitation embodied in the different forms of life of all the evolutionary inventions that nature has deployed in its geological periods ("*cohabitation entre ancêtres*"); second, as cohabitation in each body of disparate biological ancestries ("*le corps d'ascendens*"); third and last, as cohabitation of the different stages of ecological succession which alternate in the process of the formation of a primary forest and which can undergo episodes of reversion due to disturbances ("*mosaïque totale*"). "*J'appelle 'mosaïque totale'* [...] *une forêt qui dans son visage actuel manifeste tous les âges et les dynamiques de la forêt dans le même présent, et où le dernier stade, mâture ou ancien, est présent en abondance*" ["I call a 'total mosaic' [...] a forest that in its present appearance manifests all the ages and dynamics of the forest in the same present, and where the last stage, mature or ancient, is abundantly present"]. (Morizot and Mantovani 2022, p. 102). This last sort of heterochrony, Morizot continues, is the defining feature of the primary forest, and the one that gives meaning to the purpose of bringing it into European ecological reality. In other words, it is the feature that legitimizes the idea of implementing actions to reintegrate it into the landscape. Carrying out this project is an act of ecological justice. Doing so means, if not repairing, at least mitigating the damage caused by humans to the biosphere as a whole.

But do not abandon the time vector yet. It is important to emphasise that silvigenesis involves extraordinarily slow processes, at least compared to the life of a human being. And there are no means to accelerate the processes of forest generation and succession that culminate in the formation of a primary forest. At this task, our incompetence is as clear as our own incompetence at making artefacts capable of replicating photosynthesis. As it is, there is only one way to shorten the timeline: instead of starting the project on barren soil, choose a potentially conducive forest. The AFH has already scouted a few corners of Europe and identified two strategic sites: the Northern Vosges and the Ardennes, both in border regions. Meanwhile, European authorities have welcomed the project and recognized its interest in the framework of the Green Climate Pact. In a letter (09/12/2022), the executive vice-president of the European Commission for the European Green Deal, former Dutch Minister of Foreign Affairs Frans Timmermans, welcomes its trans-generational and cross-border character, and describes the project as "*un pacte de restauration de la nature entre générations*"[1] ["a pact to restore nature across generations."] Well, if the efforts are successful and the pact holds firm, if the estimates made by forest ecology specialists are not wrong, if, finally, the project is not derailed by some unfortunate event, the primary forest will have emerged as a material reality by the end of this millennium. Of course, that horizon is remotely distant from us, but only to the extent that we have become accustomed to the immediacy and expiration of our artefactual world.

It must be stressed that the processes of forest succession do not need human help. Only a forest yielding to its natural dynamics can unfold its full potential, increase its structural complexity, and develop over time until it acquires the appearance of a primary forest. However, it takes an extraordinary visionary effort to let forests abandon adolescence, a period particularly favorable to the interests of

[1] The complete letter can be found in AFH's site: https://www.foretprimaire-francishalle.org/nos-actualites/leurope-exprime-son-interet-pour-le-projet/

lucrative extractivism. As is well known, Bialowieza forest, located between Poland and Belarus, may preserve the largest span of primary forests in Europe.

In 1828 Julius Brincken, the chief conservator of Polish forests wrote a short memoir about Bialowieza. Despite being accustomed by profession to travelling through vast forest areas, he could not get over his astonishment: "*Une forêt de cette étendue, rudement élevée par une nature brute, n'est pas seulement un aspect rare en Europe, mais elle offre aussi un vaste champ à des remarques intéressants*" (Brincken 1828, p. 18). Having declared his surprise, he then eloquently outlined the contrast between two ways of characterising the specificity of this type of forest:

> *Tandis que l'historien y observe les plus bas degré de la culture où puisse se trouver un pays, ou ce qu'il deviendrait dans le cas d'une émigration universelle, le forestier y voit avec intérêt, comment sans le secours de l'homme la nature aménage les forêts, comment elle sème, élève, détruit, et comment les débris des bois morts servent de berceaux aux nouvelles générations* ["While the historian observes the lowest levels of culture a country can reach, or what it would become in the event of universal emigration, the forester sees with interest how, without the human help, nature manages forests, how it sows, raises, destroys, and how the debris of dead wood serves as a cradle for new generations"] (Brincken 1828, p. 18; English translation by the book editor).

Brincken was speaking as a forester, not as a historian, a perspective that allowed him to consider the ruinous and moribund appearance of some enclaves of the forest not as the result of the negligence of their inhabitants, but as the product of a free form of natural development. And he used the occasion, moreover, to combat a prejudice that seems as inherent to his own time as it is undoubtedly to ours. For Brinken, the forest is not a helpless creature; it is self-sufficient. Its dependence on us is fictitious. And it is rather the other way round: we depend on the forest. We are, according to Alasdair MacIntyre's famous formula, rational and dependent animals, thus vulnerable to the havoc wreaked on the biosphere as a whole, to which we have mistakenly slipped the characteristic of dependence.

Western Europe's tree cover has increased significantly in recent decades. Today, one third of its territory is covered by forests. There is no doubt that reforestation campaigns have been successful, and it seems that recent innovations introduced by laboratories and agro-technological companies could even lead to a significant acceleration in the rate of forest cover recovery. Computational data analysis techniques, together with the use of drones for aerial planting and intelligent seeds, are emerging as useful tools in the quest for a greener Europe. Despite this, the forests that populate our continent are mostly young forests, many of which exist as mere business opportunities or as *green washing* advertising formulas. There are hardly any mature forests, and only Bialowieza conserves today a vast area of primary forests. Even the Bialowieza forest is in danger of losing its status as a forest relic. In addition, many of the wooded areas classified as forests are in fact no more than monospecific plantations, and many of our real forests are showing signs of decay and massive mortality, leading to an alarming impoverishment of their ecosystem services (Lloret 2022).

The initiatives that today bring the rewilding projects together –the same ones adopted by Francis Hallé's project– take Bialowieza forest as a reference, because of its natural dynamics having hardly been disturbed since the early Holocene.[2] This is why the AFH is belligerent towards the actions of the ultra-conservative government of the Law and Justice party, which is determined to turn the forest into a lumber pantry and is ready to ruin in only a few days the patient and millenarian work of nature. In this part of the world, the well-known proverb of Seneca, that "growth is slow and ruin is swift," with which the ecologist Francisco Lloret (2022) recently epigraphed *La muerte de los bosques* ("the death of the forests"), takes on its full meaning.[3] This is a shameful example of inhumanity towards thousands of vulnerable migrants and, moreover, a blatant disregard for the forest's own ecological dynamics. At a time when the politics of wilderness insist on the need to create steppingstones and ecological corridors to ensure the connectivity of ecosystems, the only large span of forests that still survives from the early Holocene could be cut in two by a concrete wall, broken up without any consideration, and cleared to facilitate the passage of heavy vehicles. It is quite understandable that the measures taken by the Polish government should cause revulsion among scientists who promote rewilding. It is not for nothing that these measures firmly confront the ideal of the free evolution of natural processes, that they interfere in a harmful way with the ecological functions of species and that they tear apart the complex web of their relationships. As Marta Tafalla (2022) reminds us in her *Filosofía ante la crisis ecológica*, ecosystems should not be understood as static entities with taxa, but as dynamic entities inhabited by forms of life in a relationship of dependence. From this perspective, the harmful effects of the environmental policy of our European neighbor are highlighted, and the deficiencies inherent in traditional nature conservation policies, where the human being is chosen as the center of a reality that has neither center nor periphery, are exposed.

We have mentioned above two types of cities in Cartesian rhetoric. We compared each of them with two types of forest ecosystems: monospecific plantations and primary forests. We know that Descartes used the metaphor of the city to legitimize his project and declare the primacy of reason over history, of human beings over nature. Another philosopher, Ludwig Wittgenstein also used the metaphor of the city to explain his theory of "language games." Wittgenstein's metaphor could be useful for our purposes.

> Our language can be regarded as an ancient city: a maze of little streets and squares, of old and new houses, of houses with extensions from various periods, and all this surrounded by a multitude of new suburbs with straight and regular streets and uniform houses (Wittgenstein 1953, p. 18).

[2] For more information on this topic, please see: https://whc.unesco.org/en/list/33/

[3] Despite warnings from Brussels and UNESCO itself, Andrej Duda's government has not been intimidated. On the contrary, it is maintaining its opposition against European environmental legislation. The confrontation has recently (2022) grown in intensity after the Polish authorities decided to build a wall in response to the migration crisis with Belarus. A 186-km wall, topped with concertinas and fitted with thermovision systems, will extend over the border line separating the two countries over which the forest stretches.

The Wittgensteinian city offers an idea of balance between the type and (counter) type of city supplied by Cartesian logic. Whereas Descartes separates and isolates them, Wittgenstein juxtaposes them by forming spaces of contiguity and convergence. There is no discussion about the need for reforms and extensions, but not always to the detriment of the inherited works; the construction here or there of a street, a neighborhood or a ring road is admitted, but not at the cost of the destruction of the common heritage. Each language game has its own specific value, just as each element of the urban landscape has its own specific value just as each language game has its own one.

Suppose now that we use Wittgenstein's chosen terms metaphorically. Then, the "old houses" and the "tangled alleys" would correspond to the domains of wild nature such as reefs and mangroves, taiga, forest, rainforest, and desert, and the "new houses" and the "new suburbs," the "straight and regular streets" would become the domain of human life. And just one more thing: these "houses with extensions from various periods" would eventually map onto intersection spaces between these two domains. What is missing is to turn this city into the world, into the city of the biosphere, into our global village, a village, indeed, because any event has the capacity to have an impact on our existence, and also a village because it treasures the evolutionary heritage of multiple forms of life in a situation of mutual dependence and collaboration.

In this city of the biosphere, it is very important to preserve the history of the species with which we live, to give them their rightful place, to guarantee them the fulfilment of their ecological functions, to respect their ecosystems, and to gradually withdraw from places where our presence constitutes a certain threat to them. Finally, it is desirable that this global village –the village of the biotic community– should be able to accommodate wild nature and human life without today's hostility and, now that the primeval forests are disappearing at an accelerating pace, should look forward to a world ready to bring them back. This is the aim of the AFH, which should not be interpreted as other ecosystems being treated with contempt. On the contrary, because of its capacity to harbor life, plural and multi-diverse life, the primary forest here is understood a synecdoche addressing nature as a whole.

Just as an undeniable interest in humanity's past has been cultivated inside us, it is high time to turn our collective attention towards our future, where the elements of uncertainty are greater and more serious than those we find in the review of the past events. Were this new perspective adopted, Hallé's project would no longer be an unrealistic utopia and would become a multi-generational enterprise. As Alan Weisman notes in his disturbing *The World Without Us*, "the thought of rural Europe reverting one day to original forest is heartening" (Weisman 2007, p. 14). However, humans tend to become uneasy when thinking about investing their time in projects whose fulfilment they will never see. The foundation stone of this cathedral of European nature could be laid very soon. It remains to be seen whether we will have the generosity to leave our future generations the good fortune to see the project accomplished.

The compound term developed by R. Panikkar, sacred secularity, offers us new understandings about the divine, the world, and the human being. His *advaita* interpretation enriches the understanding of them seen from a relational consciousness, which unites and distinguishes at the same time, without falling into monism or dualism. On the other hand, recognizing the character of ultimacy of the temporary structures of reality, without making them absolute, leads us to propose new hierophanies where the very existence of the human being and of Earth need to be lived as sacred realities, even more so in the present context of where the possibilities of planetary ecocide are real. It is in this scenario that the words of Panikkar resonate strongly when pointing out that "the sacred-secular conception, which does not believe in 'second' chances or in 'another' world, can activate forces of salvation that are dormant in the human race" (Panikkar 1999, p. 65).

References

Adam C, Tannery P (1996) Œuvres de Descartes. Vrin, Paris

Brincken J (1828) Mémoire descritif sur la forêt impériale de Bialowieza. Chez N. Glücksberg, Warsaw

Hallé F (2021) Manifeste. Pour une forêt primaire en Europe de l'ouest. Actes du Sud, Paris

Lloret F (2022) La muerte de los bosques. ¿Qué es un bosque? ¿Cómo funciona? ¿Por qué están colapsando algunos bosques en todo el planeta? Arpa, Barcelona

Morizot B (2020) Raviver les braises du vivant. Actes Sud/Wild Project, Paris

Morizot B, Mantovani AO (2022) S'enforester. Mythologie et politique de la forêt d'Europe. D'une rive à l'autre, Paris

Panikkar R (1999) La intuición cosmoteándrica: las tres dimensiones de la realidad. Trotta, Madrid

Servigne P, Stevens R (2015) Comment tout peut s'effondrer. Petit manuel de la collapsologie à l'usage des générations presents. Seuil, Paris

Tafalla M (2022) Filosofía ante la crisis ecológica. Una propuesta de convivencia con las demás especies: decrecimiento, veganismo y rewilding. Plaza y Valdés, Madrid

Weisman A (2007) The world without us. Thomas Dunne Books, New York

Wittgenstein L (1953) The philosophical investigations. Blackwell, Oxford